Modeling and Simulation of Everyday Things

Modeling and Simulation of Everyday Things

Michael W. Roth
Hawkeye Community College, Waterloo, Iowa

CRC Press
Taylor & Francis Group
Boca Raton London New York

CRC Press is an imprint of the
Taylor & Francis Group, an **informa** business

CRC Press
Taylor & Francis Group
6000 Broken Sound Parkway NW, Suite 300
Boca Raton, FL 33487-2742

International Standard Book Number-13: 978-1-138-48544-0 (Hardback)
978-1-4398-6937-6 (Paperback)

Visit the Taylor & Francis Web site at
http://www.taylorandfrancis.com

and the CRC Press Web site at
http://www.crcpress.com

Visit the eResources: https://www.crcpress.com/9781138485440

This book is dedicated to many people who have had an impact in my life. First, there are the two best teachers I've ever had. The first is my Dad, Henry Roth, who got me into science and cultivated a true love of learning and teaching through anatomy lab excursions in that '64 Rambler, museum trips and rock hunting weekend mornings. The second is my graduate school mentor, Bogdan Kuchta, who nurtured me, cultivated a love for computer simulation research and encouraged me when things got tough. Then there are family members. My mother, Elizabeth: I know you don't remember when you went with me to check my grades after my first semester in college but I certainly do; thanks for your investment in my success and so much more. My wife, Diane, and children, who each came into my life right when they needed to: Jacqui, Amanda, Becca, Thomas, Christopher, and Sara all have heard about this project and gave of themselves as I worked on it. And my cousin, Sally, provided a nice peaceful venue when I was working on the chapter on visualization. My friend and band leader, Kent Aschenbrenner, and collaborator and engineering comrade, Tom Yops, gave me inspiration, and my Uncle Mikey taught me how to laugh, which not only helps in the rough times but also helped be build my current set of classroom jokes. The unconditional love of pets was a huge positive, courtesy of Lucky, Misty Luna, Jewel, Friendly, Paula in Pink Rauchstimme, and NellieBella Fiorella. There are my close collaborators who appear as co-authors (in addition to Bogdan): Carlos Wexler, Lucyna Firlej, Rui He, Paul Shand, and Paul Gray. Then there are the many, many students whose interests helped broaden my research agenda and whose skills contributed to it. They can be found either explicitly acknowledged or as co-authors on publications. I value the opportunity to be a mentor and engage in student-driven collaboration, so in many cases their ideas are the ones that came to life. And of course I would be remiss to not mention the unwavering support, encouragement, and friendship from my wonderful publisher, Lou, and throw in some Divine intervention. I simply don't see how the project could have started or reached completion otherwise. And finally I raise a toast to Velocity Bike and Bean in Florence, KY, as well as Cup of Joe's in Cedar Falls, IA, for providing wonderfully enriching and relaxing venues to work on this project.

Again, thank you all. You own part of this.

Contents

SECTION II Models of Everyday Things

SECTION III Beyond Everyday Phenomena

SECTION IV A Glimpse into More Advanced Computing

9. Parallel Computing, Scripting and GPU's   321

Appendix A: Integrated C++ / Python Simulation of
Guitar Sounds   339

Index   345

Preface

This book is intended to sound like a conversation. It is mostly dictated by the motivation to engage the reader, encourage creativity, and also elicit a chuckle now and then. I want it to sound like I'm trying to talk to you and work with you as you learn how to construct, code, debug, validate, and interpret results from computer models and simulations of things we run into every day. The content comes directly from my research with students as well as my having taught Modeling and Simulation of Physical Systems in a Professional Science Masters' class. So it's got proof of concept from "here to the end of the block" and I want you to take heart in that if the going gets a bit rough. The book is designed for (i) persons who don't necessarily have experience in computer coding or physics, (ii) nonphysics and nonscience majors in community colleges, colleges, and universities, (iii) students in Professional Science Masters' or other nonphysics graduate programs, and (iv) high school students in science courses. It's meant to support, encourage, and empower the reader to do some decent simulations of routine things with smatterings of what's beyond in order to challenge you to grow and expand. When your first few models come to life at your fingertips, I think you'll find it very gratifying.

Because I want this text to sound and act like a conversation, I want to emphasize that readers should feel free to contact me with any questions they might have because, especially with the first edition, I want to make sure things run smoothly for those who have invested in it. There are homework problems that invite communication as well and there is, of course, the proviso that interactions with students mustn't transcend class ethics and rules.

There are also things this book is clearly not intended for. First, it isn't intended to be a contemporary programming language or platform primer (there's enough of those out there already). It's about the construction of the models and understanding the algorithms. This means that if you're going to work for Disney or a gaming development company, what you learn from this book will help you accomplish some very cool things, which you will then go and learn to express in the language that your field uses. So don't be alarmed at the use of FORTRAN, UNIX, or C++. They're not the point. There are many languages and they're changing; you will almost certainty have to translate and it will become second nature for you. This text is also not meant to be a reference for numerical methods, integration algorithms, or historical artifacts, although all those things are of deep value. Rather, it is meant to introduce you to useful models and tools, generate an awareness of challenges, and give you a starting point from which to delve deeper into areas of interest. I hope the references listed will give you a good sense of where to go for further information.

It's been a pleasure to write this book. I have gotten to remember all the wonderful students who helped make my research program what it is and make me who I am as a physicist and teacher. I've gotten to make my simulations better, organize my thoughts, spend a lot of time in coffee shops, and research a lot of interesting topics. The project has followed my life through a great many significant changes so, in a deep sense, it has become an extension of myself. I truly hope you use it well and enjoy it heartily.

Online Supplementary Materials

Readers, please note that the e-book editions of this book contain color versions of most of the images used. Also, downloadable color figure files have also been made available, together with programming files and appendix materials, online at the publisher's website (see the tab marked 'Downloads/Updates'): https://www.crcpress.com/9781439869376.

Author

Michael W. Roth serves as Dean of the School of STEM and Business at Hawkeye Community College in Waterloo, Iowa. Prior to that, he has held faculty positions at a variety of community colleges and universities in Colorado, New Mexico, Texas, and Iowa, and was most recently department chair at Northern Kentucky University's Department of Physics, Geology and Engineering Technology. He has a passion for teaching and research involving undergraduates across all levels of undergraduate study, and has participated in course and program development and assessment. He has involved a large and diverse group of students in his computational physics modeling and simulation-based research program, and has published numerous articles and presented at conferences with them in the fields of condensed matter surface physics, bullet impact, groundwater flow, snow remediation, solar system formation, and planetary impact. He is a collector of antique science books and laboratory equipment.

SECTION I

Getting Your Feet on the Ground

CHAPTER 1

Building Your Basic Tool Box

1.1 Introduction: When are Computer Simulations Useful?

The aim of this textbook is to cultivate a working knowledge of how to construct models of things we encounter in our everyday lives. If you haven't done much computing, or you don't have much of a physics background, it's not a problem because this book is designed to guide you through the modeling process and offer the tools needed to build your proficiency in modeling things from scratch. It's not a computational physics book replete with numerical recipes; neither is it a computer programming book that stresses high-level computer architecture and programming elegance, nor is it a physics book. Although such references may be interesting and useful, this text provides a springboard to venture into more specific areas. It's an apprentice card of sorts, and its purpose is to get you to successfully model things—really that's it. In a sense, I view it as being full of thoughts and ideas that would be exchanged in an incredibly extended series of coffee shop meetings, with rigor, equations, and example code thrown into the mix when needed. I hope it nurtures the amateur, yet challenges those seasoned in modeling with new thoughts and ideas. There certainly are more sophisticated software packages, but the work you do here will be your creation and in a sense you will shape a world and get to watch the rules unfold in front of you. All well and good, but the question still remains: why should people simulate things on computers?

It could be that you're taking a course for a degree, and you need to learn about computer modeling and simulation as part of your requirements. Although a perfectly legitimate reason, simulations can also be motivated in other ways, and we will address some of them in this book. They will not be contrived ones like "Your car has a flat tire and you need to drive home, and so you whip out your cell phone and do a quick molecular mechanics simulation to make sure the tires will not overheat before you get there." Instead, they will be useful simulations and models that fill a unique spot because they operate on a simple set of rules that play themselves out in complicated and sometimes extravagant detail. For example, Newton's second law tells us how force, mass and acceleration are related. We're no strangers to that law—we've known it for centuries now and discovered it intuitively when we were children. Most people also know about gravity and how to solve gravity—driven problems on pencil and paper where there's weight and maybe air resistance. But now consider a model where we have to determine if the surface roughness of a soccer ball affects its flight—this is a very difficult problem with subtleties necessary to take into account to make it relevant and useful, usually impossible to solve with just pencil and paper.

So how would I like this text to get you to model things? I would like it to encourage nonscientists to leverage their familiarity and proficiency with computers to help understand everyday things better, equipping your toolbox for simulations from the ground up and perhaps most importantly, enticing us all to "lift the hood" and try our hand at constructing abstract models. Consider real scenarios where it would be very useful to conduct a computer simulation. Some situations are personal, others may occur in a professional setting, and all are precisely the kind discussed in this book, where we build from the ground up.

Suppose that you invested in a house and you've been there for 20 years; you just found out that its cinder block foundation is leaking. Repair costs are more than you can afford in a lump sum but you also know that as time progresses the leak will continue to do more damage. The question is whether to take a loan out to fix the house, or if you should save money and fix it in 5 years. How does the interest you pay on the loan compare to increased repair costs if you wait until you can afford to fix the leak using your own savings? The scientific issue is a diffusion through porous medium problem and you could utilize a finite element computer simulation to estimate the water seepage and hence the extent of leakage and damage over the next 5 years to see whether or not the cost to remediate is likely to exceed the interest of the loan. Sometimes, *real life processes are too slow to provide useful information* but computer simulations can help us understand how they might unfold.

Enter a mechanical engineer who had a family member badly hurt in a car accident. In this particular case the fenders of the car crumpled as they should have. That is, they did their job of taking energy away from the cars' motion and dissipating it so the collision was not so dangerous as it would have been otherwise. However, it was very cold that day and the collision angle was such that the doors cracked in certain spots and crumpled inwards, causing injury to the driver. As would be expected, the engineer is now on a mission to make cars safer in cold weather collisions, where metal alloys are more likely to crack and fail. Their first thought is to prevent the doors from fracturing at low temperatures by either modifying the metal alloy composition or perhaps adding an anti-fracture coating to the inner surface of the door, much like is already done for glass in cars. In either case, the physical issue is an understanding of material fracture and fracture propagation as a response to stress. A Material Point Method/Equation of State simulation or possibly a large scale Molecular Mechanics method could be used to understand how the material fractures and how the fracture propagates, ultimately causing material failure. Sometimes, *real life processes are too fast to provide useful information* and computer simulations can slow them down in a sense, affording better understanding and quite possibly refined control of them.

Then there is the single mother working at a factory outside the small city she lives in, who has to spend an hour in traffic on the way to work and an hour and a half on the way home. Although regular (steady) traffic can cause long commute times, it turns out that in her case traffic jams are the problem. It fits our intuition that perhaps relocating stop signs, rerouting traffic or offering driving restrictions could help solve the problem, but the much simpler act of retiming the stop lights would be ideal and much less trouble than any of the other options mentioned. In order to understand how the stop light timing should be changed, the system would need to be modeled somehow. The closer the cars get the stronger their drivers brake, so the cars can be thought of as being connected by springs and rush hour traffic can be thought of as a resonating mechanical system. Stop lights then act as "hands" that push or pull the system at certain times. If the lights turn certain colors in certain timing, it could be that the traffic jams can be dissipated and commute times can be dramatically reduced, much like if one pushes a child in a swing randomly and not in synchronicity with the swing. Because of the complexity of the system and the interactions involved it is easy to see that a scaled down mechanical model simply isn't realistically achievable. However, a dynamical computer simulation could capture all the physics necessary to understand the resonant phenomena at hand and the dampening needed to restrain it. Sometimes, *real life processes don't have reasonable experimental analogs* and computer simulations are needed to better understand them.

Astronomy is a field where asking and addressing (not always answering!) scientific questions affords an appreciation for nature's beauty and also helps us understand our own planet in broader context. Consider a beautiful planetary family in our own Solar System: Uranus and its five natural satellites. The obliquity of the axis of rotation of the planet is yet unexplained: the planet rotates so that it is tilted nearly completely on its side relative to the other planets. Moreover, the same is true for the orbital plane of its rings as well as for its satellites. It's almost like a small section of the solar system got reoriented somehow. There are ideas that a collision happened early in the formation of the Uranian system or that perhaps novel mechanisms of angular momentum transfer (they would be novel to us but not to nature; we're just trying to catch up) took place in the distant past. Designing a scaled-down experiment to investigate such a problem would prove exceedingly complicated because there would have to be a three-dimensional many particle model with no local gravity from Earth possessing only inverse square forces. Since electrostatic forces are inverse square—just like gravity—it's possible that such a model could be contrived but then electrodynamic forces—magnetic and others—from the moving charges would be present. What's more, gravity is too weak to use in the lab. Workable models of astrophysical systems in three dimensions are essentially not tenable; unlike the example in the previous paragraph, the only workable experimental analog to the system is the system itself. Sometimes, *real life processes have experimental analogs that can't be constructed* and computer simulations can help understand detailed processes of the physical behavior of the system at hand.

Then there's the college graduate embarking on a new business venture. The loan is taken out, office space has been rented and employees have been hired. Overall things went as well as could be expected for the first year but during the first winter the building had a snow drifting problem that impeded easy access to it and that caused light blockage, leakage in the springtime and energy loss. She needs to make a decision on how best to divert the snow accumulation away from the building and has a few options: snow fences, deflection fins or roof extensions. The company's operating budget constrains the situation such that a particular method can be implemented only once and so the first guess can't just be a good one; it's got to be great. The physical issues are wind flow and snow accumulation around building structures and computer simulations can help the CEO make a very good guess as to what kind of a structure to put up. Sometimes, *real life processes are too expensive to be attempted on a trial and error basis* and computer simulations can help predict their effects and determine a workable, or even optimal choice.

There never seems to have been a shortage of crime scenes throughout human history. Since natural laws of physics govern both noble and ignoble activities, there are outcomes—clues—left in the wake of crimes and when they are understood in the context of natural laws, then activities during the crime can be much better understood. What does the blood spatter pattern tell us about the type and direction of bullet injury? What was the most likely direction of travel of the bullet(s) that killed JFK in 1963? Was it possible for just one bullet to have done it? Was it optically possible for a victim to see an assailant from the room he was in given the placement of mirrors in the house? Sometimes, *real life processes are too dangerous or too hurtful to be attempted* and computer simulations can help.

On the night of April 14, 1912, an event in the icy north Atlantic indelibly etched in our minds that the machines we construct were not invincible. There have been many studies about the sinking of the Titanic in an attempt to understand the details of her demise. An interesting feature of the wreckage is that the bow and stern have starkly contrasting appearances, suggesting that the two sections of the ship experienced dramatically different pressure effects between the interior and exterior. Finite element computer simulations are able to reasonably well explain how the ship's bulkheads filled with water, how she sank, the separation of her bow and stern, and why they have the strikingly different appearances they do on the ocean floor. To simulate the sinking in a laboratory with a real model almost everything could be scaled down but not material strengths, and so unrealistic and spurious (out of nowhere) forces would have to be introduced in order to deal with the stress created as half the ship sank in the water. And a real challenge is that in many instances the passage of time itself has to be scaled, which is entirely nontrivial. To alleviate such problems, or to at least minimize them, a very large model would have to be used. Sometimes, *real life processes are too large to conveniently observe* and computer simulations can help us understand them so future processes can be made safer or easier. On top of the size and scaling issues, natural as well as human-made disasters fall into another category for simulation. Sometimes, *real life processes happen only once—or at least we certainly hope so—*and computer simulations can help replay them in detail so as to aid our understanding of them.

When molecules and atoms are placed on surfaces they interact more strongly and in ways much different from how they do in three dimensions. Just imagine how a traffic jam or crowded mall would be different for us if we could easily move in all three dimensions! In experiment, X-rays, neutrons or electrons are thrown at ("shined" on) the system and by studying how they emerge from it, information about the structure of the system can be deduced. However, there are limits to our experimental understanding: we can see steps, terraces, rows and clusters on the nanoscale but we miss detailed aspects of the dynamics of atoms and molecules such as exactly how they may be deforming, rotating, or diffusing. Molecular Dynamics computer simulations can be used to simulate a nanoscale system evolving in time and can give us an idea of the details of the physical processes and motions present in the system, as shown in Figure 1.1 [1]. If the system is

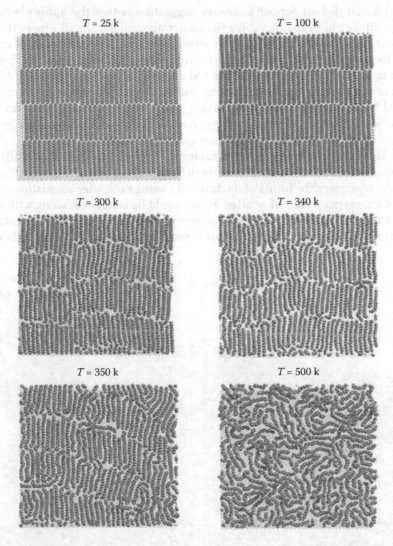

Figure 1.1. Final snapshots from a Molecular Dynamics computer simulation of a long hydrocarbon, $C_{24}H_{50}$ (tetracosane) on graphite. The simulation gives us fairly reliable information that an experiment can't: as temperature is increased, the ends of the molecules flip up (red atoms); then the molecules remain stiff but the rows start to shift; and, at last, the system melts. Also, the effects of motion on two distinctly different time scales are visible (Reprinted with permission from L. Firlej, B. Kuchta, M.W. Roth, M.J. Connolly, C. Wexler, Structural and phase properties of tetracosane (C24H50) monolayers adsorbed on graphite: An explicit hydrogen molecular dynamics study, *Langmuir*, 24(21), 12392–12397. Copyright 2008, American Chemical Society.) [1].

near equilibrium and we're just interested in its average behavior, Monte Carlo simulations are very useful, and the differences between Molecular Dynamics in Monte Carlo simulations will be discussed in Chapter 8 of this text. Sometimes, *real life processes are too small and detailed to observe* and computer simulations can help.

Egypt has a desert climate but the Sphinx shows signatures of atypical erosion (Figure 1.2). Dr. Robert Schoch's creative suggestion is that the Sphinx was around when the climate in Egypt afforded a significant amount of rainfall—more than about 5,000 years ago and that there is water erosion present! Knowing rates of erosion at various locations on the structure, a backwards-time simulation could be constructed and run to see if the Sphinx could have had another form a long time ago. Moreover, in cases of explosions or violent collisions, paths taken by fragments of the system can be traced back to a common point of origin, which can be very useful in any forensic analysis. Sometimes, *our understanding of real life processes is illuminated by running them backwards in time* and computer simulations can help.

An architect is helping build a bridge to replace the one that collapsed in Minneapolis. Instead of constructing models and doing stress tests on them, they would like to explore the limits of its durability using computer simulations. Physical effects of exaggerated natural weather events could be simulated, along with artificial forces and torques, and even extreme changes in material properties with extremes in temperature. In such situations, it is much more practical to study the behavior of a

Figure 1.2. Stone at the base of the Sphinx showing what appears to be water erosion. Is it possible that the Sphinx has been around a lot longer than traditionally believed? (Photo courtesy of Dr. Robert M. Schoch. For further investigation see http://www.robertschoch.com/sphinx.html.)

computer model you trust than to build real models, even if scaled down. Sometimes, *our understanding of real systems can be enhanced by investigating the effects of introducing unrealistic conditions* and computer simulations can help.

At the theatre or in the gaming world, special effects have become very realistic over the last 30 years, and certainly more accessible. Humans may not find physics and mathematics popular subjects in general but nonetheless we apply tenants of such fields with stunning effectiveness when we press down on the brake of a car in coming to a safe stop, or when we throw the ball through the hoop during the last minute of the game or when we run to catch up with somebody we want to speak with. Indeed, we're good at doing science implicitly. It makes sense then, that the more realistic physical behavior in movies and games is, from dynamics to hydrodynamics and optics, the more engaging and entertaining they will be. Computer simulations can do a wonderful job of helping interject realistic physics into games and motion pictures, and people from many disciplines have made their careers creating more realistic simulated environments. And there's scientific entertainment as well. Some time ago, at Sandia National Labs, the sound of a Parasaurolophus, a crested dinosaur, was simulated. Just like in studying how a wooden guitar cavity resonates and supports the sound of its strings, the shape of the air cavity inside the dinosaur's skull was coded into a computer and the wave equation for sound was solved inside it to see what sounds were supported. Then, an audio simulation program combined the individual supported notes from the wave equation to simulate the actual sound the animal made. So, even if we're not out to solve a problem, *we may just want good entertainment* and computer simulations can help.

There is a plethora of other types of situations where simulations are useful, such as automobile and aircraft design (both constructive and accident), flight and driving simulation, weather forecasting, financial and investment forecasting, noise cancellation, water and air pollution management, process engineering, robotics, population and city growth, and even simulation of computers that can't be built yet, known as emulation.

1.2 How Much Should a Simulation be Trusted?

It's good to trust your own work. As far as computer simulations go, the basic idea is that reasonable rules and sound algorithms make for robust models; when they work well in known situations they can then be trusted in many new situations. It probably won't surprise you to know that computer simulations and models have limitations outside of which their validity becomes questionable. And, in talking about the usefulness of any scientific tool it is perhaps most useful to understand situations where the tool is not much help at all, because, just like driving on an icy road, a keen awareness of the limitations is simply good practice and promotes healthy perspective.

First, computer models don't help when there is a fundamental aspect of the physical model missing. For example, simulating a baseball game without effects from air will result in dramatic miscalculation of the ball's trajectory. Simulating a bowling ball thrown down an aisle with a ball that can't rotate but just slides misses out on the beautiful interplay between kinetic friction, rolling friction and transfer from translational energy to rotational energy as the ball begins to spin. In addition, simulations aren't much good if there is an element of the physical system that is misrepresented. Running a heat transfer simulation with a negative sign reversed could result in hot boundaries cooling the system down or some other unphysical effect. Running an aerosol pollution simulation with a mistake could result in a workplace contaminant obediently rushing back to its source, not going away. And the mistake could be much more subtle. It could be that the simulation of lift on an airplane runs fine, except for the fact that the buoyant force—almost unnoticeably small—is downwards, not upwards. You would notice your mistake when helium balloons sank to the ground like lead but you wouldn't notice it dramatically for planes or people.

Almost all the situations where computer simulations don't help can be avoided by a process termed *validation*, where the computer simulation is checked against known results at each significant stage of its development. So you would run a baseball simulation without gravity, and check to make sure it works correctly. Then add gravity and run it again and check it out, then add drag resistance and run it again and check it, then add the Magnus force from spinning and run it again, and of course check. The idea is then that if none of the checks bounce and the simulation yields results consistent with known results it can be trusted in new territory. So then you could perhaps add a prevailing wind to the program or code in an altitude change or maybe incorporate some temperature-dependent elasticity of the ball and you will be able to trust what comes out of your simulation. In each of the different types of models presented in this text, specific validation sequences will be suggested in the narrative or homework problems so that the simulations you construct are helpful and work well; its part and parcel to building your model from the ground up.

Even when validated, computer models that are too simple also don't help very much, even if they are perfectly validated. After all, isn't the reason that we're not doing pencil and paper (analytical) modeling so that we *can* include a bunch of interesting and relevant details that make the problem intractable otherwise? Such things don't make the problem significantly more difficult to solve. For example, suppose we are studying the effect of bullet insult to the human arm and are interested in how the humerus fractures when somebody is shot. If we model the arm as one single cylindrical object with average properties of muscle and bone, we can certainly validate our program but we will be missing out on the effect that the overlaying muscle itself has on the bone and what the difference in elastic properties between muscle and bone mean. And of course if we just leave the muscle out, then we're simulating skeletons getting shot. And if we leave the bone out, yeah... we're not simulating chicken wings... you get the picture. But, because we're doing simulations, we could add effects from muscle shape and tone, bone density issues, bone marrow issues, and even calculate how much a bullet heats up the wound area.

If you pursue computer simulations for some time, or even have to work with an in-depth model even once, your skill in modeling various things will develop further and it will be easier to create simulations that have an appropriate balance between simplicity and completeness. You'll get a natural feeling as to when your model has captured the important physical aspects of the system, and when you are making it too cumbersome. And, in tandem with getting that gut feeling as some of your model is complete or at least complete "enough," you will nurture a sort of detective feeling about the validation process.

1.3 Who First Used Them and Why They Came About

Various cultures have recognized the importance of mathematics and numbers since antiquity. Up until the time true computer simulations were used our computational ventures were mainly numerical in nature. It is probably a combination of the search for accuracy and speed throughout history that humans felt prompted to construct computing devices. Admittedly going back to the abacus would be a bit overkill and so it is likely best to visit a very charming period where a confluence of mechanics, mathematics and science gave rise to bulky computing devices that were actually well ahead of their time.

The field of celestial mechanics incorporates mathematics that allows one to predict the positions of stars and planets in the sky (including our recent August 21, 2017, solar eclipse!), and in the late 1700s to early 1800s observational astronomy was in full swing. In order to make accurate measurements, astronomers needed to match calculated positions of planets and stars to actual ones out to several decimal places. Moreover, the mathematical functions used two obtain predicted numbers involve logarithms and trigonometric identities such as sine and cosine. Around 1812, a *difference engine* (Figure 1.3) was conceived (and a few years later built) by Charles Babbage. The difference engine was a very large mechanical computing device which was hand powered. It used tumblers and cylinders to approximate trigonometric functions and logarithms with polynomial approximations, thus making it very versatile. It was perfect for its time—it was programmable with punch cards and utilized a mill (CPU) which stored memory in pegs and rotating drums. With more development it was able to produce its results on stereotype plates and print them out directly so as to eliminate fallible humans in transferring the mathematical results to the printed page, which was undoubtedly important when getting results to contemporary royalty. It could even plot curves and it had a return bell. It passes the Turing Test [2–5], which is in a sense a measure of machine intelligence and sets indistinguishability between the responses given by the computer and an actual human as its standard.

So, if you are a machine and your responses to problem statements are completely indistinguishable from those a human would give, then you are said to be *Turing complete*. Incidentally, there are other interesting and fundamental theoretical thoughts on computing such as the halting problem [6] that can be found in many references that the reader is encouraged to seek out.

With the construction of the difference engine Babbage quickly realized that it would be possible to construct a much more general-purpose machine. In in the early 1800s, Babbage devoted his life to designing such a machine named the *analytical engine* [7–9]. The analytical engine was well ahead of its time, but unfortunately it was never built by Babbage. It was a digital mechanical

Figure 1.3. Babbage's Difference Engine. (Adapted from https://commons.wikimedia.org/wiki/User:Geni.)

Table 1.1. Parts of and Operations by the Analytical Engine Alongside Their Modern Counterparts

Babbage's Analytical Engine	Contemporary Computer
Accidental sign	Sign bit
Analyst	Programmer
Attendant	Operator
Axis	Buffer register
Barrel	Microcode
Card	Instruction
Card, operation	Arithmetic instruction
Card, combinatorial	Instruction, control transfer (test/jump)
Card, number	Instruction, load immediate
Card, variable	Instruction, load/store
Carriage	Carry propagation
Curve drawing apparatus	Plotter
Column (of Rack)	Memory (RAM) cell
Cycle	Loop
Mill	Arithmetic and Logical Unit (ALU)
Rack (of Columns)	Random Access Memory (RAM)
Run up	Overflow or sign bit condition code
Stepping down	Right shift
Stepping up	Left shift
Store	Memory (RAM) array
Turn of the handle	Clock cycle

Source: John Walker, http://www.fourmilab.ch/babbage/contents.html [7].

computing device, an absolutely charming device to anybody interested in the brass and mahogany days of scientific instrumentation. It clearly showed roots in science, industrialization and aesthetic appeal, with parts and operations having direct modern computing analogs, as mentioned in Table 1.1.

There has been interest in constructing the analytical engine (Figure 1.4). Around the year 2000 a museum in London commissioned building an analytical engine directly from Babbage's plans. Aside from some minor design errors, which were corrected in the 2000 construction, the machine worked flawlessly. In fact, many of the computers built during the 1940s were not so advanced as the analytical engine was. Ada Lovelace, the daughter of the poet Lord Byron, corresponded extensively with Babbage. In 1842–1843 Lovelace translated Luigi Menabrea's memoir of the analytical engine into the English from its native Italian. In her translation, she added a completely new section on calculating Bernoulli numbers with the analytical engine. Had the analytical engine existed, the method and algorithm would have run correctly. In light of her accomplishments she is by many accounts the first computer programmer. It would be interesting to also see if she was credited with the first bottomless pot of coffee (Figure 1.5).

The great computing machines of the 1800's carried out *numerical analysis*, where a specific mathematical problem—a relationship—is solved without requiring reference to any physical thing or process. The mathematical engines helped numerical analysis to get a foothold and advance in science, but numerical analysis is rather different from simulating or modeling, both of which involve creating an abstract concept of some physical system. Hence, there had been no computer models or simulations actually done. One could argue that just like with chemistry there were a couple curious high school whiz kids who tried some computer simulation out in their garage after school but the chances are likely slim.

Not surprisingly the first *bona fide* computer simulations were carried out during World War II. The problem at hand was a very important one: nuclear chain reactions. Now there are three types of easily observable particles that make up atoms: protons, neutrons and electrons. Protons are heavy positively charged particles and neutrons are heavy neutral particles, both of which reside in very close proximity in atomic nuclei. They carry most of the atom's mass. Electrons are negatively charged particles about 1800 times lighter than neutrons, and they exist in orbitals about the nuclei. For nuclear fission to take place, particles of some type of have to run into a large parent nucleus of the fissionable material or materials. The nucleus then breaks apart into daughter nuclei and energy is liberated; that's just one event however. For a nuclear chain reaction to propagate there have to be progressively more nuclei broken apart, liberating energy for use and more neutrons to keep the chain reaction going. The chain reaction process is actually expressed in the popular riddle: if you had a choice of getting paid $1,000,000 per day for one month or being paid a penny on the first day, two pennies on the second day, and doubling every day thereafter, for a month, which would you choose? It is very tempting to take $1,000,000 per day but you may recognize the penny progression doubling each day as a *geometric progression*; you will be owed the national debt well before the end of the month! Nuclear reactions progress like the payment of pennies, and that's where neutrons come in to the picture. Because nuclei are positively charged, the particles that run into them to sustain the reaction should not be repelled by nucleus and should be heavy. Of all three particles, neutrons fit the bill: they are our pennies. It is also important to note here the porta-

Figure 1.4. Reconstruction of Babbage's Analytical Engine. (Courtesy of Bruno Barral.)

```
for (time>=birth; &&
time<over; time++)
{drink.coffee();
if(cup_empty==true) then
{fill.cup();
}}
```

Figure 1.5. What might have been a typical scene in the life of the first computer programmer, underscoring our association with bottomless pots of coffee and computing.

bility of computer simulations and models. Even though no one has enough money to pay you enough pennies to mimic the neutron growth in a nuclear reaction, it is easy to see that, if they could, the computer programs for the two vastly different systems would be almost identical. So even though it is easy—maybe even natural—to think that one particular algorithm is applicable only to something like nuclear physics, medicine, or some branch of engineering, looking just a little bit closer will almost certainly show the relevance of the heart of the problem to other issues, many of which you will find relevant to everyday life.

It was very important to the scientists in the Manhattan project that the devices constructed worked with a minimal number of mistakes and trials. It was not feasible to construct prototypes of the weapons they were trying to build; the cost, as well as the danger involved, were overwhelming. In fact, people in the project had discussions where they speculated what would happen if the chain reaction were sustained outside the influence of the weapon: could our atmosphere or maybe even a large portion of our planet detonate? One obvious choice the mathematicians (Jon Von Neumann and Stanislaw Ulam) had was to perform analytical calculations on the trajectories of the neutrons and nuclei. The problem at hand, however, was that the mathematics used allows direct (analytical, or pencil-and-paper) solution of the problem when two inter-acting particles are present. Clearly the nature of the nuclear chain reaction called for more particles in the model. However, with more particles approximations had to be made which made the calculation too simplified. So the choice with analytical methods was either to have a *tractable* (doable) problem with too few particles or do it with more particles with an oversimplified model. Neither conceptualization offered a workable solution to the problem. Computer simulations, however, were an attractive option because the interactions of many particles could be programmed in at once, and many of the simplifying assumptions were not needed. The model utilized was a Monte Carlo simulation [10] of 12 hard spheres [11] and was very successful in exploring and making predictions of neutron scattering.

With the remarkable success of the solution of the neutron problem, modeling and simulation of physical systems became a very active field of science right after World War II in appropriate tandem with the development of computers. There are many kinds of computers that entered the scene: some had binary arithmetic; some decimal. Some are mechanical and electric others just electronic; some were not programmable, others were using cards, punched paper tape or 35 MM film stock, and some used cathode ray tube or magnetic memory. In the decade to follow World War II, the robustness of computing devices lagged scientific inquiry, and generally it was viewed that realistic problems were too difficult and lengthy to solve on computers. In the 1960s IBM came on the scene with its Gordon simulator and there were a wide range of computer simu-lation packages and languages developed such as FORTRAN. In the 1960s, computers were fairly large and bulky but now were able to start tackling meaningful simulations and models. Development of larger computers continued to the 1970s; the 1980s saw the dawn of the personal computers and personalized operating systems. Then with the event of the Internet in the early 1990s, computer simulations took on much more relevant context, and things have been progressing ever since then. In fact, there's a formula for understanding how computing power has advanced throughout civiliza-tion, which will be discussed in the following section. In addition to the formula to be discussed, there are many word pictures around describing the dramatic advances in computing technology in recent decades (Figure 1.6). The state of the art in computing ability and tools is quite dynamic in nature and is driven by our constant desire to push computing limits.

1.4 What's the State of the Art? What Limits Have Been Pushed?

In order to adequately discuss pushing limits and computing it is important to understand the limits themselves: the state of the art. You can easily get on the Internet and look up factoids about the biggest computers, the largest simulations, the ones that have run the longest, and the latest hot spots in the field; in fact it's a possible homework problem. And, I think it's best to defer to contemporary media because it is likely that by the time you read something mentioned in this text it will be either superseded or outdated. With that said, there is a very interesting empirical relationship between computing power as the years pass, called Moore's law. In a 1965 paper, Gordon E. Moore [12] showed that computing power as viewed by the number of transistors on a CPU chip doubles every 2 years (Figure 1.7) [13]. On logarithmic plots such as the one in Figure 1.7, straight lines indicate purely exponential behavior and the slopes tell you how fast the models are growing or decaying. It is remarkable that Moore's Law places the growth of computing power between two very rapidly growing areas in science: particle accelerators underneath and fusion reactors above [13]. Moore's Law is rather accurate and has been used as a guideline in social, economic, and technological arenas. The trend in computing growth cannot continue forever, and there are many discussions that estimate the limit of Moore's Law. Since transistors on chips are becoming smaller, there are limits of atomic size transistors which use quantum tunneling to communicate (quantum computers). When looking at purely idealistic considerations such as energy contained in a certain volume of matter, ultimate estimates of 10^{50}

Figure 1.6. The *Osborne Executive* computer (1982) set aside an iPhone (2007). The more modern counterpart computes 100 times faster and is certainly less prolific in dimensions.

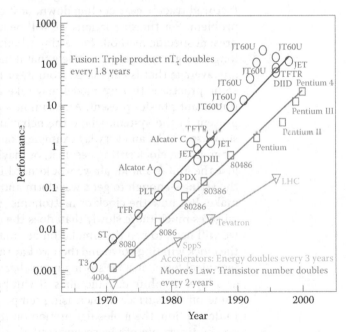

Figure 1.7. Moore's Law, graphed on a logarithmic axis. The linear behavior indicates exponential behavior and note that it places computing power growing roughly as fast as two historically hot spots in science. (Courtesy Laboratory for Plasma Physics.)

operations per second can be obtained compared to an Intel Pentium 3.2 GHz processor with 3.2×10^9 operations per second. Other elements of computer technology also follow similar laws: hard disk storage, network capability, and pixels per dollar on cameras.

1.5 The Simulation's New Clothes

Computer simulations and models are inherently artificial, and are therefore always limited in their applicability to the real systems being represented. In fact, there are two obviously adjustable aspects of computer models the cause them to fall short of reality: the fact that they don't run forever and the fact that they're not infinite in size. I imagine that for the same reason people climb mountains and try to run faster than anybody previously, individuals from various branches of science and technology have tried to push their simulations into a spot where the limitations have little to no effect. And what's more interesting about computing is that the playing field is dynamic because the state of the art of computing tools is also advancing with time in a fairly predictable way and it is the interplay between pushing the limits and the state of art that drives us towards better and more efficient computing devices.

Indeed, as the first limitation suggests, computer models don't run forever. In constructing models, we have to choose how long to run the simulations involved. The length of a simulation could express itself as the amount of simulated time in a diffusion problem dealing with pollution, or it could be simply the number of steps in a steady state problem dealing with the temperature distribution of a newly manufactured tractor gear cooling down, or it could be number of steps in a Monte Carlo problem. For the simulations we will be building from the ground up, this text will present specific methods for testing whether or not the simulations we built have run long enough, meaning intuitively that it has settled down and gives good statistics on any average that is calculated from it, or has accomplished its dynamical goal. Now some processes that we model may take a long time to run in order to capture the important physics present. An example would be a model with different time scales present in the system, where the behavior of the system has, say a very fast and a slow aspect. As an everyday example, consider simulating a guitar being played and a pendulum clock sitting beside it, or maybe a pendulum metronome. In considering just the guitar sound, all we would need is a few milliseconds of simulated time for the string—enough to get a waveform and an idea of the component sounds it would make. But now the clock or metronome introduces a different time scale because it vibrates much more slowly than does the guitar, and to capture the behavior of both we will need to run the simulation out much longer. Now add having to worry about the clock winding down and then we have a third time scale to deal with! Or for a more technical feel, consider storing natural gas on the surface of a carbon structure such as a layer graphite or nanotubes. It will be important to track how many molecules come off the surface of increasing temperature; such a process is called desorption. In desorption, the molecular motion on the graphite is important, and that happens on the timescale of picoseconds, 10^{-14} seconds. The desorption processes, however, happen on a much larger timescale, say milliseconds or seconds. To get around such situations, sometimes simulations with multiple time steps are used but if you're looking for detailed dynamics and the best answer possible, if you can do it, is to run the program for a very long time, seeking capture of multiple timescales with onetime step. In the work we did for long molecules on graphite [1] (Figure 1.1), the vibration of the molecules happens much faster than when the rows of molecules start to slide but only one time step is used.

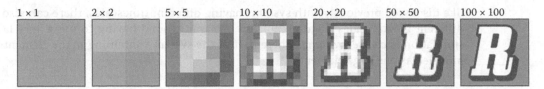

Figure 1.8. Example of resolution. A greater number of smaller pixels (ranging from 1×1 to 100×100 left–right) give clearer definition of the system, but how much is enough is very problem and system specific. Each type of simulation in this text will be presented with guidelines to help with such choices.

The second limitation mentioned is the finite size of the model, which has profound effects on its results. Whether we're talking about maximizing the number of particles that a volleyball is created out of in a Material Point Method simulation that explores the ball's elasticity, or maybe maximizing the number of boxes we partition space into for a really cool finite element tornado simulation, or if are trying run an atomic or molecular simulation and get results close to the thermodynamic limit (having infinitely many particles), the common drive in a perfect world is toward large-scale simulations. Although specific methods will be suggested for optimizing simulation size, the general rule of thumb is that your simulation is big enough when doubling its size has a negligible effect on the results. And really it's about resolution in large part—we are representing continuous things we find in real life with discrete things and there is no "back of the book" answer as to how big "big" really needs to be (Figure 1.8). Or in the case of periodic boundary conditions, we repeat the system on all sides in itself and then consider the middle as if it were one big system. If you could imagine a picture on a wall repeated on all sides of itself as opposed to one big picture, you could see that a repetition

Figure 1.9. Large-scale simulation of hexane on graphite, showing two distinctly different length scales: the molecule–molecule spacing and then the width of domains in the superlattice.

Figure 1.10. Another artificial effect of periodic boundary conditions. In this case, it is easy to tell when the simulation should be stopped for a single bullet shot.

introduces artificial patterns that you wouldn't find in real life (Figure 1.9), or perhaps there is dynamical behavior introduced from periodic boundary conditions that clearly should not exist (Figure 1.10), and those artificialities can dramatically affect the results of the computer simulation. Can you identify the effects in Figures 1.9 and 1.10 that should not show up, as well as the actual system we intend to model? Just like discussed previously with systems having different timescales, there can also be different length scales (Figure 1.9), and one way around having to use a very large simulation is *course graining*, where part of the system containing many elements is modeled as one large element.

1.6 Computer Modeling is a Very Interdisciplinary Field

Numerical analysis, computer simulation and computer modeling are all very different things. Although they all integrate diverse concepts, I view them as being in a hierarchy of sorts and, not surprisingly, the sequence in which they historically appeared is cosmetically similar to the level at which they integrate diverse interdisciplinary concepts.

Of the three efforts mentioned above, numerical analysis was the first to appear on stage and to actually be useful. Numerical analysis involves solving, or attempting to solve, a particular mathematical problem that is not easy to solve analytically (with pencil and paper). Historically, numerical analysis became useful in astronomical calculations but with reference to things we bump into daily its relevance could be found in a tax calculation, savings projection, a grade point average calculation or maybe something more technical like a stock market or population projection employing the derivative or an integral from calculus. Once there was a high school science fair project where a winning team calculated the volume of a Hershey's Kiss. Although the shape of the kiss arises directly from physical processes there's no closed form mathematical function describing the curve of the surface of the candy. So they had used numerical analysis to curve fit and essentially defined an expression for a curve that did express the shape of the candy's surface. Once the curve was found the volume of revolution was formed by the curve's being imaginarily rotated about an axis and then the solid volume was cut into slices and calculus—actually an integral—was used to add up the small volumes of all the slices to get the total volume of the Hershey's Kiss. It seemed to be a great problem in numerical analysis; I wouldn't classify it as a simulation because there are no dynamics involved, no motion, no melting and no elasticity, and it certainly would not qualify as a true model of any kind. By the way, if you understood the process of slicing the kiss up into very small slices and adding all their volumes to get the total then you understand the basics of integral calculus! Problems that involve only numerical analysis are characterized by usually having a specific purpose or close-ended question that needs to be answered. Certainly, the problems they address can still be very important; we may not normally think of the volume of a piece of candy as being significant but think about how many are made throughout the manufacturing process and about the cost of materials and manufacture analysis then it becomes more relevant. Those indented glass bottoms of wine bottles are equally as important in manufacture. Normally there are books that contain recipes for numerical analysis and it's easy to look up various techniques [14].

Computer simulations are the next step up the ladder. They normally incorporate numerical analysis but take the techniques a step further, leveraging a purely mathematical technique to characterize the behavior of a physical system, perhaps through time or through space. Computer simulations are typically differentiated from simple numerical techniques because the former involves actually tracking the system as it changes and also involves a bit more thought in its construction. As an example, say I wanted to model the behavior of a pendulum. For starters I could numerically solve the ordinary differential equations describing pendulum behavior and calculate its position and rotational speed in time. Sounds fairly straightforward but there's more to consider than just the numerical work. What is the best way to do the calculation? What things do I have to think about? I have to worry about the length of the pendulum because I know that will affect its period of oscillation. I also have to worry about the mass distribution: is the mass in the pendulum spread out more uniformly like a baseball bat or more at the bottom like in a traditional clock? The mass distribution tells me if I have a physical pendulum or a simple pendulum. Next, I have to consider friction. Is there a lot of friction at the pivot point and is there resistance from the air I need to consider? This tells me if I have a dampened pendulum. If the pendulum is in a clock or maybe on a platform on a ship and it sways back and forth there's a driving force in the pendulum's reference frame and so I have to properly consider any driving force and I may have... well... a driven pendulum. So similar differential equations work if I have a simple pendulum, a damped pendulum, a driven pendulum or maybe a combination... but I had to decide what I was simulating. I could also go back to Newton's laws for example and solve the equations a motion for a planet orbiting the sun and simulate what would happen if say the universal gravitational constant changed or some other physical parameter in this situation were to vary. I can consult charge and current conservation rules and simulate an electric circuit, calculating how much current flows through a given resistor as a function of time. And it's not just limited to pure physics. I could simulate a chemical reaction such as burning or condensation. I could simulate the number of cells in a growing embryo or in a tumor. I could simulate how much fluid flows through a pipe and how much flow loss is incurred by sediment collecting on the pipe's inner wall. I may be interested in simulating how a parking lot fills up or how people interact, so then I need to figure out how to represent likes, dislikes, trends for choices, fights, and relationships with formulae and ultimately numbers. Simulations are very useful and much can indeed be gained by doing them but sometimes what we do on the computer has to be even more involved than a simulation if we want to mimic and better understand real systems.

A main goal of this text is to entice you and draw you in to computer simulation as well as modeling and to facilitate your proficiency and constructing physical models. A computer model is much deeper than a simulation, because a model places the physical system in context with many diverse external factors. Models are often characterized by a host of open-ended questions they raise. So, in large part the success of any model lies in its ability to incorporate enough features from its environment to be relevant but not too many features to be unsolvable.

After all, since nothing in life is isolated the models mustn't be either; we often say that good science doesn't happen in a vacuum. (Note: household appliances are very interesting to simulate and they can be reduced to combinations of essential physical elements.) So let's *swing back to* the pendulum simulation previously discussed as a springboard to talk about what it would look like as a model. As a model, the question would not be about simulating a straight up pendulum but could look something like this: let's simulate a grandfather clock, or maybe a tree blowing in the wind or perhaps a tall building with a counterbalance to stabilize it as it sways from the wind hitting it. You probably can already imagine how many different fields of science, history, politics, and society to come into play. For the grandfather clock we might want to know the size of the case because it can limit the pendulum's motion. We may want to know the age of the clock because that could tell us qualities of the spring and the hands, which may be important. We'll need to know exactly how the pendulum is constructed and shaped—whether it is solid or hollow—whether it is more uniform or very lopsided in its mass distribution, so the manufacture will be very useful there. We'll need to know if it is a military clock. As far as the tree blowing in the wind we'll need to know various aspects of biology and botany. For example, what is the size and material strength of the branch that's in the wind? The season of year will tell us about air properties, as well if we have a lot of ballast or not on the tree, along with its elastic properties. Is it in an orchard where workers pick fruit? As for the building with a counterbalance, what are its dimensions? Do the terrorist attacks of September 11, 2001, influence its construction? Should the structure be something stable enough for a plane to fly through it? Are we using material in the World Trade Center sites? What other construction has gone up around the building site? Are there railroads, bus stations or airports nearby or other things that could affect vibration? Are there natural fault lines that could cause quaking? Such considerations would have to go into a good model. So I can simulate a car moving on a slippery surface but can I intuitively think of importing elements in a model of an accident? What would be needed? We can brainstorm here. First off, I would need to know the size of the car. This is relevant too because if the car is hit off-center rotational motion will take place. Also, you need to have information on the kind of tires because then I can have an idea of the friction properties between the tires and the road, as well as how they displace water and work in rain. I then need to know the positioning of the passengers in the car to help me get an idea of the center of mass, which will in turn tell me how readily the car rotates given where it was hit. I clearly need to know the weather conditions on the day of the accident. Then I'll be able to get an even better idea of the properties of friction between the car and the road. It would also be nice to know the timing of the placement of stoplights in the vicinity of the crash as well as speed limit signs. Such things can help me account for driver reaction times and distractions. And if any of the drivers make unsubstantiated claims then I need to know perhaps the time of day or if there was a football game in town, or maybe if schools were in session because then I can rely on statistics and trends to predict what a likely scenario is. So, interestingly enough, when we move from a simple pendulum to things that are relevant and contextual in our lives, we are now modeling. Your ability to incorporate ideas from many different and seemingly unrelated disciplines will determine your success with constructing, running, and interpreting results from computer models.

1.7 What Types of Models are Most Important for Everyday Things?

There are many kinds of computer simulations, and a quick Internet search will yield a list with copious amounts of examples (it's a suggested homework problem!). I want this text to facilitate your constructing, running, and interpreting the results of computer simulations and models of things we encounter in the everyday world. With that purpose in mind, there are mainly two large categories of computer simulations and models to be discussed here. The good news is that since the models are taken from things and processes in real life, their algorithms reflect real life processes; they are mostly fairly intuitive and have components with real analogs.

The first general category of computer models discussed here are probably the most applicable: *deterministic models*. Such algorithms are a sequence (usually in time) where the initial conditions are specified and then each successive step depends in a predetermined way on the previous one. And as you might imagine, deterministic models have many interesting subcategories that can be relevant for our endeavors here. The first subcategory includes *continuous deterministic models*. In such models, important variables of the system are incorporated into the program and then it is advanced in simulated time. They are classified as continuous because the simulated time flows at regular intervals from start to finish and there are no events taking place in the physical system that are introduced by the algorithm that happen instantaneously. In other words, it's just a scaled time sequence... a compressed or expanded version of some physically real thing or process.

For students and people learning how to code and construct models, *classical deterministic simulations* are by far the easiest to learn and the most intuitive. *Classical Mechanics* was formally developed by Newton but DaVinci and Galileo gave the field a great head start. It is a field that deals with everyday types of phenomena such as motion, sound, heat, oscillations, and so much more. There are copious examples of deterministic models relevant to the work presented in this book, including sports, traffic jams, waves in musical instruments, haemodynamics, heat transfer, planetary motion, and so many others.

Another kind of deterministic model is termed as a *discrete event* model. In discrete event models, there are events that happen at regular or random intervals that are instantaneous which we cannot describe as existing on a continuum. Moreover, the algorithm of the program using the simulation determines where the events take place. An example of a discrete event model would be an individual traveling on the road and having to make a choice as to which way to go at a fork in the road. Another is the Eden model of cell growth for metastatic cancer. The algorithm for the Eden model is fairly straightforward: select a site for a cell to appear and then randomly select an adjoining site for the next cell to appear. You can include effects of oxygen, nutrition and other things that affect cell growth. Researchers have used the model with some success in two and three dimensions.

In both discrete and continuous deterministic models, time is advanced in the simulation (unless it is reversed in the case of backward time simulations) in small steps or intervals. The intervals can be uniform or can vary. As we will see in the examples given in this book, time intervals appear in deterministic computer models because time derivatives from calculus have to be broken apart and approximated as ratios of differences and the equations then manipulated so as to be updated with time. Space derivatives can also appear in differential equations, and computer models can be categorized by how space is partitioned. If it is broken up into uniform segments and subsequently used for derivatives, it's a *finite difference* simulation. An example of a finite difference simulation would be if you have an aerosol contaminant at the workplace and need to simulate how it infuses the office building. There's likely no spot in the simulation where we need an increased number of the grid points because it is an area of particular interest, and so the office space is broken up into the grid with uniform size blocks and then the diffusion equation is solved on the grid. If however, space is broken up into a nonuniform mesh, then each partition or element in the simulation carries with it various physical properties and the simulation is called a *finite element* model. Suppose you're studying ground water flow in a particular region. Most of the spatial changes, that is the steepest derivatives (slope of the water table, in the case of open aquifers) happen near the wells whether they are for pumping or injection. Since you have regions in the model where derivatives are much sharper than in other places, it is wiser to have finer meshes near the steep points. Hence ,space would be partitioned nonuniformly so there would be a larger number of smaller sized elements near the parts of the model where most of the action is physically.

Deterministic simulations can also be categorized by how they track the particles involved in the model. *Lagrangian* techniques involve actually tracking the trajectory of particles in the simulation through time. In contrast, *Eulerian* methods are used to solve the physics (in the form of various discretized equations) on a background grid and to look at a snapshot of the behavior of all the particles at one time. For example, consider a fluid mechanics model. Such models are useful in airplane wing design, automobile design, building design, and in far less practical and theoretical areas such as magnetohydrodynamics and astrophysics. In such cases, the Lagrangian technique would be used to track particles that represent the physical system as they flowed around barriers and encountered boundaries, but the Eulerian technique would look at what are called the streamlines of the fluid—a snapshot of the flow at a given time. Engineers often use wind tunnels to study the aerodynamic behavior of various manufactured objects, and in such cases streamlines are directly accessible in the lab.

There are also interesting simulations which fall into the class of *deterministic chaos*. We encounter chaos every day, and not just when it feels like we're going crazy. Chaos is actually a physical concept, and it means sensitivity to initial conditions. So, a system can be deterministic in that all physical rules apply and it evolves in a definite, prescribed way as time advances, but where it ends up is very hard to predict—and highly dependent on—how it started out. That is, changes in how it ends is almost impossible to predict, given changes in where it started. A good example of a chaotic system can be found in a place I spent long hours in high school and college, intently studying physics: the pool hall. In a pool game, there are six possible outcomes for the pocket a ball can land in. So consider a lone queue ball on the table. (Note: later you will see that this is quite a dry joke for UNIX users). If you hit it with enough energy it will eventually land in a pocket. Now consider very slightly changing the angle you strike the ball at. Given enough energy it may land in a different pocket, so the outcome is dramatically different for a very small change in the initial conditions. Another example of a chaotic system is a pendulum. When a pendulum swings back and forth with low energy it does just that; it swings very well. Now imagine cargo on a ship that is able to swing much like a pendulum. As this ship sways back and forth, a driving force is experienced by the pendulum. If the energy of the system is low the pendulum swings regularly but when the pendulum clears the top of its reach then it behaves as a chaotic system, ending up doing things dramatically differently depending on slight changes in initial conditions. There are many ways to analyze the behavior of chaotic systems and computer simulations can afford much insight into them. In fact, one can calculate driving forces and damping that will help the cargo not swing so much.

So the computer models we're talking about here in this text are motivated by real-life things. Even though deterministic models and simulations track with our linear intuition, there's another type of simulation that is very useful in many situations: *stochastic simulations*. Stochastic systems are those that exhibit or are driven by random processes. There can be many types of random events to simulate. For example, when looking at flipping a penny, or leaves falling from a tree, or gas atoms and molecules hitting the side of a wall to create pressure, events are random but occur with roughly equal probabilities. In such cases, we can learn a lot by taking equally weighted averages within the simulation much like when you need to figure out a grade point average (GPA) when your classes all have the same credit hours. Some systems, however, are biased towards lower energy states and spend more time expressing them. The power of Monte Carlo methods is to allow unequally weighted averaging easily, much like when you have a one credit and a four credit class and need your GPA. The Monte Carlo method is named after the famous gambling capital [10,11] and in fact had its origins in the first computer simulations of neutron scattering in World War II. Although developed by many individuals, Metropolis, Rosenbluth, and Rosenbluth are widely credited as being the originators of Monte Carlo simulations [11]. In Monte Carlo simulations, the system is not stepped through real time as it is for deterministic models. Instead, the system is changed slightly during the program and then the likelihood of the system taking on that particular change is calculated. Normally, the likelihood is calculated by seeing how an exponential function involving the change in energy of the system compares to some random number between zero and one; the method is discussed later in the advanced portion of this text. So instead of a time sequence, we're generating what is called a Markov chain of the most likely configurations the system is going to take, where successive frames are not linked as strongly as they would be in deterministic simulations. So what is done with the configurations we get? Actually we calculate maybe their energy or maybe accumulate some distribution function, and then average directly over the configurations or distributions... and it's great because it's already a weighted average based on how the configurations were selected. As you might imagine, since it's not a direct time sequence, Monte Carlo methods are not the best choice when looking at vibrational modes, fracture, or any kind of dynamics but are wonderful, for many reasons we will see later, at calculating equilibrium properties for systems not undergoing significant changes with time.

There are many other kinds of simulations and models used in physics, engineering, mathematics, and industry but it is the representation of real things—everyday things—that really shine through in stochastic and deterministic simulations.

1.8 When Do You Build Your Own Tools and When are Black Boxes the Best?

My desire is to not only show you the value of simulating things but to also entice you to cultivate a passion for actually thinking the model through and constructing a successful simulation. Therefore my philosophical point of view—in fact the purpose of this text—would prompt me to argue against your using black boxes. Certainly writing your own code is the more enriching endeavor in terms of learning, thinking the model through, and considering what you need to calculate. And what's more, when bugs happen—not if—but *when* bugs happen,it will be far easier for you to manipulate the code and to know what part to go to for fixing and debugging. Yet, much to my chagrin, the words sang by Kenny Rogers ring true: "you've got to know when to hold 'em and know when to fold 'em." Certainly, just like we don't build our tools before we work on cars, there must be times when using black boxes is the best decision; there must be situations where we have to use somebody else's code or computer model to get something done. How do we discern those times?

One situation that arises favoring black boxes is in the courtroom. Let's say you're doing a bullet trajectory model for a crime, and your models are going to help decide whether a person goes to prison for 35 years or walks free. The model you employ has to be accepted as valid; it then holds scientific as well as legal credence. In such instances, codes personally written are often not accepted in courts but commercial packages (often times obtained as freeware) are provided by code developers for such purposes. After all, credit should be given where it is due and people have made their careers by writing, checking, and optimizing various simulation tools. I may be able to do physics but it doesn't make sense for me to think that I could outperform dedicated groups. In addition, you could be writing your own code and you want to check it. Although it may be useful to validate your code against experimental data there are many situations where it is much more practical and prudent to validate your model against other models. A perfect example of this that comes to mind is oil flow in a field. In calculating subsurface liquid flow of any kind (oil, water, fluid contaminants), one has to know details of the hydrogeologic properties of the soil and surface as well as subsurface media. Even if such properties are known, in many real-life cases the properties are not uniform because there's rocks, inclusions, lenses, fractures, and many things that alter the flow in irregular ways. So it is highly unlikely that any subsurface fluid flow model would reproduce what's happening in the field. However, the model being used must agree with other such models or even analytical solutions in entirety or limiting cases, as appropriate. So if you're writing your own code, having a black box to validate against is very useful. Another situation involves optimized code. Suppose you're working on a project or trying to solve a problem that contains very large models, as you might encounter in a weather simulation. In such situations, runs may take on the order of months or years to complete. Even though it is very good for you to be able to program and understand the model, there may be other tools available that have been optimized and run much faster and more efficiently. What's more, you really wouldn't lose out so much because you still have to write the input and you have to analyze the output, both of which efforts are very specialized and can often involve as much thought as constructing the simulation itself. Finally, you might find yourself at a job where, say the electric system of a specific tractor or car needs to be simulated before being built and the company you work for has the software all ready to go. It is a good decision to go forward, utilizing their software as opposed to trying to write your own. Granted, after some time you might want to make preprocessors or postprocessors; however, for specific applications in industry that have been developed by teams already familiar with the products being manufactured, black boxes can prove to be an invaluable tool and starting point.

PROBLEMS

1.1. It is likely that between the time this book was written and the time it was published, the state-of-the-art of computing has changed. Conduct an Internet search of the extremes in computing and computer simulation and modeling: what are the biggest, fastest, longest, etc. simulations around?

1.2. Problem 1.1 dealt with extremes but now look for trends in simulation. Search around for hot spots and areas of vigor in the field. Can you identify any patterns or trends?

1.3. Locate yourself with respect to Moore's Law (Figure 1.8). Problem 1.1 dealt with extremes but now search around for the state of the art in computing. What is an average processor speed of a PC you could buy at the store?

1.4. It is very likely that, by the time you are using this book, Moore's Law will be extended past 2015. Conduct an Internet search and modify all three lines in Figure 1.8 so it is current.

1.5. Even though it is likely that you will mainly need deterministic and stochastic models and simulations, conduct an Internet search for the various kinds that are out there, from mundane to exotic. Identify types that pique your interest and also that you might use for a project or solution to a problem.

1.6. Consider a computer simulation that you might envision being involved in. Discuss the model and the assumptions you would make in constructing it. With what things is it accurate and where does accuracy fail? How will you determine what things matter and which don't?

1.7. The term "bug" has a quite physical origin. Look it up, and comment as to how such an intrusion could hamper computing today. What other physical events could hamper your computing and how could you either circumvent them or proactively prepare?

References

1. L. Firlej, B. Kuchta, M.W. Roth, M.J. Connolly, C. Wexler, Structural and phase properties of tetracosane (C24H50) monolayers adsorbed on graphite: An explicit hydrogen molecular dynamics study, *Langmuir*, 24(21), 12392–12397, 2008.

2. A. Turing, On computable numbers, with an application to the Entscheidungsproblem, *Proceedings of the London Mathematical Society*, Series 2, 42, 230–265, 1936.

3. A. Turing, Machine intelligence, in Copeland, B.J, *The Essential Turing: The Ideas that Gave Birth to the Computer Age*, Oxford University Press, Oxford, ISBN 0-19-825080-0, 1948.

4. A. Turing, Computing machinery and intelligence, *Mind LIX* (236), 433–460, October 1950. doi:10.1093/mind/LIX.236.433, ISSN 0026-4423, http://loebner.net/Prizef/TuringArticle.html, retrieved 2008-08-18

5. A. Turing, Can automatic calculating machines be said to think?, in Copeland, B.J, *The Essential Turing: The Ideas that Gave Birth to the Computer Age*, Oxford University Press, Oxford, ISBN 0-19-825080-0, 1952.

6. M. Sipser, Section 4.2: The halting problem, in *Introduction to the Theory of Computation* (Second edition). Mac Mendelsohn (ed.), Boston, Massachusetts, PWS Publishing, 173–182. ISBN 053494728X, 2006. https://s3.amazonaws.com/academia.edu. documents/35002661/_Sipser__2006__Introduction_to_the_Theory_of_Computation__ Second_Edition.pdf?AWSAccessKeyId=AKIAIWOWYYGZ2Y53UL3A&Expires=15 14477989&Signature=A56jkMg5cQoF6zu9S0QHpVCtl%2F0%3D&response-content-disposition=inline%3B%20filename%3DINTRODUCTION_TO_THE_THEORY_OF_ COMPUTATIO.pdf

7. John Walker, http://www.fourmilab.ch/babbage/contents.html

8. Babbage's Calculating Engines Being a Collection of Papers Relating to Them; Their History, and Construction, Charles Babbage Edited by Henry P. Babbage, Cambridge University Press, 1889, 2010, 2012, Cambridge, NY https://monoskop.org/images/4/40/ Babbage_Charles_Calculating_Engines.pdf.

9. P. Henry, Babbage's analytical engine, *Monthly Notices of the Royal Astronomical Society*, 70, 517–526, 645, 1910 [Errata] (1910), describing his construction of a portion of the Mill and Printing Apparatus, used to compute a table of multiples of Pi.

10. N. Metropolis, S. Ulam, The Monte Carlo method, *J. Am. Statist. Assoc.*, 44, 335–341, 1949.

11. N. Metropolis, A. W. Rosenbluth, M. N. Rosenbluth, A. H. Teller, E. Teller, Equation of state calculation by fast computing machines, *Journal of Chemical Physics*, 21(6), 1087–1092, 1953.

12. G. E. Moore, Cramming more components onto integrated circuits, *Electronics Magazine*, 38(8), 114–117, 1965.

13. http://iter.rma.ac.be/en/community/Worldwide/index.php

14. http://www.nr.com/oldverswitcher.html

CHAPTER 2

///

Getting to Know the Neighborhood

2.1 Overview

In scientific computing, there are many ways to accomplish your goals and I strongly encourage you to develop your own set of skills and procedures. What I will be presenting in this chapter is an introduction to some commonly used techniques, filtered through my eyes. In the end, you will be doing your own thing but it is my hope that what I offer here will help you piece it all together. We start from square one and walk together through a beginning session with UNIX. After that, there is a list of Visual editor (Vi) commands, as opposed to a session. I organized things in such a way because I feel that a session of random edit commands would have the same effect as a conversation with your friend and each sentence you speak is about a different topic. There are some things not even coffee can fix but if you want a session you certainly can make one out of the commands. After the operating system and editing tools we then dive into C++ coding, compiling and execution, both correct and incorrect. After that, I will try to show you that working with another machine, language, or operating system other than discussed here is fairly straightforward so long as you have a basic skill set that can be mapped from one to another.

In addition to understanding computing tools in the neighborhood, we also have to make a stop regarding the current state of work in the scientific community. Specifically, it is very important for you to understand where your project sits in the context of what has already been done as well as the hot topics of what people are doing. Viewing your work in broader context can help you understand the history of its development and almost certainly will inform the construction of your own simulations. Even for the most independent researcher or student, a reality check as far as where your work stands is a good thing.

2.2 The UNIX Operating System

If you look on the Internet, at the library, or in a standard computer science textbook, you will find copious amounts of information on the timeline and development of the UNIX operating system. (References 1 and 2 for example) In summary, it seems to support the notion that scientific discovery and technological advances are driven more by out of the way ideas rather than driving purpose coming to fruition at the end. The text you are reading now is not intended to be encyclopedic, but rather to be relevant in getting you to the point where you can do nice modeling and simulation. Hence I don't cover here what you can easily find elsewhere (wow, a good Facebook quote right there) and I suggest that, in your spare time, you check out some background information about UNIX.

UNIX is the system of choice for scientific computing applications, which is not surprising knowing that it started out at Bell Laboratories about 1969. Of course, you don't have to be running UNIX and have a supercomputer in your basement to do nice simulations, but if you're going into any aspect of the professional world it is very useful to have dabbled in it. As a scientific tool, UNIX comes equipped with a range of compilers and file editors that are designed to interface with a large assortment of packages and languages commonly seen in scientific settings.

Especially if you have little or no experience with the UNIX operating system, working through the next section of this book is likely a very good idea. We start with how to communicate with the UNIX machine and log on, and then progress through a session illustrating important basic UNIX tools. This text is intended to be an introductory treatment of UNIX but references exist for more extensive treatments. (References 3 and 4 for example)

2.2.1 Remote UNIX Machines: Logging On and File Transfer

It's hard to dance with nobody leading; communication with a UNIX machine is of fundamental importance. You can "hit up" a UNIX machine from almost anywhere... your home, a favorite coffee shop, or even a hospital waiting room; all you need is an account on the machine with your *username* and *password*. The machine can be a mainframe that a university or corporation houses or could simply be your computer at home running LINUX. Wow, so if it's your own machine, why study remote access? Isn't that like sending snail mail to your roommate? It's because I want you to learn to be able to access your machine from anywhere and so long as you have an IP address and your network is correctly configured it will work well. A common interfacing freeware tool is PuTTY [5]. If you install it on your local computer and then run it (Figure 2.1), you can ask for a "terminal" screen to appear like that in Figure 2.2 and you can enter all the required information to connect. The Host name is the name of the remote UNIX machine and may look something like *mainframe.physics.uni. edu*. You can also enter your *username* and subsequently you will be prompted for your *password*. If all has gone well you will find yourself in the home directory of your account on the remote UNIX machine. If you don't like the PuTTY color settings offered by default you can change them by double clicking on the terminal icon in the upper left hand corner and exploring various options. It is often the case that you will need to transfer files between local and remote machines, and one tool (of many) that can help you accomplish file transfer is SSH (Secure SHell) [6] which will call up into a window as shown in Figure 2.3. The login procedure is very similar to that of PuTTY and, after having logged in with SSH you can transfer files and folders to and from your desktop. Within SSH you can also call up a terminal having a slightly different flavor from PuTTY.

2.2.2 UNIX Shells

The *shell* "makes its living" interfacing between you and the computer, interpreting the commands you issue (typing the command and then <return>). There are several different types of shells in UNIX but the most widely used is the Bourne Shell, designated by the path /bin/sh. When you log on to your UNIX account you will be in your home directory and will see a prompt, usually a % or $. Prompts are very customizable and, even though a cosmetic issue, adjusting yours could provide you with some fun experience when we get to file editing. So the first thing you should do in the shell upon the first login on a new account is change your password. It could be that if your account is new it is likely that somebody has recently seen the new password that was assigned to you and you want to offer no room for password compromise. To change a password, issue the command "passwd" and follow the on-screen commands for changing your password.

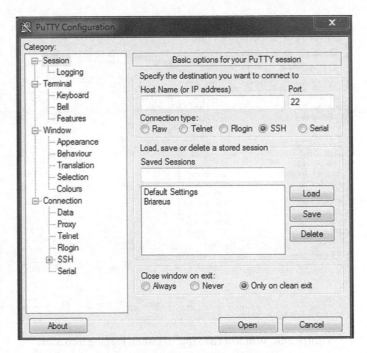

Figure 2.1. PuTTY front-end screen.

Figure 2.2. The terminal screen that appears when logging into PuTTY.

2.2.3 Directories

Upon logging on, the shell should start you in your home directory, which is space dedicated to you on the UNIX system for your files. For example after I log on to *mainframe.physics.uni.edu* I see a prompt that looks like this:

```
roth@mainframe:~$
```

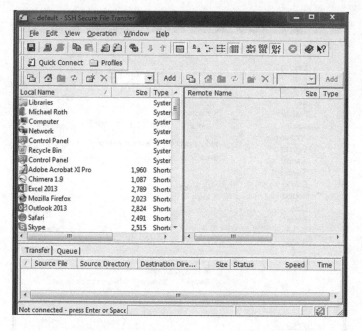

The "$" designates whatever command line prompt you have, which may be some other symbol that makes cents (wow).

You can find out the name of your current working directory by issuing *pwd (print working directory)*; The response that follows on the next line will be the name of your home directory, where the name following the last slash should be your *username.* I get

```
/user/roth
```

The directory structure can be conceptualized as an inverted tree. From your home directory, create a new subdirectory named "coffee" by issuing *mkdir coffee (make directory coffee).* To check to see that you successfully created the directory, issue *ls.* I get

```
roth@mainframe:~$ coffee
```

Figure 2.3. Window that appears when logging into SSH.

Some systems will show directories and various files in different colors and, like the command prompt such things are customizable and following through could prove to be a very good exercise.

There are many different options for commands in UNIX and you can access the system Manual by...well...issuing the manual command, which is *man command (manual lookup for command with name "command").* The computer responds with the name of the command, synopsis, description and syntax for various command options.

The space bar can be used to advance through *man* results when there is more than one page, and typing "*q*" gets you out and back into the Bourne Shell.

Later in this section we will be going through directory and file removal commands. Please be exceedingly careful with any remove and delete commands in UNIX. Unlike in DOS and Windows, many UNIX systems do not have undelete and recovery options available.

Since coffee should never remain unused, let's move into the directory we just created by issuing *cd coffee (change directory into coffee).* The prompt will tell us if we were successful in the directory change

```
roth@mainframe:~/coffee$
```

and we can also issue pwd:

```
/user/roth/coffee
```

No matter where in the directory structure you are, you can always get back to your home directory by issuing "*cd*" or "cd ∼":

```
% cd
```

or

```
%cd ∼
```

If you returned to your home directory, then go back to the *coffee* directory now.

2.2.4 Files

Files live within directories, but since you just created the directory, nothing will be listed because the directory is empty. Create your first file using the Vi text editor by issuing *Vi hazelnut (visual editor hazelnut). Vi* can be used to enter or change any kind of text, letter, essay, report, or program. We'll cover Vi in detail in the next section. For now simply type "*i*" to enter a typing mode. Then type a sentence; I suggest "*I will simulate many cool physical systems in the near future.*" Then press "Esc" followed by "ZZ" to save and quit out of *Vi.* Yes, "quit out" is being used as a verb meaning to quit and exit a particular computer application of *Vi.* Now when you list your files, you will see file "hazelnut" listed. You can view a text file by issuing cat *filename (concatenate file with name filename)*, and issuing cat hazelnut gives me

```
I will simulate many cool physical systems in the near future.
roth@mainframe:∼/coffee$
```

You can also use *cat* to display multiple files together on the screen. If you have a file that is longer than your 24-line console window, use instead "*more*" to list one page at a time or "*less*" to scroll the file down and up with the arrow keys. The above commands work well with ASCII (American Standard Code for Information Interchange) files but not binary ones, in "machine language." If you try to display binary (nontext) files on your console—the attempt to print the nonprintable control characters might alter your console settings and render the console unusable.

Now copy file "hazelnut" to "darkroast" by issuing *cp hazelnut darkroast (copy file "hazelnut" to file "darkroast").* The *cp* command works with any two file names with acceptable names: *cp file1 file2.* By doing this you have created a new file named "*darkroast*" which is a duplicate of file "*hazelnut*". Issuing the file listing command *ls* reveals the presence of both files:

```
darkroast hazelnut
roth@mainframe:∼/coffee$
```

Now rename the file "*darkroast*" to "*mocha*" by issuing *mv darkroast mocha (move file "darkroast" to file "mocha"),* which is a special case of the *mv* command *mv file1 file2.*

Listing the files still shows two files because you haven't created a new file.

```
hazelnut mocha
roth@mainframe:~/coffee$
```

If you "*cat*" the second file, you'll see the same sentence as in your first file.

```
I will simulate many cool physical systems in the near future.
roth@mainframe:~/coffee$
```

Renaming a file is just a special case of moving a file and keeping the result in the same directory as the original file. The *mv* command allows you to actually move files and not just rename them. Let's explore moving files by issuing the following commands:

```
mkdir wholebean (make directory "wholebean")
mv mocha wholebean (move file "mocha" into directory "wholebean")
ls wholebean lists the contents of the directory wholebean.
```

You can list even more information about files by using the "*−l*" option when issuing "*ls*": Issuing `ls  -l` *(long list)* gives me

```
total 8
-rw-r--r-- 1 roth roth 63 Aug 6 23:45 hazelnut
drwxr-xr-x 2 roth roth 4096 Aug 7 09:47 wholebean
```

The listings are of the general form

```
drwxrwxr-x 2 username group 512 May 22 17:11 some_name
```

where "*username*" will be your username and "group" will be your group name. Among other things, *ls −l* lists the creation date and time, file access permissions, and file size in bytes. The letter "d" (the first character on the line) indicates the directory names.

Next issue *cd wholebean*. This changes your current working directory to the "sub" subdirectory and *pwd* confirms your new location:

```
/user/roth/coffee/wholebean
```

Issuing "*cd..*" backs you up one directory and *pwd* again confirms the change /user/ roth/coffee

The ".." always refers to the parent directory of the given subdirectory. Finally, clean up the duplicate files by removing the "roast" file and the "sub" subdirectory *rm wholebean/mocha* and then removing the subdirectory itself. *rmdir wholebean*. The rmdir command works for empty directories but in other cases (as well as for empty ones) you can also remove it by issuing *rm −f −r directory_name (force a recursive removal of directory with name directory_name).*

When referring to files you can use the absolute path or a relative path name to the file: you can also use the absolute path "*/user/roth/coffee/wholebean*" or "*~/coffee/ wholebean*" perhaps a relative path "*../wholebean*" if you were in some other subdirectory of *coffee*.

2.2.5 Commands between Machines and File Transfer: *ssh* and *sftp*

Here we will discuss communication and file transfer between two UNIX machines. Logging on to a remote machine (say, at a university or business) from your UNIX account can typically be achieved using the "ssh" command *ssh username@ remotemachine.department.university.edu* after which you will be prompted for your password.

Transfer files between machines utilizes "sftp". For example, to copy file "myfile" from the remote machine named "remotemachine", the command would be *get myfile* with the response

```
myfile 100% 2564 106.8KB/s 00:00
```

To put "myfile" on to the remote machine you would issue *put myfile*

```
myfile 100% 2564 302.7KB/s 00:00
```

To get out of sftp, just issue *"exit"*.

When you are in an s ftp session, typing an *"l"* (the letter) before a command directs it to the local host and issuing regular commands deal with the remote host. So *"ls"* lists the files in the home directory of the remote host, *"cd"* changes directories in the remote host and *"lls"* and *"lcd"* accomplish the same things locally.

2.2.6 Other Useful Commands

The current date and time are printed with the following command:

date (what's the date)
Sun Aug 07 17:39:04 CDT 2011

You can find out who is logged on by issuing *who (who's on)*:

```
roth pts/1 2011-08-07 14:48 (vpn078.uni.edu)
```

To determine how long the machine has been up, issue *uptime*:

```
14:54:30 up 88 days, 21 min
```

To find out which of your processes are running, issue *ps–u username*:

```
ps-u username
ps PID TTY TIME CMD
27638 ? 00:00:00 sshd
27639 pts/1 00:00:00 bash
27664 pts/1 00:00:00
```

To see who is running what processes on the system issue *top*:

```
top - 14:54:30 up 88 days, 21 min, 1 user, load average: 1.01, 1.01, 1.00
Tasks: 135 total, 2 running, 133 sleeping, 0 stopped, 0 zombie
Cpu(s): 24.8%us, 0.2%sy, 0.0%ni, 74.8%id, 0.2%wa, 0.0%hi, 0.0%si, 0.0%st
Mem: 2060528k total, 1386692k used, 673836k free, 278916k buffers
Swap: 3229024k total, 1028k used, 3227996k free, 720704k cached
```

```
PID USER PR NI VIRT RES SHR S %CPU %MEM TIME+ COMMAND
27691 roth 20 0 10864 1200 884 R 0 0.1 0:00.04 top
```

A summary of UNIX shell commands is shown in Table 2.1.

2.2.7 More Advanced Commands

Issuing *locate textstring* will return the full path where the specified text string *textstring* appears in a directory or file name.

The *grep textstring* command (from *global/regular expression/print*) is a very powerful text search utility that displays full paths and files containing the specified text string. Many people think affectionately of *grep* as "get repetition".

sed (stream editor) is a text transformation utility and is so powerful that entire tutorials are devoted to it. I suggest that you look up some uses of *sed* on the Internet and try it out. *awk* combines the functionalities of *grep* and *sed* and is a product of the surnames of its inventors. As with *sed* I strongly suggest that you look up some information about it and try it on for size. Please follow through if for no other reason, because I sed so. A summary of miscellaneous and advanced UNIX commands is shown in Table 2.2.

Table 2.1. A List of Basic UNIX Shell Commands for Your Reference

% vi myfile	*Text edit file "myfile"*
% ls	*List files in current directory*
% ls —l	*Long format listing*
% cat myfile	*View contents of text file "myfile"*
% more myfile	*Paged viewing of text file "myfile"*
% less myfile	*Scroll through text file "myfile"*
% cp srcfile destfile	*Copy file "srcfile" to new file "destfile"*
% mv oldname newname	*Rename (or move) file "oldname" to "newname"*
% rm myfile	*Remove file "myfile"*
% mkdir subdir	*Make new directory "subdir"*
% cd subdir	*Change current working directory to "subdir"*
% rmdir subdir	*Remove (empty) directory "subdir"*
% rm —f —r subdir	*Force a recursive removal of directory subdir*
% pwd	*Display current working directory*
% date	*Display current date and time of day*
% who	*Who's on?*
% man command	*Display man page for "command"*
% man —k "topic"	*Search manual pages for "topic"*
% exit	*Exit a terminal window*
% logout	*Logout of a console session*

Source: http://www.ks.uiuc.edu/Training/Tutorials/Reference/unixprimer.html.

Table 2.2. A List of Miscellaneous and Advanced UNIX Commands for Your Reference

% date	*Text edit file "myfile"*
% who	*List files in current directory*
% top	*Long format listing*
% uptime	*View contents of text file "myfile"*
% more myfile	*Paged viewing of text file "myfile"*
% less myfile	*Scroll through text file "myfile"*
% cp srcfile destfile	*Copy file "srcfile" to new file "destfile"*
% mv oldname newname	*Rename (or move) file "oldname" to "newname"*
% rm myfile	*Remove file "myfile"*
% mkdir subdir	*Make new directory "subdir"*
% cd subdir	*Change current working directory to "subdir"*
% rmdir subdir	*Remove (empty) directory "subdir"*
% rm -f -r subdir	*Force a recursive removal of directory subdir*
% pwd	*Display current working directory*
% date	*Display current date and time of day*
% who	*Who's on?*
% man command	*Display man page for "command"*
% man -k "topic"	*Search manual pages for "topic"*
% exit	*Exit a terminal window*
% logout	*Logout of a console session*

Source: http://www.ks.uiuc.edu/Training/Tutorials/Reference/unixprimer.html.

2.2.8 Online Help

You can get online help from the "man" pages ("*man*" is short for "*manual*"). The information is terse, but generally comprehensive, so it provides a nice reminder if you have forgotten the syntax of a particular command or need to know the full list of options. Display the man page for "*cp*" by issuing: *man cp*.

You may scroll down by pressing "enter" and you may exit the manual by typing "q".

2.2.9 Logging Out

It is very important to log out of your account whenever you are done using it, especially if you are on a public machine. To close a shell window in a graphical environment, you can type:

exit

Logging out from a graphical environment will require clicking on the appropriate icon. Logging out of a remote session can be done by either using the "*exit*" command or by typing "*logout*".

2.3 The Vi Editor

2.3.1 Starting the Vi Editor

I view the introduction of the Vi editor to be slightly different form the introduction to UNIX in the previous section, and so here I do not offer a session but rather a "get 'er done" approach accompanied by getting right into the commands. I encourage you to play around with the commands mentioned here and to begin to get a feeling for which ones you might find regularly useful. Issuing "*vi filename*" will initiate the process of editing or creating a file, and issuing "*view filename*" will open the file in a read-only mode.

"*:set all*" displays various options, and I strongly encourage you to explore the list of those available for your particular Vi session. For example, issuing "*:set nu*" will display line numbers on the screen, and its compliment "*:set nonu*" suppresses them.

2.3.2 A Couple "in a bind" Commands

The following commands in Table 2.3 can be very useful when undesirable things happen that you would like to undo. And could potentially save you a considerable amount of time and work.

Table 2.3. "In a Bind" Commands that are Good to Have Handy

vi -r "filename"	*Retrieves the latest version of a file before a system crash and opens it in vi*
u	*Undoes the latest command*

Source: http://www.ks.uiuc.edu/Training/Tutorials/ Reference/unixprimer.html.

Table 2.4. A Summary of Character Move Commands in Vi

h	*One space to the left*
j	*One line down*
k	*One line up*
l	*One space right*
w	*One word right*
B, b	*One word back*
e	*End of the current word*
E	*End of word*

Source: http://www.ks.uiuc.edu/Training/Tutorials/ Reference/unixprimer.html.

2.3.3 Moving Around in Vi

It is very important to know how to move around in Vi and it can be accomplished at several different levels. Hence, it seems to make sense to start talking about the smallest moves that can be made and to increase successively in levels from there. Table 2.4 summarizes how to move around by one character at a time in Vi.

If moving around by characters is not sufficient, then you can also jump around by lines, columns and pages using the commands in Table 2.5 and even by screens and windows as shown in Table 2.6.

So far, I have gone through how you can relocate the cursor, or "move around" in Vi, but as you might imagine, we actually have to use the tool as a text editor. As such, it is useful to be able to move around by *text structures*, not just lines of code or characters. Table 2.7 contains some commands that are useful for jumping around based on sentences and paragraphs.

Once you are proficient at moving around by text structures, you can actually search for characters and strings to as shown in Table 2.8.

Now that you can move around proficiently and locate text strings, you can add text, replace, and change text, delete text as well as move and join text; relevant commands are contained in Tables 2.9 through 2.12 respectively.

Table 2.5. A Summary of Moving Around by Lines, Columns, and Pages

nG	Go to the nth line of the file (G goes to first line)
0	Go to beginning of the current line
$	Go to end of the current line
^d	Down half a page
^u	Up half a page
^B	Backwards one page (^nB for n pages)
^F	Forwards one page (^nF for n pages)
n|	Column specified by count
^E	Scroll forwards one line (^nE for n lines)
^Y	Scroll backwards one line (^nY for n lines)
^M	First character on next line

Source: http://www.ks.uiuc.edu/Training/Tutorials/ Reference/unixprimer.html.

Table 2.6. A Summary of Moving Around by Screens and Windows

^f	Forward one screen
^b	Backward one screen
H	Upper left of the screen
M	Middle of the screen
L	Last line on the screen
^D	Forwards half a window (^nD for n windows)
^U	Backwards half a window (^nU for n windows)

Source: http://www.ks.uiuc.edu/Training/Tutorials/ Reference/unixprimer.html.

Table 2.7. A Summary of Commands for Moving Around by Text Structures

(	Beginning of sentence
)	Beginning of next sentence
{	Preceding paragraph
}	Next paragraph

Source: http://www.ks.uiuc.edu/Training/ Tutorials/Reference/unixprimer.html.

Table 2.8. Commands That are Useful for Locating Text Strings

/string	Searches for text string
?string	Searches for text string
n	Repeat last / or ?
/	Repeats last text search
N	Repeat last / or ? in reverse
F char	Backwards line search
T char	Search current line backwards for character
f char	Search current line for character char
;	Repeat last f, F, t or T
,	Repeat last f, F, t or T in reverse

Source: http://www.ks.uiuc.edu/Training/Tutorials/ Reference/unixprimer.html.

It is also possible to manipulate files within Vi (write, delete save, etc.) as shown in Tables 2.13 and 2.14 contains miscellaneous commands.

Finally, it is very important to gracefully exit Vi because you created a new file you are going to save, or a long series of file edits hang in the balance. Table 2.15 shows various commands for making a curtain call on your Vi session.

2.4 A Working Introduction to C++: Basic Coding

2.4.1 General Comments

I learned the value of organization in research as well as in my thoughts only after I began graduate school. In computer-oriented fields a lack of organization can translate into the

Table 2.9. Commands for Adding Text in Vi

u	*Undo*
a	*Begin general insert mode (na automatically repeats insert)*
A	*Append text at the end of the current line*
i	*Insert text to the left of the cursor (ni repeats insert)*
I	*Insert text at the beginning of the current line*
o	*Open a line below current one and insert text*
O	*Open a line above the current one and insert text*
p	*Places deleted text after where the cursor is*
P	*Places buffer text before where the cursor is*
:1,2 co 3	*Copy lines 1-2 and place them after line 3*
:r filename	*Inserts contents of filename after current line*

Source: http://www.ks.uiuc.edu/Training/Tutorials/Reference/unixprimer.html.

Table 2.10. Commands for Replacing and Changing Text in Vi

r	*Replace the character you are on with the one you type*
~	*Change case of the letter you are on*
R	*Overstrike*
Y	*Yanks selected text and places it in a buffer*
y	*Cursor yank*
yy	*Yanks current line*
ny	*Yanks so many lines*
C	*Change text to the end of the line*
S	*Change an entire line*
c	*Change*
cc	*Changes current line*
r	*Replace one character under the cursor (nr for n characters)*
s	*Substitute one character under the cursor*
s/p1/p2/options	*Substitutes p1 to p2 (s/p1/p2 does once; s/p1/p2/g does all)*
cw	*Changes current word*
:g/search/ s//replace/gc	*Case-sensitive replacement of every occurrence of "search" with "replace". The "gc" option asks for confirmation and I encourage you to look into what g and gi do.*

Source: http://www.ks.uiuc.edu/Training/Tutorials/Reference/unixprimer.html.

Table 2.11. Commands for Deleting Text in Vi

x	Delete character
X	Delete character before cursor
d	Cursor delete
ndw	Delete remainder of word right of cursor
ndb	Delete word to left of cursor
ndd	Delete current line
D (D$)	Deletes from cursor to end of line
Y	Yanks selected text and places it in a buffer
y	Cursor yank
yy	Yanks current line
ny	Yanks so many lines

Source: http://www.ks.uiuc.edu/Training/Tutorials/ Reference/unixprimer.html.

Table 2.13. File Manipulation Commands in Vi

:wq (:x) (ZZ)	Write to the file and quit
:w filename	Writes contents to a file
:e	Edits a new file.
^\, Q	Exit Vi and enter ex editor. Undone by :vi
^G	Show current file name and status
:n1,n2w newfile	

Source: http://www.ks.uiuc.edu/Training/ Tutorials/Reference/unixprimer.html.

Table 2.15. Commands for Properly Exiting Vi

:q (:q!)	Quits the Vi editor
:wq (:x) (ZZ)	Write to the file and quit
ZZ	Exit Vi and save changes
^\, Q	Exit Vi and enter ex editor. Undone by :vi

Source: http://www.ks.uiuc.edu/Training/Tutorials/ Reference/unixprimer.html.

Table 2.12. Text Moving and Joining Commands in Vi

J	Joins the current line and next line (nJ for next n lines)
< linenum	Shifts lines up to linenum one space left
<<	Shifts current line to the left
> linenum	Shifts lines up to linenum one space right
>>	Shifts current line to the right
z	Puts current line on top of the screen (z. for center and z- for bottom)
:4,5 m 6	Move lines 4–5 and put after line 6

Source: http://www.ks.uiuc.edu/Training/Tutorials/ Reference/unixprimer.html.

Table 2.14. Potentially Useful Miscellaneous Commands

^L	Redraws your screen in case it has become disorganized
J	Joins the current line and next line
^R	Redraws screen removing false lines
'l	Cancels partially formed command
^^	Go to last file edited
!command	Executes UNIX shell command "command".

Source: http://www.ks.uiuc.edu/Training/ Tutorials/Reference/unixprimer.html.

feeling of one's drinking out of a fire hydrant, which generally should be avoided. Although the order of statements in a C++ program is very flexible, it is good to develop a standard protocol for organization's sake. When I teach my physics students about what friction does, I suggest that they view the world without friction and imagine what will not happen; the same idea applies here. Once you have a program running, you can alter, remove or add one thing at a time and see what, if anything, your program does not do correctly. Later in this chapter we will explore the deep value found in making our programs work incorrectly with, I hope, humor interjected. And, as mentioned in previous sections of this chapter there are a host of external references that I recommend if you want a more exhaustive treatment [7–10]. And, if you are interested in advanced C++ data structure information and utilization I recommend Main and Savitch [11].

2.4.2 Playing with Arithmetic and Basic Algorithmic Elements

The first thing I do is to call standard libraries in headers, whether I need them in the program or not. The library calls below take care of most of the modeling and simulation applications I have ever done in C++. On some machines the libraries may have a ".h" extension; you will have to inspect compilation results to see if your header calls match file names on your machine.

First call important libraries through various headers.

```
#include<cstdlib>
#include<iostream>
#include<fstream>
#include<math.h>
#include<iomanip>
#include<stdio.h>
#include<conio.h>
using namespace std;
```

cstdlib: General purpose functions, including dynamic memory management, integer operations, random numbers, and external communication.

iostream: Establishes input and output using streams called by cin and cout.

fstream: Establishes file manipulation using streams.

math.h: C++ numeric library that establishes trig functions, square roots, exponentials, etc.

iomanip :define needed manipulato functions related to input and output.

stdio.h : sets up standard input/output for screens, keyboards, terminal and external devices.

conio.h: Establishes console input and output. Not present in UNIX applications.

using standard namespace: supports fstream. Can be dangerous in more advanced programming applications.

Start the program by declaring an integer expressing the status of the program.

```
int main()
{
```

Set up the input and output protocol; name and open input and output files.

```
fstream input_file;
input_file.open("example_input.txt");
ofstream output_file;
output_file.open("example_results.txt");
```

Declare scalar variables.

```
int n;//Number of samples in each average
int seed;//Random number seed
```

```
double sum,num2,fullsum,sqsum,ransum;
float ran_num;
```

Now declare an array (in this case a one dimensional array, or a vector).

```
float powersum [4];//Sums for the 2nd, 3rd, 4th and 5th powers of numbers.
```

Read the input file and seed the random number generator.

```
input_file>>seed;
srand(seed);
```

Prompt the user to enter the number of samples and read the answer from the screen.

```
cout<<"\n Enter the number of samples: ";
cin>>n;
cout<<"\n";
```

Initialize the scalar sums.

```
sum=0.0;
fullsum=0.0;
ransum=0.0;
```

Now initialize the sums of various powers using a *for loop*. This should be done for all arrays.

```
for(int i=0; i<4; i++)
{
powersum[i]=0.0;
}
```

Start looping through trials.

```
for(int i=0; i<n; i++)
{
```

Calculate a random number between 0 and 1000.

```
ran_num=(rand()%1000);
```

Convert it to a number between 0 and 1 and calculate its running sum.

```
ransum=ransum+ran_num/1000.0;
```

Now calculate a random number between −1 and 1 as well as its running sum

```
num2=2.0*(ran_num/1000.0-0.5);
fullsum=fullsum+num2;
```

and calculate the running sum for 2nd through 5th powers of num2.

```
for(int j=0; j<4; j++)
{
powersum[j]=powersum[j]+pow(num2,j+2);
}
```

Use a conditional to calculate positive running sum.

```
if(num2>=0.0)
{
sum=sum+num2;
}
```

Close the loop involving number of steps.

```
}
```

Next write out to the screen.

```
cout<<"The random number average is equal to:
"<<ransum/float(n)<<"\n";
cout<<"The positive average is equal to: "<<sum/float(n)<<"\n";
cout<<"The sum is equal to: "<<fullsum/float(n)<<"\n";

for(int i=0; i<4; i++)
{
cout<<"The average of power "<<i+2<<" is equal
to:"<<powersum[i]/float(n)<<"\n";
}
```

Write to the output file.

```
output_file<<"The random number average is equal to:
"<<ransum/float(n)<<"\n";
output_file<<"The positive average is equal to: "<<sum/float(n)<<"\n";
output_file<<"The average is equal to: "<<fullsum/float(n)<<"\n";

for(int i=0; i<4; i++)
{
output_file<<"The average of power "<<i+2<<" is equal
to:"<<powersum[i]/float(n)<<"\n";
}
```

Close all files for reading and writing.

```
input_file.close();
output_file.close();
```

For a Windows-based system hold the screen up so it doesn't close and disappear with the following *getch* command, and for a UNIX-based system it will need to be commented out:

```
getch(); when used or //getch(); when commented out.
```

Signal the main program to end by returning an integer value to it

```
return(0);
```

and close.

```
}
```

In the previous section we examined simple coding in C++ where we looked at how we can deal with numbers within loops. Here, now I want to tie in the coding we talked about (no really I talked and you were a captive audience, simply reading) to more formal mathematical algorithms and constructs, including those from calculus.

2.4.3 Getting a Bit More Formal: Sequences and Series

Now I will write down the formal mathematical expressions of what the programs in the previous section are doing. Writing code is actually part of your homework! Consider the C++ code from the previous section. It calculates various sequences and sums of numbers related to the loop index. A sequence S_N is just a set of numbers that follow a certain order:

$$S_N = \{x_1, x_2, x_3, \ldots x_N\}. \tag{2.1}$$

There are finite and infinite sequences. There may be a formula describing the sequence and limits that can be taken for say $N = 0$ or N very large ("tending towards infinity"). For a review on sequences there are a good many calculus texts "out there" you can check in to (See Reference 15). In the code for Section 2.4, the sequences we are generating with this loop involve finite sets of powers of random numbers x_i:

$$A_N = \{x_1, x_2, x_3, \ldots x_N\}.$$
$$B_N = \{x_1^2, x_2^2, x_3^2, \ldots x_N^2\}. \tag{2.2}$$
$$E_N = \{x_1^5, x_2^5, x_3^5, \ldots x_N^5\}.$$

We don't actually write the sequences out, but where would you put a write statement in the program to accomplish such a thing?

The code we are talking about also deals with series, where we simply sum the members of a sequence. The series corresponding to the sequences mentioned above are

$$AS_N = \sum_{i=1}^{N} x_i = x_1 + x_2 + x_3 + \cdots + x_N$$

$$BS_N = \sum_{i=1}^{N} x_i^2 = x_1^2 + x_2^2 + x_3^2 + \cdots + x_N^2 \tag{2.3}$$

$$ES_N = \sum_{i=1}^{N} x_i^5 = x_1^5 + x_2^5 + x_3^5 + \cdots + x_N^5$$

2.4.4 Derivatives and Integrals

I thoroughly encourage my students to program as much as they can directly, unless the calculations they are involved in have to do with a legal (or other type of) issue where validated and accepted software must be used. As such, we need to discuss how to program derivatives and integrals.

As discussed later in this book, a derivative is nothing other than geometrically the slope of a tangent line to a curve or physically the rate of change of a quantity with respect to another at a given point on the curve, which meets the mathematical requirements of differentiability [12]. The slope is a special one where the points are infinitely (arbitrarily) close together but we can't have the computer calculate such a thing, we therefore use a finite difference quotient to express a derivative. Then $\frac{dy}{dt}(t)$ (which is not a fraction and can't be pulled apart as one) can become

$$\left(\frac{\Delta y}{\Delta t}\right)_n = \frac{(y_{n+1} - y_n)}{\Delta t}. \tag{2.4}$$

This is called a forward difference and there are many other ways to calculate derivatives that will be touched on in the next section. We can go on and on, so second derivatives are derivatives of derivatives and a central difference expression of $\frac{d^2 y}{dt^2}$ (again not a fraction) can become

$$\frac{(y_{n+1} - 2y_n + y_{n-1})}{2(\Delta t)^2}. \tag{2.5}$$

I say "can become" because there are many ways to calculate derivatives depending on the problem construction, system parameters and such that we will touch on below [13–17]. As homework you are asked to code in derivatives and examine error with the actual answer as you change the size of your system.

An integral is geometrically the area under a curve between two points and physically the sum of a quantity where the system being summed over had to be divided into an infinite number of infinitesimally small pieces. We visualize integrals in the next chapter but they are nothing other than sums.

The *composite trapezoidal rule* for numerical integration approximates the integral of a function $f(x)$ on a closed interval $[a,b]$ divided into n equally spaced sections of length Δx as

$$\int_a^b f(x)dx \approx \frac{1}{2}\sum_{i=1}^{n}\{f(x_{i-1}) + f(x_i)\}\Delta x = \frac{(b-a)}{2n}\sum_{i=1}^{n}\{f(x_{i-1}) + f(x_i)\}.$$

(see References 13–17 and Chapter 3 for a visual). Then

$$I = \int_a^b f(x)dx \tag{2.6}$$

becomes

$$I_n = \sum_{i=1}^{N} \frac{1}{2}(f(x_n) + f(x_{n+1}))\Delta x$$
(2.7)

Where $x_1 = a$ and $x_N = b$. As you may have guessed there are a great many methods of integration. For homework, I will ask you to code up some integrals and evaluate the error from the actual answer.

Monte Carlo methods in numerical integration are named after the famous gambling utopia and involve random numbers in one way or another. First, consider the integral $\int_a^b f(x)dx$, which is geometrically interpreted as the area between the curve $f(x)$ and the x-axis. Now imagine a rectangular box which exists on $[a,b] \times [0,c]$ with c being greater than the maximum of $f(x)$ on $[a,b]$ (Figure 2.4).

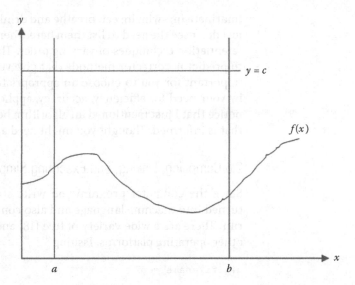

Figure 2.4. Diagram for Monte Carlo integration of $y = f(x)$ on $[a,b]$.

Imagine "throwing mud" at the (x,y) plane (within the box only) so it hits at random positions within the box only. After many mud hits, let's say there are n_t spots above the curve and n_b spots below the curve. The area under the curve (and hence $\int_a^b f(x)dx$) is given by the simple ratio $n_b(b - a)c/n_b + n_t$. The key here is to *randomly* throw points within the box.

There are many different situations where you don't have an actual answer but you still have to make sure that the derivatives and integrals you are calculating are behaving well and are correct. In those cases you need to (1) perform self-consistency checks, (2) examine how the code behaves for special cases and (3) look at how your answers change as you make the system sample larger and larger. You guessed it; there are homework problems you can do that help you with such things and are part of a larger consideration (and another section of this book) where I underscore the importance of and suggest ways to critically evaluate your work and place it in context.

For further details on visualizing derivatives and integrals as well as homework problems please see Chapter 3.

2.5 How Do I Choose a Good Algorithm?

The best way to choose a good algorithm is to stay away from the bad ones. (Laughter subsides...) There are a wide range of numerical, differentiation, and integration algorithms that can be chosen, depending on the nature of the system you are simulating [13–17]. As in the spirit of this text I want to introduce you to scientific computing and numerical methods, and empower you to thoughtfully apply them. If you can look up (marinate in; swim in; eat, breathe and drink...) a wide range of numerical algorithms I just don't see the need to list them here. There are backwards, central, and forwards differentiation techniques of varying order. The same applies for integration, in addition to predictor-corrector methods that give very high accuracy. However, it is extremely important for you to choose an appropriate algorithm that is stable and is informed by your need for efficiency, accuracy, application, and so forth. The astute reader will notice that I just mentioned an algorithm being informed but it is actually your choice that is informed. Thought you might need a laugh—I surely got one!

2.6 Compiling, Linking, and Executing Simple Programs

Since the computer program you write starts out as written words, it needs to be turned into machine language and also connected with various libraries it requires to run. There are a wide variety of free [18] and commercial [19] compilers in UNIX and other operating platforms. Issuing

```
g++ filename.cpp
```

invokes the g++ compiler and issuing

```
ls
```

should show a new file named a.out which is the executable. Compilers have many options; for example you can issue

```
g++ filename.cpp>mychoice.out
```

which creates an executable with name mychoice.out
     or there is an optimizing option which may help longer programs to run faster:

```
g++ -O3 filename.cpp>mychoice.out
```

I thoroughly encourage your exploring compiler options.
     To execute (run) the program, just issue

```
./a.out
```

and it will run in the foreground. If it is a longer program that you would like to run in the background so you can log out, you can issue

```
nohup./a.out &
```

To check on any jobs running in the background, issue

```
ps-u username
```

or

```
top
```

If you don't have any explicit input/output directives in the file you can direct it yourself by issuing

```
./a.out<inputfile>outputfile
```

or

```
nohup./a.out<inputfile>outputfile &
```

Just remember that when you are dealing with screen directed input to start a program (as for the one in the previous section) running the program in the background is challenging at best because you have to remember the questions or prompts it is going to give you and answer them on your own in sequence if there are more than one.

The input file for the program in the previous section has one line containing the integer "333" for the random number seed. When executed with ./a.out, the first screen output is

```
Enter the number of samples:
```

To which I responded with

```
10000
```

and obtained the following output to the screen:

```
The random number average is equal to: 0.50402
The positive average is equal to: 0.253822
The sum is equal to: 0.008039
The average of power 2 is equal to:0.33192
The average of power 3 is equal to:0.0029373
The average of power 1 is equal to:0.109775
The average of power 5 is equal to:0.00176488
```

Issuing ls reveals the presence of a new output file and examination of it with more, less or

Vi reveals the same numerical answers, as it should.

2.7 Examples of What Can be Done Wrong: Compile Errors, Execution Errors, and Bugs

Even for seasoned programmers, compiling and executing programs can offer a whole new experience in discovering errors. It can be a journey through frustration and an exercise in perseverance. In this text, my aim is to give you a feeling for what kind of things you can run into when things don't go quite as planned. We will cover compile, linking and execution errors, as well as bugs. And by the way, I encourage you to research why certain problems are called "bugs;" you may find the historical anecdote interesting and entertaining.

2.7.1 Compile Errors are Mistakes which Prevent Compilation of the Program

1. The first compile-time error that comes to mind is that you can mistype the compile command, in which case the machine will simply respond by telling you that the *command is not recognized.*

2. If you leave out the number sign (#) in the header, the compiler will interpret the line of code as a command and you will get back about a page full of "invalid operands" errors.

3. Omitting the call to fstream results in

```
random.cpp: In function âint main()â:
random.cpp:15: error: âinput_fileâ was not declared in this scope
```

and a similar result is obtained for omitting ofstream.

4. Forget to declare a variable in a particular program will give

```
random.cpp: In function âint main()â:
random.cpp:34: error: ânâ was not declared in this scope
```

5. Leaving out the number in declaring size of an array gives

```
random.cpp: In function âint main()â:
random.cpp:26: error: storage size of âpowersumâ isn't known
```

6. You can always add an extra decimal point to a number but the following will happen

```
random.cpp:38:5: error: too many decimal points in number;
```

I know you will observe that you didn't need college for that one.

7. Misspelling a variable or referencing one that wasn't declared gets you

```
random.cpp: In function âint main()â:
random.cpp:39: error: âfulllsumâ was not declared in this scope
```

8. Leaving out a semicolon from a line of code results in

```
random.cpp: In function âint main()â:
random.cpp:46: error: expected â;â before â}â token
```

9. One too many or few brackets

```
random.cpp: In function âint main()â:
random.cpp:106: error: expected â}â at end of input
```

10. Missing a semicolon in a for loop

```
random.cpp: In function âint main()â:
random.cpp:43: error: expected â,â or â;â before âiâ
random.cpp:43: error: expected â;â before â)â token
```

11. Typing a colon instead of semicolon in a for loop

```
random.cpp: In function âint main()â:
random.cpp:43: error: expected â;â before â:â token
random.cpp:43: error: expected primary-expression before â:â token
random.cpp:43: error: expected â)â before â:â token
random.cpp:43: error: expected primary-expression before â:â token
random.cpp:43: error: expected â;â before â:â token
```

12. Forgetting to declare the variable type of the loop counter

    ```
    random.cpp: In function âint main()â:
    random.cpp:43: error: âiâ was not declared in this scope
    ```

13. Referencing a loop counter variable other than the one intended

    ```
    random.cpp: In function âint main()â:
    random.cpp:43: error: âiâ was not declared in this scope
    random.cpp:49: error: âjâ was not declared in this scope
    ```

14. Wrong number of parenthesis in a math expression

    ```
    random.cpp: In function âint main()â:
    random.cpp:43: error: âiâ was not declared in this scope
    random.cpp:59: error: expected â)â before â;â token
    ```

15. Leaving off quotes from text you intend to write out to the screen or a file

    ```
    random.cpp:78: error: stray â\â in program
    random.cpp:78:69: warning: missing terminating " character
    random.cpp:78: error: missing terminating " character
    random.cpp: In function âint main()â:
    random.cpp:43: error: âiâ was not declared in this scope
    random.cpp:78: error: âTheâ was not declared in this scope
    random.cpp:78: error: expected â;â before ârandomâ
    ```

16. Using a single arrow instead of a double arrow for output direction

    ```
    random.cpp: In function âint main()â:
    random.cpp:43: error: âiâ was not declared in this scope
    random.cpp:78: error: invalid operands of types âconst char
    [40] â and âdoubleâ to binary âoperator<<â
    ```

2.7.2 Linker Errors are Mistakes which Prevent Compilation of the Program

1. You can misspell a header or ask for one that doesn't exist

    ```
    random.cpp:2:18: error: iosream: No such file or directory
    random.cpp: In function âint main()â:
    random.cpp:33: error: âcoutâ was not declared in this scope
    random.cpp:34: error: âcinâ was not declared in this scope
    ```

2.7.3 Bugs and Execution Errors that Prevent a Compiled Program from Running

"Bugs" usually are reserved for programs that compile and either run but give an incorrect answer, or hang and produce no results. Execution errors, on the other hand are those errors which prevent a program from running. I would have classified them into either category in this section but I didn't because it is a homework problem! Some types of bugs are discussed below but, as with the previously discussed errors the list is not exhaustive. Not surprisingly they can certainly exhaust even the most experienced programmers.

1. If you type something like "3A3" in input,

    ```
    The number is read up to the letter and that's it.
    ```

2. Now try E33 in input, and

```
0 is assumed but it really could be anything
```

3. Forget to initialize variables in the program, and the machine may or may not assign the correct value for the number.

4. If the random number percentage disagrees with what you divide out in the program we covered previously for example

```
The random number average is equal to: 0.0497512
The positive average is equal to: 0
The sum is equal to: -0.900502
The average of power 2 is equal to:0.814242
The average of power 3 is equal to:-0.739231
The average of power 4 is equal to:0.673805
The average of power 5 is equal to:-0.61659
```

5. If you declare ransum as integer in the program we previously went through, for example, it gets truncated to zero:

```
The random number average is equal to: 0
The positive average is equal to: 0.249479
The sum is equal to: -0.00130906
The average of power 2 is equal to:0.333577
The average of power 3 is equal to:-0.00188413
The average of power 4 is equal to:0.199939
The average of power 5 is equal to:-0.00171958
```

6. Code a loop as for (int i=0; i<n; i), and

```
The program will hang and do nothing
```

7. Likewise, if a for loop is miscoded as for (int i=0; i=n; i++)

```
The program will hang and do nothing
```

8. Forget brackets enclosing loop and the program we covered previously as an example runs through only once

```
The random number average is equal to: 4.3e-05
The positive average is equal to: 0
The sum is equal to: -1.4e-05
The average of power 2 is equal to:1.96e-06
The average of power 3 is equal to:-2.744e-07
The average of power 4 is equal to:3.8416e-08
The average of power 5 is equal to:-5.37824e-09
```

9. Ask for more array elements than there are declared and you will get one of the most difficult errors to track down:

```
Segmentation fault
```

10. Coding an array sum over itself powersum[j]==powersum[j]+pow(num 2,j+2) in the program we went through as an example is not good practice:

```
The random number average is equal to: 0.496921
```

```
The positive average is equal to: 0.248417
The sum is equal to: -0.0061578
The average of power 2 is equal to:0
The average of power 3 is equal to:0
The average of power 4 is equal to:0
The average of power 5 is equal to:0
```

11. Miscoding a conditional such as if(num2=0.0) using the previous program as an example gives:

```
The random number average is equal to: 0.496921
The positive average is equal to: 0
The sum is equal to: -0.0061578
The average of power 2 is equal to:0.336749
The average of power 3 is equal to:-0.00345799
The average of power 4 is equal to:0.202568
The average of power 5 is equal to:-0.00232249
```

12. If there is mistyping of the input file, you might get a very well executed result that is incorrect. There's nothing like driving 100 miles/hr in the wrong direction!

13. Unlike Pascal, C uses the / operator for both real and integer division. It is important to understand how C determines which it will do. If both operands are of an integer type, integer division is used, else real division is used. For example:

```
double half = 1/2;
```

This code sets `half` to 0 not 0.5! Why? Because 1 and 2 are integer constants. To fix this, well, that's a homework problem.

14. Infinite loops will run *ad infinitum* unless there are execution limits set by the machine you are using. Sometimes, they can present exactly like a hang, and so debugging techniques can definitely help you to identify the problems and fix them. There are many lists of errors available for programmers and Reference 20 is just an example.

2.7.4 Are There Best Practices in Debugging?

Perhaps there are best practices for debugging but if they exist they are sitting right next to best practices for winning a poker game. There are general rules but no specific advice I have for you other than to make the program write out copious amounts of information at strategically located places. The way I view it, providing you with a script to debug the programs you write would be like scripting out your activities on a visit to a new and exciting place you have never been before. Note: mine would be the skeleton hall in the Paris Museum of Natural History, a remark completely tangential to the discussion here. In many senses, being a successful program debugger is like being a chef who cooks wonderful food by throwing the recipe away.

My recommendation to learn debugging actually is much like cooking: just start constructing code by working on something. It would be extremely rare for no errors to appear and so you will naturally run into them and have to track them down before you take next steps. You have to cultivate a detective instinct in order to identify, locate, and a repair all kinds of mistakes—something that just takes time, some of us more than others. Many of the homework problems in this chapter are designed to help you begin to do just that. I have found that the greatest challenge students face in debugging is reaching the patience threshold. So, when it you feel your basic human instinct creating the urge to sell the computer at a pawnshop and use the money for your favorite dinner, it may be best to leave the problem and do something else for a couple hours as it runs through your mind. Another suggestion would be to get together with people who are working on similar problems and have a debugging session. You can work on your own code or maybe you could even make bugs for each other. Debugging, and coding for that matter doesn't always have to involve a computer lab, if it can involve a coffee shop, some social time, and some fun.

2.8 Doing it Without a Supercomputer: Computing on Macs and PC's

Computing on a PC or Mac feels identical to using a mainframe but there can be large differences in the compilers and processors and there exists quite a range of them with various options and features. Getting a sense for what is out there is…drum roll…a homework problem! A word of caution about the references provided for compilers and processors is that they are changing quite rapidly so you should always apprise yourself of the most current situation in the field before progressing. If you compare the results below from a PC with the results at the end of section 3.2 on a mainframe, you will notice some differences. Under what conditions would such deviations be unacceptable or acceptable? You guessed it, that is a homework problem for you to think about!

Enter the number of samples: 10000

```
The random number average is equal to: 0.504019
The positive average is equal to: 0.253822
The sum is equal to: 0.00803898
The average of power 2 is equal to:0.33192
The average of power 3 is equal to:0.0029373
The average of power 4 is equal to:0.198335
The average of power 5 is equal to:0.00176488
```

Generally speaking, for homework and classroom applications a PC and mainframe will both be equally useful. The differences would come if you have to run, for example, research applications needing large scale or massively parallel simulations, which a mainframe would favor. On the other hand, if you travel or have limited Internet access, then running Linux or Windows on a PC or Mac may be most prudent for you.

For whatever calculations you do, it is important to understand the variations in results across the different machines so I strongly suggest that you do initial simulations with many different devices and compare results to make sure that you understand how your results compare in the larger perspective. Once again, there is no written script I am going to give you; it involves investigation and understanding on your own.

2.9 Mapping Your C++ Knowledge to Other Computing Languages

I am going to guess that is easier than you think to map your knowledge of C++ to another language. I think that it's true for all languages and that the most important part of mastering diverse coating is understanding the algorithm and calculations you're using. In fact, that's why I'm not too worried about the languages I'm presenting examples in throughout the book here. There are many and they are changing. You will doubtless have to translate the algorithms into whatever languages of interest to you at the time you are simulating. I picked FORTRAN for the other language to talk about here because it is widely used in the scientific community. The FORTRAN code I present is uncommented and it is a homework problem to comment the code in detail.

Obvious cosmetic differences are that FORTRAN always starts off with the program name and nominally in the sixth column and there are no semicolons. The loops are numbered and the numbers as well as code continuation lines exist in columns 1 through 5. There are also slight syntactical differences between the two languages and I will let you get a feeling for all that by looking at this next example and related code provided, Averages.f.

Start off with a program name and always start nominally in the 6th column.

```
      Program Averages

      integer n
      integer seed
      real sum,num2,fullsum,sqsum,ransum
      real ran_num;
      real powersum(4)
```

Open files for I/O. See homework about using certain file numbers.

```
      open (1,file='example_input.txt')
      open (2,file='example_results.txt')

      read(1,*) seed

      write(*,*) "Enter the number of samples: "
      read(*,*) n

      sum=0.0
      fullsum=0.0
      ransum=0.0
```

Notice loop construction.

```
      do 100 i=1,4
      powersum(i)=0.0
100   continue

      do 200 i=1,n
      ran_num=rand()
      ransum=ransum+ran_num
      num2=2.0*(ran_num-0.5)

      fullsum=fullsum+num2

      do 300 j=1,4
      powersum(j)=powersum(j)+num2**(j+1)
300   continue

      if(num2.ge.0.0) then
```

```
      sum=sum+num2
      endif

200 continue
```

And here notice how a number in column 5 serves as a continuation of the line above.

```
      write(*,*) "The random number average is equal to: ",
     1ransum/float(n)
      write(*,*)"The positive average is equal to:",sum/float(n)
      write(*,*) "The sum is equal to: ",fullsum/float(n)
      do 500 i=1,4
      write(*,*) "The average of power ",i+2," is equal to:",
     1powersum(i)/float(n)
500 continue
      write(2,*) "The random number average is equal to: ",
     1ransum/float(n)
      write(2,*)"The positive average is equal to:",sum/float(n)
      write(2,*) "The sum is equal to: ",fullsum/float(n)
      do 600 i=1,4
      write(2,*) "The average of power ",i+2," is equal to:",
     1powersum(i)/float(n)
600 continue

      close(1)
      close(2)

      return

      end
```

The results from compiling and running the program are below.

```
Enter the number of samples:
10000

 The random number average is equal to: 0.50182772
 The positive average is equal to: 0.25175485
 The sum is equal to: 3.65340267E-03
 The average of power 3 is equal to: 0.33458865
 The average of power 4 is equal to: 3.09360796E-03
 The average of power 5 is equal to: 0.20202026
 The average of power 6 is equal to: 3.14053753E-03
```

2.10 Critically Thinking about Your Work: Relevance, Applicability, and Limits

As far as errors and limitations, the bad news is that every computer simulation is "wrong" because they are limited in space and time. It's true that we're only approximating things or stepping things through time in fairly large chunks as opposed to infinitesimally small ones. We can have finite size effects where the size of a computer simulation affects the results because it's not a truly infinite system. Also, our models include natural phenomena that we need but not everything. For example, in our baseball simulations we don't incorporate length, mass and time dilation described by special relativity. Validation of the simulation is very important, that is making sure it gives acceptable answers in a simulation where we already have real things.

Personally, one of the most difficult skills I have had to develop is self-criticism but it behooves us all to do so. If we are harder on our work than others it means that we will be sharing and applying well thought out and executed science and technology.

2.11 Your Work in the Broader Context of the Scientific
and Technological Community

When you complete a project and want to disseminate the information, it is always important to have an understanding of the context of your results. There is no real prescriptive way to gain that, but the more work you do, the better perspective you will have on how it fits into the body of knowledge. One way I suggest is to search on the Internet for projects having some of your key words in the title or document. And you can determine what papers reference those, which is really useful. In scientific fields, the number of people who reference to your journal article or technical report is an important measure of how relevant the work is. Just as in *Strawberry Fields*, it usually takes nearly forever. At the end of the day, if others are able to use your work to advance their own research projects or use it to help people, then your work is effective and in some sense complete.

There is also a scientific consciousness that I encourage you to exercise. For example, Einstein was a pacifist but said that if people didn't fight Hitler we might as well lay down. And then of course there is the Manhattan project. It's a very delicate balance but the barometer I use is that when I am describing to my grandchildren what I did with my career (hopefully *am doing*, as I'm one of those guys that simply doesn't have retirement on the radar), I can tell them that I helped others and made the world a better place to be in if even by a small measure. Will your legacy sing a positive song?

PROBLEMS

NOTE: The sample programs provided are meant to be good starting points for the homework and so before you begin any of the homework I suggest that you "play with" (run and execute for a few different cases) them so as to gain a basic familiarity.

2.1.   Consider the sequence $S_n = (2n^2 - 5)/(5n^2 + n)$, $n = 1,2,3,\dots$. Now, code and run an algorithm that calculates the sequence. Comment on your result as n grows very large.

2.2.   Same as Problem 2.1 but with the sequence $S_n = 1/(5n^2 + n)$, $n = 1,2,3,\dots$.

2.3.   Consider the series $S_N = \sum_{i=0}^{N} r^i$. Code and run an algorithm that calculates the sum and comment on the result as n gets very large for $r = 0.8$ and $r = 1.01$. This is an example of a geometric series, so look up the result for an infinite number of terms and comment on the quality of your results. How will you determine how close "close" is?

2.4.   There are many integration algorithms. Consider the function $f(x) = x^2$ on [1,3]. OK, thanks for considering it. Now, compare the results from three different integration algorithms. First, use rectangles where the left edge determines the height of each rectangle. Second, change things so that the height of the rectangles are determined by the right hand side of each rectangle. Third, use the composite trapezoid rule. Which algorithm(s) overestimate the integral and which one(s) underestimate it? Comment on your results when the number of rectangles is very large—I suggest $n = 1$, 10, 1000 and 10,000 rectangles.

2.5.   Write a program which uses Monte Carlo integration to calculate $\int_0^{2\pi}(1+\cos x)dx$. Have your program write out the value of the integral after every power of 10 steps (1, 10, 100, 1000, etc.) up until you think the integral has converged. Please plot the value of the integral vs. step number to convince yourself of convergence.

2.6.   Monte Carlo integration works in 3D, only the region of integration is now an area (the square $[0,1] \times [0,1]$ in this case), and we deal with throwing random points in *boxes* (volumes), not areas. All we need is that the bottom of our box is at $z = 0$ and the top is somewhat above the maximum value of the integrand. Compute $\int_0^1\int_0^1(x^2 + y^2)dxdy$ as a *volume ratio* using Monte Carlo integration. Try to visualize this if you get stuck!! Write out the value of your integral every 1000 steps and construct a plot similar to the one in 2.4 and 2.5 to convince yourself of convergence. Hint: it may help you to calculate the exact (analytical) value of the integral.

2.7.   Now repeat 2.5 but for $\int_0^{3\pi/2}\sin^5 xdx$. Compare with the analytical result.

2.8.   For the function $f(x)e^x + 7$ on [0,3] use a forward, backward, and central difference approaches to calculating df/dx. Briefly compare and contrast the results and discuss how they compare as the number of intervals grows very large.

2.9.   Classify the issues presented in Section 2.6.3 into bugs and execution errors, and justify your reasoning. Give brief accounts of related examples you have encountered, if any.

2.10.   Code "if(j=3)" into a conditional statement in a program and describe the results.

2.11.   Create a program where you divide by zero. Discuss the results and see how many unique errors you can generate with a zero division.

2.12.   Again dwelling on the sweet taste of execution failure, compare and contrast the errors NaN and INF.

2.13.   Let's plan to fail! Use the incorrect postfix and prefix operator (i++ as opposed to ++i) in a loop and discuss the results.

2.14.   Return a value in a void function and discuss the results.

2.15. Create an infinite loop in as many ways as you can, and provide proof that the computer program is running. How is the behavior of an infinite loop different from when the program hangs?

2.16. Use the code provided in this chapter (or create your own) that runs but has mismatched "{" or IF-ELSE statements. Discuss the results along with a proposed debugging solution for someone who suspects they might have an error like this in their work.

2.17. To get some informative and entertaining background history, research and briefly discuss the origin of the term "bugs" in computer programming.

2.18. Get together with a few classmates in the computer lab or a coffee shop with good Internet. Give another individual or group reasonable assignments for writing code to simulate various physical systems. Regardless of the time frame for the work carried out, upon completion share the results with the group.

2.19. Get together with a few classmates in the computer lab or a coffee shop with good Internet. Using working computer simulations, create five bugs in your program and then pass it on to another individual or group to debug.

2.20. Set out to mess up! Construct code where you confuse the name of an array with the contents of the first element. Discuss the outcome.

2.21. Fix bug 13 in Section 2.6.3 and discuss your results.

2.22. Create your own code or use code from this book to study the differences in computation results across various computing machines/platforms. Carefully discuss the levels of acceptable and unacceptable numerical results and justify your viewpoint.

2.23. Now you get to pick up the pieces left by... me! Comment the FORTRAN code provided in Section 2.6.

2.24. Research various PC's available to you. Compare processor speeds, hard drive space, and memory. Which options are best for you, and why? Which options are acceptable or unacceptable, and why?

Acknowledgment

For tables 2.1–2.15, I would like to acknowledge Emad Tajkhorshid, PhD Group Leader, Computational Structural Biology and Molecular Biophysics Group, University of Illinois at Urbana Champaign, for allowing use of various UNIX tutorials found here.

References

1. J.R. Levine, *UNIX for Dummies*, Margaret Levine Young, IDG Books Worldwide, Inc., Foster City, CA, 1993.

2. J. Peek, G. Todino, J. Strang, *Learning the UNIX Operating System*, ISBN 0-596-00261-0, 5th edition, January 2002.

3. J. Larson, S. Teach, *Yourself Unix in 24 Hours*, ISBN: 0672337304, 5th edition, 2015.

4. Unix Primer—Basic Commands In the Unix Shell, http//www.ks.uiuc.edu/Training/ Tutorials/Reference/unixprimer.html, Theoretical and Computational Biophysics Group, University of Illinois at Urbana-Champaign, Chicago, IL.

5. PuTTY: a free SSH and Telnet client, https://www.chiark.greenend.org.uk/~sgtatham/ putty/

6. Secure Access for Today and Tomorrow, www.ssh.com, Copyright © 2017 SSH Communications Security, Inc.

7. P.J.Deitel, Harvey Deitel, *C++ How to Program* (10th Edition) (Paperback), Pearson, Essex, England, 2017.

8. K. Jamsa, *Jamsa's C/C++/C# Programmer's BiblePaperback*, Delmar (Thornton Learning), Albany, NY, December 18, 2001.

9. J. Hubbard Schaum's, *Outline of Programming with C++* (Paperback), Mc Graw-Hill, New York City, June 6, 2000.

10. S.B. Lippman, J. Lajoie and B.E. Moo, *C++ Primer* (5th Edition), Addison-Wesley, Westford, MA, 2015.

11. M. Main, W. Savitch, *Data Structures and Other Objects Using C++*, 4th edition, Paperback, Addison-Wesley, Lebanon, IN, 6, 2010.

12. R. Larson, R.P. Hostetler, B.H. Edwards, *Calculus of a Single Variable*, 7th edition, Cengage Learning, Boston, MA, 2001.

13. S.S. Ray, *Numerical Analysis with Algorithms and Programming*, Chapman and Hall/CRC, Boca Raton, FL, May 17, 2016.

14. C.F. Gerald and P.O. Wheatley, *Applied Numerical Analysis* (7th Edition), Pearson Education, Boston, MA, 2004.

15. W.H. Press, S.A. Teukolsky, W.T. Vetterling and B.P. Flannery, Numerical Recipes in C++: *The Art of Scientific Computing*, Cambridge University Press, New York, NY, February 7, 2002.

16. S. Chapra, *Applied Numerical Methods with MATLAB for Engineers and Scientists* McGraw-Hill Education, New York, NY, February 6, 2017.

17. B. Carnahan, H.A. Luther, J.O. Wilkes, *Applied Numerical Methods*, Wiley, New York; Sydney, 1969.

18. Free C/C++ Compilers, http://www.thefreecountry.com/compilers/cpp.shtml, Copyright © 1998-2017 by Christopher Heng.

19. Bloodshed Software, http://www.bloodshed.net/devcpp.html, Borland Software, October 2000.

20. http://ace.cs.ohiou.edu/new_users/error.html. Copyright 2014 by the School of Electrical Engineering and Computer Science, Ohio University.

CHAPTER 3

//

Visualizing Your Work and Representing Your Best Story

3.1 Introductory Thoughts

This chapter deals with one of the most important parts of any research project: data visualization. Even though the results that are obtained from a given computer simulation are *objective*, they can be viewed in many different *subjective* ways based upon the context of the application of your work. I do want to share a story that underscores the importance of visualization. When Einstein's brain was studied after he died, it was found that he had an enhanced visualization area. This is consistent with his saying that he could see his equations actually playing out in real time—for example—Brownian motion. And the ability to visualize geometry doubtlessly afforded Einstein the tools with which to develop special and general relativity (1905 and 1915, respectively) which revolutionized modern physics. It could be that the main thing about visualization is that it adds a particular relevance to the problem and allows you to interpret the results in ways other than what simple numbers can afford. I will be wondering about the significance of such things for a very long time in my career.

So anyway, let's get cracking and talk about graphs. Suppose you do a simulation and you wind up with a set of ordered pairs. That is to say, the simulation has generated (x, y) values and although that may be the pinnacle of your day in some regard it is a natural question to ask how one variable is related to the other. Normally, we say that one variable depends on the other, that is, the *dependent* variable depends on the *independent* variable. Normally, the independent variable is labeled as x and the dependent one is labeled as y. So, one of the first things to do is to plot your ordered pairs on a standard (x, y) plot. When your data is plotted you will be able to notice a few aspects

Figure 3.1. Centering, spread, and trend in a data set are exhibited by fireworks and each aspect of the data set has a physical origin in this case. (Bigstock 85787186 https://www.bigstockphoto.com/licensing-tos.html.)

of the plot visually which we can translate into calculations: centering, spread, trend, modality, and such things.

Also about the reference list... there are many citations of my own work in this chapter. The thing I want you to notice is not myself but the wonderful and diverse group of students that I have worked with and who have helped make my research program thrive.

3.2 Visualizing Two-Dimensional Data Sets

3.2.1 Data Centering: Right to the Point

As far as the first important aspect of visualization I want to cover—data *centering*—is concerned, you would simply look at the graph and try to glean where the center of the data is. For highly symmetric cases such a task is very easy, but with the lack of symmetry comes difficulty. As shown in Figure 3.1 you can likely get a very good feeling of the data centering—the point in the sky where the firework plume would balance out (if we could even do that). If you want to be more precise and calculate the parameters for data centering, you would simply find the center of the data exactly as you would find the center of mass for an object:

$$\bar{x} = \frac{1}{n}\sum_{i=1}^{n} x_i \tag{3.1}$$

$$\bar{y} = \frac{1}{n}\sum_{i=1}^{n} y_i \tag{3.2}$$

Even though the plots we are discussing are two dimensional, we can also calculate the center for the third coordinate (z) if we are dealing with ordered triplets; then the center $(\bar{x}, \bar{y}, \bar{z})$ is called the *centroid* of the data set. In the case of centering, you are visualizing your data as the point where it would balance out on the tip of a pencil.

3.2.2 Data Spread: Just How Scattered Are You?

Now for the next aspect of the data set: *spread.* You can get a good idea of the spread in your data by simply looking at your plot. There are many ways you can calculate spread/scatter and the two most popular ones are average deviation Δ

$$\Delta = \frac{1}{n}\sum_{i=1}^{n} |\bar{x} - x_i| \tag{3.3}$$

and standard deviation σ

$$\sigma = \sqrt{\frac{\sum_{i=1}^{n}(\bar{x}-x_i)^2}{n-1}} \tag{3.4}$$

We can now calculate the spread/scatter in your data set even when the plot of your data set is not necessarily symmetric. The *average deviation* basically tells you the average of how far away the data is from the average. The standard deviation is a bit more complicated and provides a measure of the square of the distance between the data points and their average. The standard deviation is the square root of a quantity known as the *variance* which, of course provides the same type of information. It is a homework problem for you to think about what average and standard deviations would look like in special cases. As shown in Figure 3.1, you can get an idea of how spread, or diffuse, your data set is. As with centering, you can also go to the next level and look at the distribution of data points about the center, if the measures of deviation presented here don't provide enough information for you. Looking at distributions and more advanced calculations you can also get a handle on the *modality* of your data set, that is how many groupings, concentrated areas, or peaks it has.

3.2.3 Data Trends: How Correlated is Your Data?

The third aspect of two-dimensional data set visualization I want to discuss is *trend*. Trend is a very important aspect of data sets and will give you an idea of how strongly y depends on x. Knowing about trends in your data set is important because it gives you a sense of correlation or even predictability with your results. Graphically, a trend shows up as a tendency of your plot to look linear. In fact, using linear regression we are fitting our data to the straight line $y = ax + b$ that minimizes variance between the line and actual data points:

$$a = \frac{\sum_{i=1}^{n} y_i - b\left(\sum_{i=1}^{n} x_i\right)}{n} \tag{3.5}$$

$$b = \frac{n\sum_{i=1}^{n} x_i y_i - \left(\sum_{i=1}^{n} x_i\right)\left(\sum_{i=1}^{n} y_i\right)}{n\sum_{i=1}^{n} x_i^2 - \left(\sum_{i=1}^{n} x_i\right)^2} \tag{3.6}$$

The preceding equations give you an idea of the degree of *linear* trends in your data as shown in Figure 3.1 and if you want to fit higher order trends so that you recognize curvature you will then need to go to more advanced expressions [1]. With curve fitting you are in essence visualizing your data set as an averaged curve which expresses collective trends in slope, curvature, and other higher order quantities. If two variables are negatively correlated, an increase in one means a decrease in the other but if they are positively correlated then they both change with the same sign.

3.2.4 Straight Line Plotting: A Powerful Data Analysis Tool

So now back to straight lines: it is very important to be able to cleverly plot data as a straight line because not only do the slope and intercept have different and unique meanings, but they can be used to extract different aspects of the data set that speak to different physical properties of your system. For example, consider the following equation that expresses the acceleration a of a block having mass m being slid across the floor with kinetic friction included:

$$a = \frac{F - \mu_k mg}{m} \tag{3.7}$$

Note: I tend to step lightly around the issue of friction, as it rubs so many people the wrong way... So anyway... μ_k is the coefficient of kinetic friction, g is the acceleration of gravity at the Earth's surface and F is the applied external force. If we are clever, which I think we are, we can recast Equation 3.7 to look like a straight line

$$a = \frac{1}{m}F - \mu_k g \tag{3.8}$$

such that the plot of acceleration a vs. applied external force F has slope 1/m and intercept $-\mu_k g$.

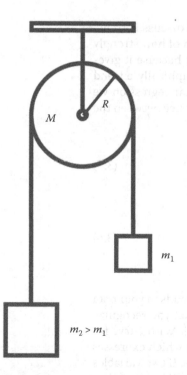

This is great because if we were to measure accelerations and applied forces and graph them then we would expect the results to follow the linear trend prescribed by Equation 3.8. Then, if we actually *measure* the physical quantities that appear in the equation (say F and a) and make a graph of them, we can get the mass of the block from the slope and pull friction out of the mix via the intercept. Wow! Friction then will not affect our measurements of mass because we are able to account for it by visualizing that it simply shifts the linear plot downwards. Pretty cool, right? This linear plotting method works for a wide range of physical systems. Here is the equation for the acceleration of an Atwood's machine shown in Figure 3.2—what used to be called a dumbwaiter in hotels. I never try to make it a point to insult people at restaurants, and always try to tip:

$$a = \frac{(m_2 - m_1)g - \tau/R}{(m_1 + m_2)} \tag{3.9}$$

It is a homework problem to derive Equation 3.9, identify the variables and address some plotting issues.

The two-dimensional plots we have been talking about will tell you a wealth of information about the systems you are studying, but I strongly encourage you to look up other types of plots that apply to really cool and interesting systems such as those exhibiting chaos as well as fractals [2].

Figure 3.2. A simplified diagram of Atwood's Machine (dumbwaiter).

3.3 Visualizing Three-Dimensional Data Sets

The next type of plot you might generate are from ordered triplets, where one variable depends on two others. A common example of this is a contour elevation map but another would be the temperature of a stovetop. In such situations, we can use a couple different graphical techniques to see the data nicely. The first is graphing a surface so whatever the dependent variable is would be expressed as height. Figure 3.3 in fact shows the surface of the potentiometric head for a groundwater system [3]. The potentiometric head can be thought of as the height the water column rises to in a well when tapped into the confined or unconfined groundwater system. Sometimes, it is better to actually draw contour lines, which are paths along which the height of the surface is constant. Contour lines are especially useful when we need to calculate gradients, which express the direction of greatest change in the values of the contour lines as we move along the surface (aka uphill and downhill). Figure 3.3 also shows a contour map (without explicit contour lines) of the same groundwater head data. Where the lines are close together the surface is steeper and where they are farther apart it's shallower. It can also be useful to add color as a visualization tool,

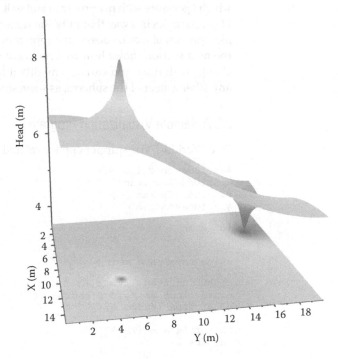

Figure 3.3. Potentiometric head surface and contour plots for a groundwater system with an injection and pumping well (M.W. Roth, C.M. Wilson, M.Z. Iqbal, *Hydrol. Process.*, 17(3519), 2003.) [3]. Grayscale reveals differences in elevation and the electronic version is in color.

so red could represent the highest points on the surface and blue the lowest. It is quite important to note that this text may be printed in grayscale, with figures appearing in electronic versions. As such, you can utilize grayscale to help visualize aspects of the data being presented and defer to the discussion or electronic figures when you need further information or details. Whatever kind of three-dimensional plot you choose to make depends on your data set and what you are trying to extract from it and the context of usage.

3.4 Making Pictures and Movies

Now suppose you are running a simulation and the results actually reflect what the system looks like as stepped through time or maybe just stepped through various configurations as in a Monte Carlo sequence as outlined in Chapter 8. In this case, normally you have outputs comprised of coordinates of particles or various entities comprising your system that need to be tracked through time. I have many such simulations and I use a tool called POV Raytracer. It is great freeware and the link is www.povray.org. The way I incorporate visual rendering in my simulations is that I store coordinates for what the system should look like at certain time steps that I specify. I incorporate the coordinates into a POV file which I generate with my program and so it writes out the header and then the coordinates of the particles in a way that POV Raytracer understands them. I can pick shapes of things like spheres, planes or boxes that represent parts of the system. The example program in the next section shows how to write out a stacked array of spheres and have their colors change with time. You can also modify it to have brightness change with time or size or any other aspect of the spheres, as suggested in the homework section.

3.5 A Sample Visualization Program

(See Working_viz_snippet.cpp for related code)

```cpp
#include<cstdlib>
#include<iostream>
#include<fstream>
#include<math.h>

int main()
{

using namespace std;

const int nx=10,ny=10,nz=10,n=1000,nfiles=10;
const float abox=10.0,bbox=10.0,cbox=10.0;
const float rad=0.5;

char filename [7];
int iball;

float x[n],y[n],z[n];
float red,green,blue;

for(int ifile=1;ifile<=nfiles;ifile++)
{

sprintf(filename,"%d.pov",ifile);

iball=-1;
for(int i=0;i<=nx-1;i++)
{
for(int j=0;j<=ny-1;j++)
{
for(int k=0;k<=nz-1;k++)
{
iball=iball+1;
x[iball]=abox*float(i)/float(nx);
y[iball]=bbox*float(j)/float(ny);
z[iball]=cbox*float(k)/float(nz);
}
}
}

ofstream myfile;
myfile.open(filename);

myfile<<"#version 3.6"<<endl;
myfile<<"#include "<<"\""<<"colors.inc"<<"\""<<endl;
myfile<<"global_settings"<<endl;
myfile<<"{"<<endl;
```

```
myfile<<"assumed_gamma 1.0"<<endl;
myfile<<"}"<<endl;
myfile<<"camera"<<endl;
myfile<<"{"<<endl;
myfile<<"location"<<"<"<<2.5*abox<<","<<2.5*bbox<<","<<2.5*cbox<<">"<<endl;
myfile<<"direction 1.5*z"<<endl;
myfile<<"right 4/3*z"<<endl;
myfile<<"look_at   <"<<abox/2.0<<","<<bbox/2.0<<","<<cbox/2.0<<">"<<endl;
myfile<<"}"<<endl;
myfile<<"background { color red 0.5 green 0.5 blue 0.5 }"<<endl;
myfile<<"light_source"<<endl;
myfile<<"{"<<endl;
myfile<<"0*x"<<endl;
myfile<< "color red"<<endl;
myfile<<"1.0   green 1.0   blue 1.0"<<endl;
myfile<<"translate <-30, 30, -30>"<<endl;
myfile<<"}"<<endl;
myfile<<"light_source"<<endl;
myfile<<"{"<<endl;
myfile<<"0*x"<<endl;
myfile<<"color red"<<endl;
myfile<< "0.5   green 0.5   blue 0.5"<<endl;
myfile<<"translate <30, 30, -30>"<<endl;
myfile<<"}"<<endl;

iball=-1;
for(int i=0;i<=nx-1;i++)
{
for(int j=0;j<=ny-1;j++)
{
for(int k=0;k<=nz-1;k++)
{
iball=iball+1;
red=float(i)/float(nx);
red=red*float(ifile)/float(nfiles);
green=float(j)/float(ny);
green=green*float(ifile)/float(nfiles);
blue=float(k)/float(nz);
blue=blue*float(ifile)/float(nfiles);
myfile<<"sphere {
<"<<x[iball]<<","<<y[iball]<<","<<z[iball]<<">,"<<rad<<endl;
myfile<<"texture {pigment {color rgb<"<<red<<","<<green<<","<<blue<<">}"<<endl;
myfile<<"finish{specular 1}} }"<<endl;
}
}
}

myfile.close();
}
return(0);
}
```

3.6 Four- and Higher-Dimensional Visualization: Yes, It Really Works!

Writing out visualization files in the way presented here is very useful not only because we can track the system through time but we can actually have four- and higher-dimensional representations of physical systems. In the snapshot (Figure 3.4) of a simulation of a real-life ping-pong ball cannon, we can see the ball colliding with a pop can. Moreover, there

Figure 3.4. Snapshot of a Material Point Method simulation of a ping-pong ball going through a pop can. Grayscale shades reveal pressure differences in the ball and the electronic version is in color.

is another ball above the system, which is a direct copy of the ball inside the can. This second ball allows us to examine deformation and oscillations of the ball as well as pressure waves traveling through the ball visualized as color. Then it's a four-dimensional plot. Clearly there are limitations as to how many dimensions we can actually graph but in theory we could have the pressure be color, the temperature be brightness, and we could also have speed be maybe reflectivity or something like that, but after some time it would be very hard to actually interpret the

visual. I am personally convinced that graphs with about five dimensions are all we can really effectively interpret but I implicitly encourage you to push the limits. Once we have rendered a POV file we can then look at the picture that is made but what if we have a simulation that writes out many POV files with time as in the example program, and we want to watch an animation? The old adage that a picture is worth a thousand words might be true, and if one believes that, then that a movie is worth a thousand words times the number of frames but you get more. A movie is worth a thousand words times the number of frames times plus the interaction between frames, which is the kicker that adds to the experience. It carries an aspect of the history of the simulation. As such we would like to smoothly string together or concatenate the picture files we generate and I use a nice freeware tool called Bink and Smacker (www.radgametools.org) I use for that and it is a homework problem to take the files generated by the example program and turn them into a movie.

3.7 Cross-Sensory Visualization: What If You Can't See or Hear?

One of the most wonderful aspects of teaching is that I get to not only build relationships with students but also am constantly challenged to show creativity in the methods I use. I clearly recall one of my most mortifying experiences, the account of which pushed me far away from my comfort zone but also made me more aware of the diverse needs of the students in my classroom. It was the first day of the Fall 1995 semester in *Conceptual Physics*. The lab section had assembled and everything was good to go. I was ready to show one of my favorite videotapes (remember it was 1995), which was *The Incredible Machine* (a National Geographic documentary from the mid 1970's). The first four words that were spoken—or at least that echo in the halls of my memory—were "We are visual creatures." Now, that's all well and good for those of us blessed with sight but I had a nearly completely visually impaired student there that semester. So, I sweated as the video played and spoke to the student after class. Throughout the semester we really had a good time with it all and one of my most precious moments as a teacher came when we worked through optical resolution together by having him feel a football and then put gloves on and feel it again, the point being that he now could not distinguish bumps that were previously detectable as separate without gloves. The bigger picture I walked away with was that it is important to help make physics assessable to all persons and cross-sensory visualization is central to that.

Relevant to simulation comes an example where a visually impaired person could visualize a popular computational problem: the cooling of a coffee cup when cream may or may not be added at a particular time in the example program to follow. A cup of coffee has initial temperature Ti and the ambient environment is at temperature Ts. The changes in temperature T are governed by the initial value problem below, where r is a positive constant.

$$\begin{cases} \dfrac{dT}{dt} = -r(T - T_S) \\ \quad T(0) = T_i \end{cases} \tag{3.10}$$

As a note, all three types of heat transfer are responsible for temperature changes, so cooling is actually a complicated process and there are assumptions made along with the equation's application. Further background reading on the problem can be found in Reference 4. Please review CoffeeCool.cpp for related code.

```cpp
#include<cstdlib>
#include<iostream>
#include<math.h>

using namespace std;

int main()
{
//Now declare variables
//loop index i
//Total number of steps nsteps
//Variable temperature T
//Initial temperature Ti
//Ambient temperature Ts
//Quitting temperature Tquit
//time step dt
//time in the simulation time
//total simulated time total_time
//4th order Runge Kutta constants k1,k2.k3.k4 s
//Cram addition flag creamflag
   int i,nsteps,creamflag;
   double T,r,Ts,Ti,Tquit,time,k1,k2,k3,k4;
//Declare and initialize
   double dt = 0.0001;
   double total_time=1000.0;
//User input
   cout<<"\nWhat is the initial temperature of the coffee?";
   cin>>Ti;
   cout<<"\nWhat is the temperature of the surroundings?";
   cin>>Ts;
   cout<<"\nWhat is the simulation's ending temperature?";
   cin>>Tquit;
   cout<<"\nWhat is the cooling constant?\n";
   cin>>r;
   cout<<"Enter 1 if you are adding no cream;\n";
   cout<<"Enter 2 if you are adding cream at first;\n";
   cout<<"Enter 3 if you are adding cream at T = 80C;\n";
   cin>>creamflag;
//Figure out constants in the program and write everything out
   nsteps=int(total_time/dt)+1;
   cout<<"The initial temperature is: "<<Ti<<"\n";
```

```
   cout<<"The surrounding temperature is: "<<Ts<<"\n";
   cout<<"The cooling constant is: "<<r<<"\n";
   cout<<"The time step is: "<<dt<<"\n";
   cout<<"The number of steps is: "<<nsteps<<"\n";
   cout<<"The total simulated time is: "<<total_time<<"\n";
   cout<<"Time    Temperature\n";//Loop output header;
//Initialize temperature;
   if (creamflag==1|creamflag==3)
   {
   T=Ti;
   }
   if (creamflag==2)
   {
   T=Ti-5;
   }
   time=0.0;
   cout<<time<<"      "<<T<<"\n";
//Start time loop...4th order Runge-Kutta integrator

   for(i=1; i<nsteps; i++)
   {
   time=float(i)*dt;
   k1=-r*(T-Ts)*dt;
   k2=-r*(T+k1/2.0-Ts)*dt;
   k3=-r*(T+k2/2.0-Ts)*dt;
   k4=-r*(T+k3-Ts)*dt;
   T=T+(k1+2.0*k2+2.0*k3+k4)/6.0;
   if(creamflag==3&&abs(T-80.0)<=0.01)
   {
   T=T-5.0;
   }
   cout<<time<<"      "<<T<<"\n";
   if (T<=Tquit) break;
   }
//End time loop
return(0);
}
```

So out of the simulation comes temperatures at different times. For a visually impaired person you could have a clock or metronome ticking for visualization 0f time, and a certain pitch of sound signifying temperature. The addition of cream could even be a splash or click. Using your creativity you can advance the level of your visualizationm rather easily.

3.8 Limiting Cases and Effective (Reduced) Systems

Many times in physics it is important to use special techniques to visualize your system. The first technique is looking at your system as a *reduced* or *effective* system, which means to conceptualize your system as if it were a simpler one with parameters and aspects that you can write down. The second I'm going to talk about is the use of *limiting cases*, which means to take certain aspects of your system and exaggerate them to the point where for all intents and purposes your system becomes a simpler one, the results of which you know well. Limiting cases can also be very useful in conceptualizing limits in beginning studies of calculus. I have chosen one example to deal with both techniques so let's get rolling (you'll get that one later)!

3.8.1 Visualizing a System as a Simpler (Reduced) One

Consider a block of mass m with four uniform solid wheels of mass M and radius R sitting on a level surface. There is perfect tractive friction between the tires and the road so there is no slipping, but there is some rolling friction at the axle–bearing interface of the wheel, which expresses itself as a retarding torque. Let's now write an expression for the acceleration a of the block if we apply a force F to it. We will assume that the block is already moving so kinetic frictional torque is the only torque acting and then each wheel presents as a retarding friction f. It's a homework problem to figure out why I just made the previous statement. This is a very nice problem because we now need to do a (linear) force balance as well as (rotational) torque balance on the wheels. So let's start. We can write down the linear force balance using Newton's second law $\vec{F} = m\vec{a}$, which becomes for this system

$$F - 4f = (m + 4M)a \tag{3.11}$$

We clearly have to do more work, so let's now include a torque balance $t_{net} = I\alpha$ for each wheel:

$$fR - \tau = I\alpha. \tag{3.12}$$

Using the conversion between linear and rotational motion $a = R\alpha$ and the expression for the moment of inertia of a wheel $I = KMR^2$ we can rewrite Equation 3.12

$$fR - \tau = KMaR \tag{3.13}$$

From which we can write an expression for f:

$$f = KMa + \frac{\tau}{R} \tag{3.14}$$

Plugging Equation 3.14 into Equation 3.11 we get

$$a = \frac{F}{(m + 4[1 + K]M)} - \frac{4\tau}{R(m + 4[1 + K]M)} \tag{3.15}$$

yep it's a homework problem to derive Equation 3.15!

Now if we compare Equations 3.15 and 3.8 for the block sliding on a surface we see something very neat! The cart with four wheels has the same equation of motion as the block sliding on a table

$$a = \frac{F}{\text{effective mass}} - \text{friction term} \tag{3.16}$$

only with an effective mass of $m + 4(1 + K)M$. This means that the system here with the wheels acts like a block without wheels if it had the mass and inertia of the wheels stuffed into it. I'd like for you to identify how the friction terms compare between Equations 3.15 and 3.8 as a homework problem: what is the effective coefficient of friction for the cart? Alternatively, and something that requires a whole lot more coffee and mental energy is to reframe the whole problem as a rotational one so we can write the angular acceleration as

$$\alpha = \frac{FR}{(m + 4[1 + K]M)R^2} - \frac{4\tau}{R^2(m + 4[1 + K]M)} \tag{3.17}$$

meaning now we are acting like we are pushing one wheel on an axle with the mass of the cart stuffed into it expressed as an inertia!! It is a homework problem to figure out what the effective moment of inertia for that "wheel" and frictional torque would be, and I wrote Equation 3.15 in a leading way. It can be beneficial to look at complicated systems as behaving like complete simpler systems—it helps us to compare, characterize, measure, and analyze much better. And, in the cases here, note that we can still pull the friction off separately if we were to make the appropriate straight-line graphs.

3.8.2 Exaggerating Aspects of a System: Limiting Cases

Now we can investigate limiting cases to further confirm that we have written down expressions describing behavior of the system that are reasonable. Unlike with reduced systems, we won't be making direct comparisons with other systems but will rather be reducing it to simpler ones by exaggerating various aspects of it. Consider the system described by Equation 3.15. Let's imagine that the mass of the block is much larger than that of the wheels of the cart, something like a Mack truck with pennies for wheels. Yeah, I know… just work with me here… we should first ask ourselves what we expect the acceleration to be. Without the wheels mattering, this thing should look like a block moving under an applied external force and subject to friction. Mathematically, we have a *limit*:

$$\lim_{\frac{M}{m} \to 0}\left[\frac{F}{m\left(1 + 4(1+K)\frac{M}{m}\right)} - \frac{4\tau}{Rm\left(1 + 4(1+K)\frac{M}{m}\right)} \right] \tag{3.18}$$

I used fractions in the denominator because when we say that something is "large' or "small" it is always meant to be in *comparison* to other quantities. When we evaluate the limit in Equation 3.16 we get

$$a = \frac{F}{m} - \frac{4\tau}{Rm} \tag{3.19}$$

Which is exactly of the form of Equation 3.4, so we do have our "sliding block"! Now if we take away sliding friction from the systems described by Equation 3.8 and rolling friction from Equation 3.15 we get exactly the same relationship, $a = F/m$. Why do these two different types of friction give the same overall dynamical results, and why do we get a different result if we remove sliding friction for the cart with wheels? Those questions are in the homework! At any rate, it might not be very exciting to push a block without friction but since the cart with wheels looks like the simpler sliding block in the limiting case, it is more likely that the mathematical expression describing its behavior is correct. If we are comparing our system to a mouse trap atop four grinding wheels on axles then the car's mass m would not matter and we essentially wind up pushing four wheels on axles. When we are looking at effective systems and limiting cases we are looking at part of the behavior and minimizing other parts and we have to make sure that our correspondence is relevant to the context of your work. Several limiting cases as well as effective systems are in the homework for you to consider.

Such tools are very important and useful in visualizing your data and systems you are modeling, and in enriching your simulation experience.

3.9 Visualizing Calculus Part I: Derivatives

We have talked about various techniques to visualize the outcomes of your computer simulations. Because this textbook is geared for people who may not have extensive training in computer coding or mathematics, it is very important to also talk about techniques we can use in order to visualize higher math, especially with reference to calculus. Of course what I am going to talk about is not the only perspective on picturing higher math but it may be useful to students who are beginning their venture in a calculus-based academic track. The first point of discussion I would like to have with you is to offer some techniques regarding how to visualize *derivatives*. When you study beginning calculus you will track through limits and finite differences in order to understand the mechanics of where these derivatives come from. For such treatment I refer you out to some very good mathematics texts, such as Reference [5]. For the purpose of our discussion here, however, I want to briefly introduce how we get to derivatives and concentrate on visualizing them and their physical relevance. So let's start off with something very familiar: a straight line $y = mx + b$ where m (slope) is some finite number as well as is b (intercept). Please be careful to keep the "e" in the slope because without that all you have is slop and none of us wants to work that way. Okay so anyway, imagine the slope of this line. The slope is defined to be the rise over the run, which is equal to $\Delta y/\Delta x$. Suppose that our graph is one of position vs. time or velocity vs. time. When you look at the rise over run you will realize that all we're doing is providing a change in position over a change in time—or a change in velocity over a change in time, respectively. With time being the independent variable we can say then that the slope of this line is the rate of change of the position or the rate of change of velocity, which ever you decided to take the slope of. We can generalize such a concept to plotting something like a height of a highway/the distance traveled in a particular direction, then the slope of the resulting curve would signify a spatial rate of change—not a temporal rate of change like we had before. We could also graph something like cost vs.

supply, and so we can again speak about a rate of a totally different pair of related variables whose changes affect each other. Notice that the intercept is just an offset and so the slope contains all the information about the rate of change. In real life, however, there are very few processes that are *perfect* straight lines. Imagine for now that we are changing our velocity such that our velocity vs. time is shown in Figure 3.5. Now when we talk about a single slope that is really a meaningless idea because this is not a straight line. What we can do, however, is look at a small section of a curve and get an idea of what's happening. So look at two different velocities at two different times as shown in Figure 3.5 and let's connect them with a straight line. The slope of the line still has a rise over run, which is the *average* rate of change of velocity with time over the

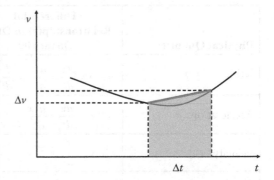

Figure 3.5. Typical velocity vs. time plot with illustrations of slope and area under the curve that lead towards concepts of derivatives and integrals.

particular time interval we picked. And yes, if we pick two points at the same vertical level then the average rate of change would be 0. The preceding idea of course also works for position vs. time graphs or any other as well provided the quantities are well behaved and the curve is continuous, which typically is the case for graphing physical quantities. OK so back to the graph… now bring the two points you selected a little bit closer and connect them, as before. You will get a new value for an average slope that may or may not be the same as in the previous case. So, let's continue making the two points closer and closer… closer yet… closer but not touching. Something very special happens if you keep this process up; we then say mathematically that we are making them *arbitrarily close.* That means we can bring them as close to each other as we want without touching. In mathematics we can make the points *infinitesimally close together* and then the average slope we calculate is actually the *instantaneous* slope of the curve. What's more, the line connecting two infinitesimally close points touches the curve at a single point and is called the *tangent line* to the curve (from Latin *tangens*, to touch). In fact, if a tangent curve walked into my office and wanted help making a resume, its career objective would be to touch a curve at one point! So the slope of the tangent line to the velocity curve we just constructed has a very special name in calculus, it is called the *first derivative* of the curve. Specifically, it is the first derivative of velocity with respect to time. So it is a rate of change but it is specifically the *instantaneous* rate of change of the dependent variable with respect to the independent variable. So the first derivative of velocity with respect to time is nothing other than acceleration:

$$a = \frac{dv}{dt} \tag{3.20}$$

So we can visualize a derivative of a curve at any point as the slope of a line that is tangent to the curve at that point. Velocity is the first derivative of position with respect to time and jerk is the third derivative of position with respect to time. Table 3.1 summarizes differential relationships between physical quantities that seem particularly useful in visualization.

We are talking here about functions that apply to physical systems and so they are said to be "well-behaved." When the curve is broken or discontinuous, or a slope is infinite then we have to bring more complicated physics and math into the picture and I strongly recommend external references if you are interested [5–7]. Derivatives are everywhere in nature and in physics and can be visualized as rates of change as well as slopes of tangent lines. The numerical methods covered in the next chapter express how to communicate them to the computer so you can calculate them.

Table 3.1. Differential Relationships between Various Physical Quantities that Lend Themselves to Visualization

Physical Quantity	Differential Relationship(s) to Other Quantities
Jerk	$j = \dfrac{da}{dt} = \dfrac{d^2v}{dt^2} = \dfrac{d^3v}{dt^3}$
Acceleration	$a = \dfrac{dv}{dt} = \dfrac{d^2x}{dt^2}$
Velocity	$v = \dfrac{dx}{dt}$
Surface area of sphere	$A = \dfrac{dV}{dR}$
Circumference of circle	$C = \dfrac{dA}{dR}$

3.10 Visualizing Calculus Part II: Integrals

In addition to talking about ways in which to visualize derivatives, we need to also take a stab at visualizing *integrals*. I always have said that calculus is an integral part of physics and of course now we will see that!! Yeah, ok back to the technical stuff for now. Just as we call on derivatives when we need a rate of change, we use integrals when we have to calculate things that require us to break a system into pieces and then recompose the calculation of it as a sum of the parts. Such is usually the case when we know how a small piece of a system behaves but not the system as a whole... there is no formula for the volume of my coffee cup but if I break it down into small cubes I can add those volumes up. Let's start out by talking about calculating the mass of a solid ball of constant density ρ. We certainly can say that the total mass is $M = \rho V$, and if we want to get specific we can say $M = (4/3)\pi\rho R^3$. Now that's not too heavy of a calculation right now, but let's up the ante. The mass–volume relationship applies actually for any system for which we know the *average* density so we can say $M = \bar{\rho}V$ and this applies to objects even for which the density varies within them. Let's say that we have a ball with varying density $\rho(r)$ and we do not know off hand what the average density is. We have no real textbook formula for the mass. But think about this though, we can break the object into small pieces. And we can do that in several different ways. Let's say for now we decided to break up the sphere into N very small shells, each of volume ΔV_i thickness Δr. We would then look at the concentric shells formed by this process and add up the mass of the shells to get the total:

$$M = \sum_{i=1}^{N}\Delta m_i = \sum_{i=1}^{N}\rho(r_i)\Delta V_i = \sum_{i=1}^{N}4\pi\rho(r_i)r_i^2\Delta r. \tag{3.21}$$

There are ways to write the sum in Equation 3.21 such that you are using the inside of the shell, outside, or middle as a reference and you will need to pay attention to such sources of error when you calculate such sums. But let's now go ahead and cut up the object into infinitely many infinitely small pieces. Then the sum becomes a *Riemann Sum* and looks like

$$M = \lim_{N\to\infty}\sum_{i=1}^{N}4\pi\rho(r_i)r_i^2\Delta r. \tag{3.22}$$

In addition to our not having to worry about the inside, outside or middle of the shells any more, something very special happens to the sum. We say that each small piece has a differential mass dm, which is an arbitrarily small mass. If we add up all the differential masses, we have to do it in a special way because we have infinitely many, and the Riemann Sum in Equation 3.22 becomes an *integral*. Then we have

$$M = \int_{Sphere} dm = \int_{Sphere} \rho(r)dV = \int_{0}^{R} 4\pi\rho(r)r^2 dr. \tag{3.23}$$

This particular integral is easily evaluated to give $M = (4/3)\pi\rho R^3$ and as a homework problem I am asking you to evaluate the integral and verify. Integral calculus is useful

Table 3.2. Integral Relationships between Various Physical Quantities that Lend Themselves to Visualization

Physical Quantity	Differential Relationship(s) to Other Quantities
Position	$\Delta x = \int_{t_1}^{t_2} v(t)dt$ or $x(t) = \int v(t)dt$
Velocity	$\Delta v = \int_{t_1}^{t_2} a(t)dt$ or $v(t) = \int a(t)dt$
Acceleration	$\Delta a = \int_{t_1}^{t_2} j(t)dt$ or $a(t) = \int j(t)dt$
Mass of sphere	$M = \int_{0}^{R} 4\pi r^2 \rho(r)dr$
Mass of circular disc having area mass density $\sigma(r)$	$M = \int_{0}^{R} 2\pi r \sigma(r)dr$

anywhere in physics where you know how the differential parts behave and you have to find the behavior of the whole, such as with calculating electric fields, gravitational fields, masses, and so much more.

Now integral calculus also ties into kinematics and motion because a change in acceleration over a given time interval is the integral of the jerk with respect to time and the change in velocity is the integral of the acceleration, and change in position is the integral of velocity. You can figure out relationships between various quantities and their integrals many times just by looking at units. Now, let's go back to the velocity vs. time graph of Figure 3.5 and let's look at it from the perspective of integrals. Consider a slice underneath the velocity vs. time graph. Provided the slice is very small its height has the value of the velocity, and width is equal to the small time interval Δt and so the small area end of that small part that we picked is equal to the small change in velocity over that small time interval. Now suppose you break up the area under the curve in some larger time interval $[t_1, t_2]$ into many rectangles. We can then add up the areas of the small rectangles to find the total change in velocity Δv over the entire time interval. As in our discussion with the sphere, we can use the left side of the rectangles, right hand side, or actually consider them as trapezoids and we will discuss some of these important ideas when we talk about numerical integration. As before, let's let the size of those rectangles get smaller and smaller. In fact, in the infinite limit the error of integration actually goes to zero. In that case, the rectangles have differential with dt and their area is not an actual area but it is a differential area $dx = v\, dt$. So, we have infinitely many differential rectangles, so we cannot just do a regular sum to add them up. We have a Riemann Sum as before which in the infinite limit becomes an integral:

$$\Delta x = \int_{t_1}^{t_2} v(t)dt \tag{3.24}$$

We say that the change in position from t_1 to t_2 is the integral from t_1 to t_2 of the velocity with respect to time. So now comes our second important way to visualize integrals: as areas under curves. And if the curve is just below the horizontal axis its contribution to the total area would be negative. Table 3.2 summarizes integral relationships between various physical quantities that are amenable to visualization, including definite and indefinite integrals for kinematic quantities.

The integrals I am showing you here are *definite* integrals; there are also *proper* integrals, *improper* integrals and *indefinite* integrals. There are *surface* integrals *volume* integrals *boundary* integrals, *double* integrals and *triple* integrals. The treatment here on calculus is meant to be a highlight reel that you can use when you are mostly simulating physical systems, but I thoroughly encourage you to read regarding the formal development of integrals and derivatives in a standard University calculus text [5–7].

3.11 Critically Thinking about How Best to Visualize Your Results

This textbook is designed to introduce you to techniques that I have found useful in trying to get students to critically think about how to model everyday phenomena. I refer you out to References [5–7] where there are copious amounts of textbooks giving certain topics a treatment; such is the case for higher math. There are elements of differential and integral calculus, Stokes' theorem, conformal mapping, the gradient operator, and things like that that are useful when you talk about solving partial differential equations that are widely used in physics 6,7. There's wonderful books and visualization tools on that stuff. I want to do something new here.

It takes a while to get a knack for excellent visualization of your results. How far in detail should you go? For example, if you're doing accident reconstruction should the wheels on your cars roll? Should the cars be painted different colors? Many such questions depend on the context of the project you're working with and the aim(s) of your work. Would you use colors to designate temperature, speed, or energy? Are any individuals you are working with either vision or hearing impaired? As you advance in your careers and develop an increasingly sophisticated sense of visualization, I strongly encourage you to play with different techniques and try new things so you can explore how best to communicate your results to interested parties. An even bigger and arguably more important challenge is to get people interested in your work and simulation in general, especially those who normally would not be.

3.12 Examples of Visualization and Presentation of Data

3.12.1 When Should Data be Presented in Tables?

There are data sets which the readership is likely to grab data from and use it to verify or extend your results. In such cases, tabular data as shown in Tables 3.3 and 3.4 is useful because having the numbers readily available saves your colleagues quite a bit of time. Potential parameters, for example are used in simulations and so in a study where we looked at the release of various noble gas species from Fullerene cages we presented potential parameters for the noble gases [8]:

We also published cage disintegration data in tables, which authors often do when there are a relatively small number of data points and/or rather large uncertainties.

Table 3.3. Parameters for the Nonbonded Lennard-Jones Interaction Potentials in Our Fullerene Cage Disintegration Study

Species	ε_{ij}(K)	σ_{ij} (Å)
He-He	10.80	2.57
Ne-Ne	36.68	2.79
Ar-Ar	120.0	3.38
Kr-Kr	171.0	3.60
Xe-Xe	221.0	4.10
C-C (type 1)	28.00	3.40
C-C (type 2)	34.839	3.805

Source: M.K. Balasubramanya et al., *J. Comput. Theor. Nanosci.*, 5, 627–634, 2008 [8].

Table 3.4. Release Rate Constants for All Endohedral Species at Representative Temperatures

Species	$K\,(T = 4100\text{ K})$, ps^{-1}	$k\,(T = 4900\text{ K})$, ps^{-1}
He	0.00359	0.0784
Ne	0.00029	0.0086
Ar	0.00024	0.0088
Kr	0.00027	0.0035
Xe	0.00081	0.0016

Source: M.K. Balasubramanya et al., *J. Comput. Theor. Nanosci.*, 5, 627–634, 2008 [8].

Note: Uncertainties are on the order of 40% of the measured value.

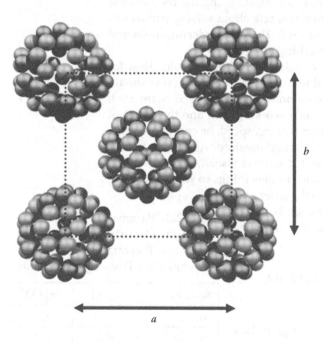

Figure 3.6. Initial conditions utilized for the Molecular Dynamics caged noble gas/Fullerene simulations. The Fullerene molecules are one layer of a fullerite crystal with $a = b = 14.4$ Å. Grayscale helps delineate the following color aspects in the electronic version. The carbon atoms in each fullerene are colored green if they are closest to its center of mass, red if they are farthest away, and blend anywhere in between. The orange atoms inside the fullerene cages are encapsulated Ne atoms (P. Tilton et al., *Mol. Sim.*, 31(11), 951–958, 2007) [9].

3.12.2 Representing Data Visually

I encourage you to inspect all the figures in this chapter, since many of them employ color. Because one really must include the topic of color usage in any treatment of visualization, our discussion here can be relevant for aspects of grayscale printing as well. Please see the note in Section 3.3 about grayscale printing of color figures. Figure 3.6 shows the initial conditions for the Fullerene cage disintegration simulations mentioned in the previous section [9]. The noble gas atom trapped in the cage is orange and the carbon atoms are colored based on their distance from the cage center: green at minimum to red at maximum. Color can be used to track the behavior of atoms or molecules by giving them a particular color based on their initial position in the simulations and then examining how the distribution changes with time (Figure 3.7a,c) [10,11]. In Figure 3.8a and b [10] open vs. closed data markers are used to emphasize the data presented for different layers. In addition, one can connect the data points with straight lines but if there is a lot of scatter then you can fit the data to a curve to convey the trend in the data and wash out the scatter. There could be enough points plotted so that trends are illustrated without any trend lines or connecting lines as in Figure 3.9. In this work [12] we are examining how the orientation of pentacene molecules depends on their height above a graphene surface and so there has to be a lot of statistics in order to see trends. When examining changes in a large number of curves for different cases, it can also be useful to present a cascade plot as in Figure 3.10 [10]. The three phases for the system are broken down by color and subtle changes in appearance of the outer peaks with increasing temperature are visible. It's also possible that it's more effective to simply plot the plots all on the same axis to emphasize subtleties in changes in similar curves such as in the pair correlation functions in Figure 3.11. Color can be used to indicate levels as in pentacene cluster height [12] (Figure 3.12),

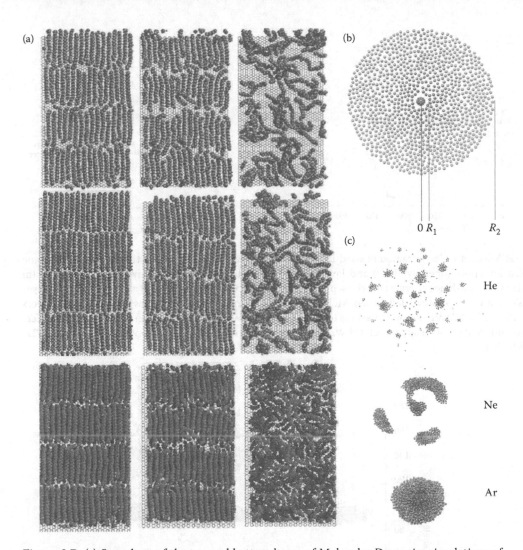

Figure 3.7. (a) Snapshots of the top and bottom layers of Molecular Dynamics simulations of tetracosane on graphite respectively at various temperatures. Can you identify the solid, intermediate smectic, and liquid phase? Grayscale helps delineate the following color aspects in the electronic version. In the upper two panels atoms colored red signify that they are significantly higher off the surface relative to their respective layer and in the bottom panel upper (green) and lower (blue) layers together (M.W. Roth et al., *J. Phys. Chem. C*, 120, 984–994, 2016, doi: 10.1021/acs.jpcc.5b09677) [10]. (b) Initial conditions for the nanoscale Molecular Dynamics rotating disc simulations. Grayscale helps delineate the following color aspects in the electronic version. The blue to green gradient color scale represents closest to farthest initial noble gas atom distance from the central fullerene. The purple line of atoms is used to help understand the system's rotational dynamics. (c) Snapshots of final configurations in our orbiting noble gas simulations at low temperature for He (top), Ne (middle) and Ar (bottom) (E. Maldonado and M.W. Roth, *J. Comput. Theor. Nanosci.*, 5(11), 1–8, 2008) [11].

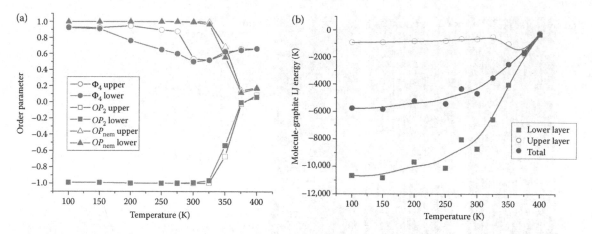

Figure 3.8. (a) Various order parameters used in determining phase transitions for the bilayer tetracosane system plotted for upper (open symbols) and lower (closed symbols) layers of the bilayer v. temperature. (b) Intermolecular Lennard-Jones energies for the separate layers, in between layers, as well as for the entire bilayer tetracosane system v. temperature (left panel). Molecule-graphite Lennard-Jones energies for the separate layers as well as for the entire bilayer system at various temperatures. In the electronic color version, curves for different layers have different colors (right panel) (M.W. Roth et al., *J. Phys. Chem.* C, 120, 984–994, 2016, doi: 10.1021/acs.jpcc.5b09677.) [10].

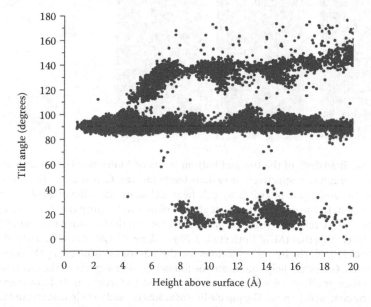

Figure 3.9. An account of how pentacene clusters with increasing height off the graphene surface at room temperature Molecular Dynamics simulations (R. He et al., *Sci. Lett. J.*, 4, 199, 2015, http://www.cognizure.com/scilett.aspx?p=200638809.) [12].

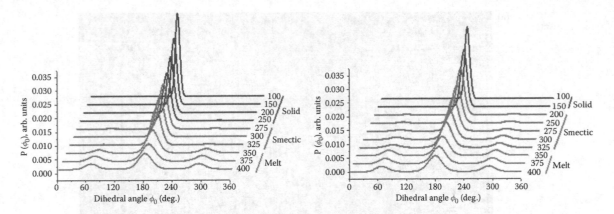

Figure 3.10. Dihedral angle distributions plotted for the upper layer (left panel) and lower layer (right panel) of the tetracosane bilayer for various temperatures. Grayscale shading delineates the different colors the curves have in the electronic version for different phases (M.W. Roth et al., *J. Phys. Chem. C*, 120, 984–994, 2016, doi: 10.1021/acs.jpcc.5b09677.) [10].

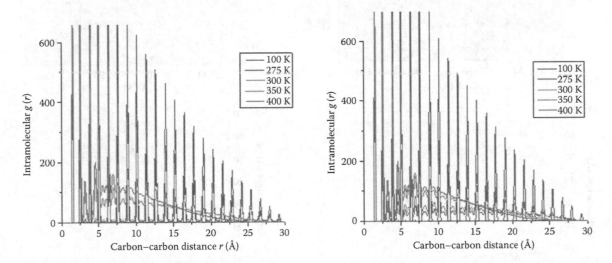

Figure 3.11. Intramolecular pair correlation functions for both layers of the tetracosane bilayer system. Grayscale shading helps delineate the different colors shown for different temperatures in the online version (M.W. Roth et al., *J. Phys. Chem. C*, 120, 984–994, 2016, doi: 10.1021/acs.jpcc.5b09677.) [10].

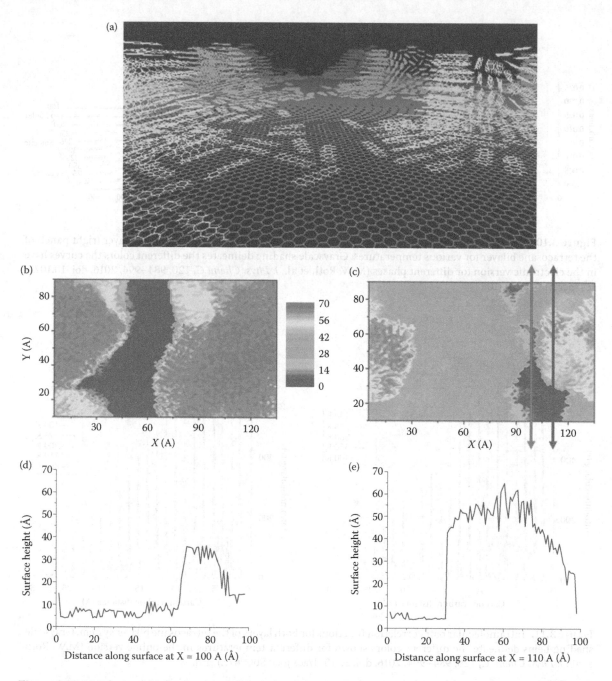

Figure 3.12. Analysis of the fascinating and complicated terrain offered when pentacene condenses out of the gas phase onto a graphene surface including visuals, height maps and terrain cross sections. The eye can use grayscale to delineate much of the information and the electronic version is in color (R. He et al., *Sci. Lett. J.*, 4, 199, 2015, http://www.cognizure.com/scilett.aspx?p=200638809.) [12].

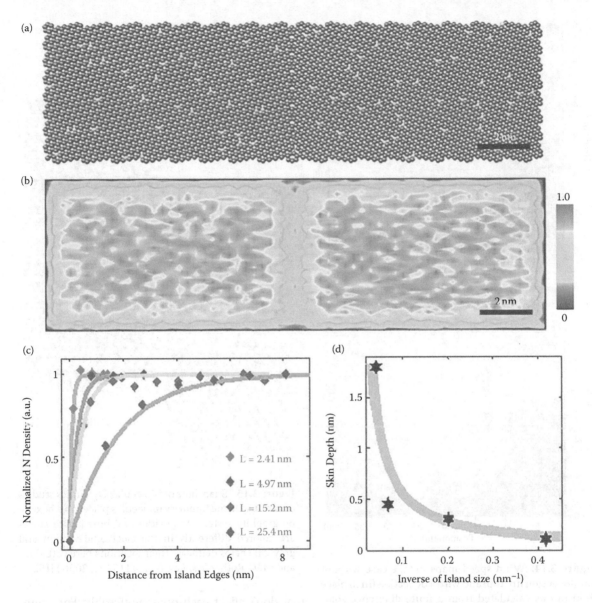

Figure 3.13. Monte Carlo simulations of dopant segregation in N-doped graphene including density profiles, density maps and skin depths. The eye can use grayscale to delineate much of the information and the electronic version is in color (L. Zhao, et al., *Nano Lett.*, 15(2), 1428–1436, 2015, doi: 10.1021/nl504875x.) [13].

amount of nitrogen dopant impurity concentration in a graphene sheet [13] illustrated in Figure 3.13 and how wind speed is affected by the presence of a static deflecting fin (Figure 3.14) [14]. Color can also be used to illustrate the extent of atomic impurities (Figure 3.15) [15] including magnetic behavior (Figure 3.16) [16]. Color can also be used to further engage the reader by cross-correlating different aspects of the system

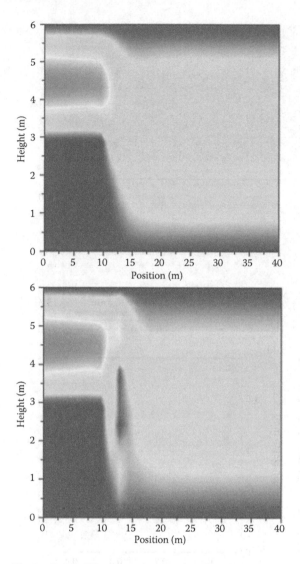

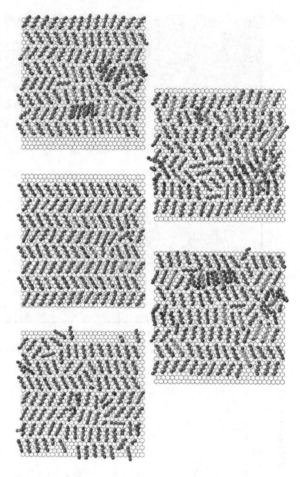

Figure 3.15. Snapshots of Molecular Dynamics studies of various alkane impurity molecules placed in a hexane on graphite system. Impurities and host lattice atoms are colored differently in the electronic version and grayscale helps delineate their locations here (C.L. Pint and M.W. Roth, *Phys. Rev. B.*, 73, 115404, 2006.) [15].

Figure 3.14. Wind speed maps for the case without any deflection fin (top) and with a deflection fin in place (bottom) as calculated from a finite difference solution to the Navier–Stokes equations. Grayscale helps delineate the color scale shown in the electronic version: the color scale ranges from red areas representing highest speeds and blue regions for zero speed and uniform color mixing in between (E. Maldonado and M.W. Roth, *Applied Mathematics and Computation*, submitted.) [14].

that don't affect each other noticeably. For example, Figure 3.17 shows subtle aspects of snowfall on a building and ties them in to how they express themselves on the roof vs. on the ground, which speaks directly to two uniquely different aspects of engineering and oversight [17]. Finally, Figure 3.18 shows court availability for a volleyball and how, if a player hits the ball on a particular color with the hit directed to the center of the ball, it will strike the corresponding color on the court [18].

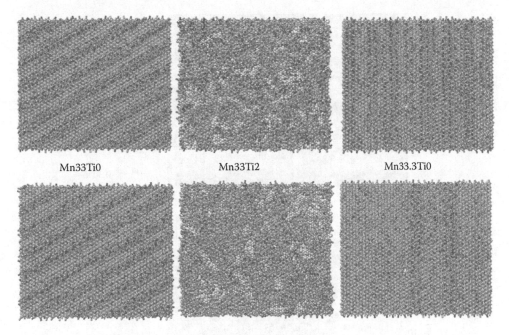

Mn33Ti0   Mn33Ti2   Mn33.3Ti0

Figure 3.16. Final spatial and magnetic configuration snapshots from Monte Carlo simulations of mixed Mn/Ti layers. Grayscale helps delineate aspects shown in the following color scale in the electronic version. Mn is purple, Ti is yellow and the magnetic moments are represented by bars placed through the Mn atoms with black (N) and red (S) ends (M.W. Roth et al., *J Phys Condens Matter.*, 28(18), 184001, May 11, 2016, doi: 10.1088/0953-8984/28/18/184001. Epub 2016 Apr 8.) [16].

Visualizing your work is as much of an art as it is a science. The more you do, the better feel you will develop for the use of color in telling your best story. Teaching has helped me immensely in understanding what techniques of visualization are better than others in various cases.

3.13 Visualizing Various Stages of Cancer Cell Growth

There are times when computer simulations can be used to visualize systems or events without the underlying physics theory being present or driving the simulation. Such efforts can carry much more value than one would think because you can get a feel, for example of the spatial extent or characteristics of a system, its behavior with time or other qualitative aspects of it. Once such application I want to discuss is visualizing cancer cell growth. There are many different stages that have been identified and in your homework I ask you to code some of the ones up I haven't covered here. Although the exercises here are purely visual (certainly in the correct chapter!) and the presentation could vary across different types of cells or population samples, the programs show construction of the membrane cell-by-cell and it's not a stretch to construct biologically meaningful tissue growth simulations from the code presented here.

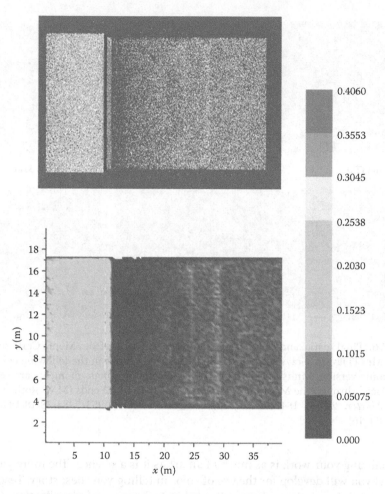

Figure 3.17. Snapshot (top) and contour map (bottom) of the snowfall depth pattern with no deflection fin present. The building's roof is the leftmost rectangle. Grayscale will reveal depth differences and the electronic version is in color.

As a baseline, we need to talk about normal cell growth, which is characterized by uniform cell size/size progression and normal organization (Figure 3.19). Please review the provided program NormalCells.cpp for related code showing step-by-step construction of the tissue.

```
#include<cstdlib>
#include<iostream>
#include<fstream>
#include<math.h>
```

Start the program.

```
int main()
{
using namespace std;
```

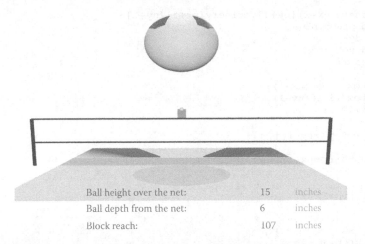

Ball height over the net:	15	inches
Ball depth from the net:	6	inches
Block reach:	107	inches

Figure 3.18. Volleyball court availability for a given block and ball position. A hit on a given marking on the ball directed towards its center will land the ball on the corresponding marking on the court. Grayscale reveals slight differences in markings and the figure is in color for teh electronic version.

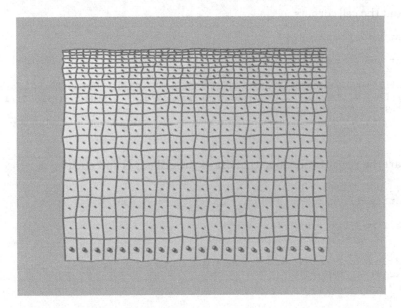

Figure 3.19. One representation of many normal cell growth scenarios within some tissue. Shade scale does not apply here but the program allows you to color the cells as you wish. In the electronic version this tissue sample is green.

Declare and initialize variables.

```
const int nx=20,ny=20;
const float abox=10.0,bbox=10.0;
char filename [7];
int ifile,ix;
int iplaced[nx][ny];
```

```
float cornerx[nx+1][ny+1],cornery[nx+1][ny+1];
float red,green,blue;
float dx,dy;
float rad,sph_rad;
float random;

dx=0.8*abox/float(nx-1);
dy=0.8*bbox/float(ny-1);
random=dx/10.0;

for (int j=0;j<=ny-1;j++)
{
for (int i=0;i<=nx-1;i++)
{
if(j==0){
iplaced[i][j]=1;
}
if(j>0)
{
iplaced[i][j]=0;
}
}
}
```

Figure out the cell corners.

```
for (int j=0;j<=ny;j++)
{
for (int i=0;i<=nx;i++)
{
cornerx[i][j]=0.1*abox+0.9*dx*float(i)+random*3.0*(0.5-(rand()%1000)/1000.0);
cornery[i][j]=0.1*bbox+(1.6*dy*float(j)+random*3.0*(0.5-(rand()%1000)/1000.0))
*exp(-0.8*float(j)/float(ny));
}
}
```

Now start the visualization loop that draws the cell edges and nuclei.

```
ifile=0;
for (int iy=1;iy<=ny-1;iy++)
{
for(int iix=0;iix<=nx-1;iix++)
{
line1:
ix=int(float(nx)*(rand()%1000)/1000.0);
if(iplaced[ix][iy]==1)
{
goto line1;
}
if(iplaced[ix][iy]==0)
{
iplaced[ix][iy]=1;
ifile=ifile+1;

sprintf(filename,"%d.pov",ifile);

ofstream myfile;
myfile.open(filename);

myfile<<"#version 3.6"<<endl;
```

```
myfile<<"#include "<<"\""<<"colors.inc"<<"\""<<endl;
myfile<<"global_settings"<<endl;
myfile<<"{"<<endl;
myfile<<"assumed_gamma 1.0"<<endl;
myfile<<"}"<<endl;
myfile<<"camera"<<endl;
myfile<<"{"<<endl;
myfile<<"location"<<"<"<<abox/2.0<<","<<bbox/2.5<<","<<12.0<<">"<<endl;
myfile<<"direction 1.5*z"<<endl;
myfile<<"right 4/3*z"<<endl;
myfile<<"look_at <"<<abox/2.0<<","<<bbox/2.5<<","<<0.0<<">"<<endl;
myfile<<"}"<<endl;
myfile<<"background { color red 0.5 green 0.5 blue 0.5 }"<<endl;
myfile<<"light_source"<<endl;
myfile<<"{"<<endl;
myfile<<"0*x"<<endl;
myfile<< "color red"<<endl;
myfile<<"0.5 green 0.5 blue 0.5"<<endl;
myfile<<"translate <-30, 30, 30>"<<endl;
myfile<<"}"<<endl;
myfile<<"light_source"<<endl;
myfile<<"{"<<endl;
myfile<<"0*x"<<endl;
myfile<<"color red"<<endl;
myfile<< "0.5 green 0.5 blue 0.5"<<endl;
myfile<<"translate <30, 30, -30>"<<endl;
myfile<<"}"<<endl;

for (int j=0;j<=ny-1;j++)
{
for (int i=0;i<=nx-1;i++)
{
if(j<=0)
{
rad=0.03;
sph_rad=0.06;
}
if(j>0)
{
rad=0.03;
sph_rad=0.025;
}
red=2.0;
blue=2.0;
green=2.5;
if(iplaced[i][j]==1)
{
myfile<<"polygon{4 <"<<cornerx[i][j]<<","<<cornery[i][j]<<">"<<"<"
<<cornerx[i+1][j]<<","<<cornery[i+1][j]<<">"<<"<"<<cornerx[i+1][j+1]<<",
"<<cornery[i+1][j+1]<<">"<<"<"<<cornerx[i][j+1]<<","<<cornery[i][j+1]
<<">"<<endl;
myfile<<"texture {pigment {color rgb<"<<red<<","<<green<<","<<blue<<">
}"<<endl;
myfile<<"finish{specular 1}} "<<endl;
red=0.0;
blue=0.0;
green=1.0;
myfile<<"sphere { <"<<(cornerx[i][j]+cornerx[i+1][j])/2.0<<","
<<(cornery[i][j]+cornery[i][j+1])/2.0<<",0.02>,"<<sph_rad<<endl;
myfile<<"texture {pigment {color rgb<"<<red<<","<<green<<","<<blue<<">
}"<<endl;
```

```
myfile<<"finish{specular 1}} }"<<endl;
myfile<<"cylinder { <"<<cornerx[i][j]<<","<<cornery[i][j]<<","<<0.01<<
">,<"<<cornerx[i+1][j]<<","<<cornery[i+1][j]<<","<<0.01<<">,"<<endl;
myfile<<rad/2.0<<endl;
myfile<<"texture {pigment {color rgb<"<<red<<","<<green<<","<<blue<<">
}"<<endl;
myfile<<"finish{specular 1}} }"<<endl;
myfile<<"cylinder { <"<<cornerx[i+1][j]<<","<<cornery[i+1][j]<<","<<0.01<<
">,<"<<cornerx[i+1][j+1]<<","<<cornery[i+1][j+1]<<","<<0.01<<">,"<<endl;
myfile<<rad/2.0<<endl;
myfile<<"texture {pigment {color rgb<"<<red<<","<<green<<","<<blue<<">
}"<<endl;
myfile<<"finish{specular 1}} }"<<endl;
myfile<<"cylinder { <"<<cornerx[i][j+1]<<","<<cornery[i][j+1]<<","<<0.
01<<">,<"<<cornerx[i+1][j+1]<<","<<cornery[i+1][j+1]<<","<<0.01<<">,
"<<endl;
myfile<<rad/2.0<<endl;
myfile<<"texture {pigment {color rgb<"<<red<<","<<green<<","<<blue<<">
}"<<endl;
myfile<<"finish{specular 1}} }"<<endl;
myfile<<"cylinder { <"<<cornerx[i][j]<<","<<cornery[i][j]<<","<<0.01<<
">,<"<<cornerx[i][j+1]<<","<<cornery[i][j+1]<<","<<0.01<<">,"<<endl;
myfile<<rad/2.0<<endl;
myfile<<"texture {pigment {color rgb<"<<red<<","<<green<<","<<blue<<">
}"<<endl;
myfile<<"finish{specular 1}} }"<<endl;
}
}
}
myfile.close();
}
}
```

Close the Program.

```
}
return(0);
}
```

Hyperplasia is often the initial stage of the development of cancer and entails an increase in cell reproduction rate (and hence cell number) but still presents somewhat normal organization. I've pulled out the sections of the above program that need to be changed in order to visualize hyperplasia instead of normal cell growth below (Figure 3.20).

```
for (int j=0;j<=ny;j++)
{
for (int i=0;i<=nx;i++)
{
cornerx[i][j]=0.1*abox+0.9*dx*float(i)+random*6.0*(0.5-(rand()%1000)/1000.0);
cornery[i][j]=0.1*bbox+(1.6*dy*float(j)+random*6.0*(0.5-(rand()%1000)/1000.0));
}
}

ifile=0;
```

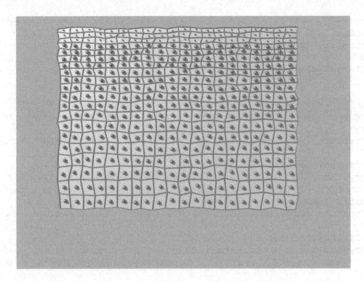

Figure 3.20. One representation of many hyperplastic cell growth scenarios within some tissue. Shade scale does not apply here but the program allows you to color the cells as you wish. In the electronic version this tissue sample is tan.

```
for (int iy=1;iy<=ny-1;iy++)
{
for(int iix=0;iix<=nx-1;iix++)
{
line1:
ix=int(float(nx)*(rand()%1000)/1000.0);
if(iplaced[ix][iy]==1)
{
goto line1;
}
if(iplaced[ix][iy]==0)
{
iplaced[ix][iy]=1;
ifile=ifile+1;

sprintf(filename,"%d.pov",ifile);

ofstream myfile;
myfile.open(filename);
for (int j=0;j<=ny-1;j++)
{
for (int i=0;i<=nx-1;i++)
{
rad=0.03;
sph_rad=0.08;
red=3.5;
blue=2.0;
green=2.5;
if(iplaced[i][j]==1)
{
myfile<<"polygon{4 <"<<cornerx[i][j]<<","<<cornery[i][j]<<">"<<"<"<<
cornerx[i+1][j]<<","<<cornery[i+1][j]<<">"<<"<"<<cornerx[i+1][j+1]<<",
```

```
"<<cornery[i+1][j+1]<<">"<<"<"<<cornerx[i][j+1]<<","<<cornery[i][j+1]
<<">"<<endl;
myfile<<"texture {pigment {color
rgb<"<<red<<","<<green<<","<<blue<<">}"<<endl;
myfile<<"finish{specular 1}} }"<<endl;
red=2.0;
blue=0.0;
green=1.0;
myfile<<"sphere { <"<<(cornerx[i][j]+cornerx[i+1][j])/2.0<<","<<(cornery[i]
[j]+cornery[i][j+1])/2.0<<",0.02>,"<<sph_rad<<endl;
myfile<<"texture {pigment {color rgb<"<<red<<","<<green<<","<<blue<<">}
}"<<endl;
myfile<<"finish{specular 1}} }"<<endl;
myfile<<"cylinder { <"<<cornerx[i][j]<<","<<cornery[i][j]<<","<<0.01<<
">,<"<<cornerx[i+1][j]<<","<<cornery[i+1][j]<<","<<0.01<<">,"<<endl;
myfile<<rad/2.0<<endl;
myfile<<"texture {pigment {color rgb<"<<red<<","<<green<<","<<blue<<">}
}"<<endl;
myfile<<"finish{specular 1}} }"<<endl;
myfile<<"cylinder { <"<<cornerx[i+1][j]<<","<<cornery[i+1][j]<<","<<0.
01<<">,<"<<cornerx[i+1][j+1]<<","<<cornery[i+1][j+1]<<","<<0.01<<">,
"<<endl;
myfile<<rad/2.0<<endl;
myfile<<"texture {pigment {color rgb<"<<red<<","<<green<<","<<blue<<">}
}"<<endl;
myfile<<"finish{specular 1}} }"<<endl;
myfile<<"cylinder { <"<<cornerx[i][j+1]<<","<<cornery[i][j+1]<<","<<0.
01<<">,<"<<cornerx[i+1][j+1]<<","<<cornery[i+1][j+1]<<","<<0.01<<">,
"<<endl;
myfile<<rad/2.0<<endl;
myfile<<"texture {pigment {color rgb<"<<red<<","<<green<<","<<blue<<">}
}"<<endl;
myfile<<"finish{specular 1}} }"<<endl;
myfile<<"cylinder { <"<<cornerx[i][j]<<","<<cornery[i][j]<<","<<0.01<<
">,<"<<cornerx[i][j+1]<<","<<cornery[i][j+1]<<","<<0.01<<">,"<<endl;
myfile<<rad/2.0<<endl;
myfile<<"texture {pigment {color rgb<"<<red<<","<<green<<","<<blue<<">}
}"<<endl;
myfile<<"finish{specular 1}} }"<<endl;
}
}
```

Neoplasia represents new and abnormal tissue growth. It presents disorganized growth as well as an increase in the number of dividing cells. A visual generated by using the code below in the program can be seen in Figure 3.21.

```
for (int j=0;j<=ny;j++)
{
for (int i=0;i<=nx;i++)
{
cornerx[i][j]=0.1*abox+0.9*dx*float(i)+random*3.0*(0.5-(rand()%1000)/1000.0);
cornery[i][j]=0.1*bbox+(1.6*dy*float(j)+random*3.0*(0.5-(rand()%1000)/1000.0))
*exp(-0.5*float(j)/float(ny));
}
}

ifile=0;
for (int iy=1;iy<=ny-1;iy++)
{
```

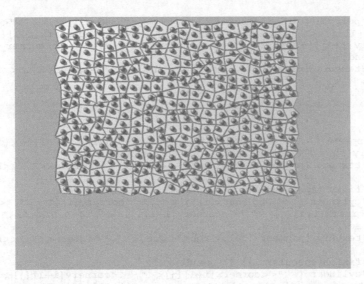

Figure 3.21. One representation of many neoplastic cell growth scenarios within some tissue. Shade scale does not apply here but the program allows you to color the cells as you wish. In the electronic version this tissue sample is salmon.

```
for(int iix=0;iix<=nx-1;iix++)
{
line1:
ix=int(float(nx)*(rand()%1000)/1000.0);
if(iplaced[ix][iy]==1)
{
goto line1;
}
if(iplaced[ix][iy]==0)
{
iplaced[ix][iy]=1;
ifile=ifile+1;
sprintf(filename,"%d.pov",ifile);
for (int j=0;j<=ny-1;j++)
{
for (int i=0;i<=nx-1;i++)
{
if(j<=4*ny/5)
{
rad=0.03;
sph_rad=0.06;
}
if(j>4*ny/5)
{
rad=0.03;
sph_rad=0.025;
}
red=2.5;
blue=2.0;
green=2.5;
if(iplaced[i][j]==1)
{
```

```
myfile<<"polygon{4 <"<<cornerx[i][j]<<","<<cornery[i][j]<<">"<<"<"<<
cornerx[i+1][j]<<","<<cornery[i+1][j]<<">"<<"<"<<cornerx[i+1][j+1]<<",
"<<cornery[i+1][j+1]<<">"<<"<"<<cornerx[i][j+1]<<","<<cornery[i][j+1]
<<">"<<endl;
myfile<<"texture {pigment {color rgb<"<<red<<","<<green<<","<<blue<<">
}"<<endl;
myfile<<"finish{specular 1}} }"<<endl;
red=1.0;
blue=0.0;
green=1.0;
myfile<<"sphere { <"<<(cornerx[i][j]+cornerx[i+1][j])/2.0<<","<<(cornery[i]
[j]+cornery[i][j+1])/2.0<<",0.02>,"<<sph_rad<<endl;
myfile<<"texture {pigment {color rgb<"<<red<<","<<green<<","<<blue<<">
}"<<endl;
myfile<<"finish{specular 1}} }"<<endl;
myfile<<"cylinder { <"<<cornerx[i][j]<<","<<cornery[i][j]<<","<<0.01<<
">,<"<<cornerx[i+1][j]<<","<<cornery[i+1][j]<<","<<0.01<<">,"<<endl;
myfile<<rad/2.0<<endl;
myfile<<"texture {pigment {color rgb<"<<red<<","<<green<<","<<blue<<">
}"<<endl;
myfile<<"finish{specular 1}} }"<<endl;
myfile<<"cylinder { <"<<cornerx[i+1][j]<<","<<cornery[i+1][j]<<","<<0.
01<<">,<"<<cornerx[i+1][j+1]<<","<<cornery[i+1][j+1]<<","<<0.01<<">,
"<<endl;
myfile<<rad/2.0<<endl;
myfile<<"texture {pigment {color rgb<"<<red<<","<<green<<","<<blue<<">
}"<<endl;
myfile<<"finish{specular 1}} }"<<endl;
myfile<<"cylinder { <"<<cornerx[i][j+1]<<","<<cornery[i][j+1]<<","<<0.
01<<">,<"<<cornerx[i+1][j+1]<<","<<cornery[i+1][j+1]<<","<<0.01<<">,
"<<endl;
myfile<<rad/2.0<<endl;
myfile<<"texture {pigment {color rgb<"<<red<<","<<green<<","<<blue<<">
}"<<endl;
myfile<<"finish{specular 1}} }"<<endl;
myfile<<"cylinder { <"<<cornerx[i][j]<<","<<cornery[i][j]<<","<<0.01<<
">,<"<<cornerx[i][j+1]<<","<<cornery[i][j+1]<<","<<0.01<<">,"<<endl;
myfile<<rad/2.0<<endl;
myfile<<"texture {pigment {color rgb<"<<red<<","<<green<<","<<blue<<">
}"<<endl;
myfile<<"finish{specular 1}} }"<<endl;
}
}
}
myfile.close();
}
}
```

Finally, invasive growth presents a tumor and a visual generated by using the code below in the complete program can be seen in Figure 3.22.

```
for (int iclear=1;iclear<=15000;iclear++)
{
blank=(rand()%1000)/1000.0;
}

for (int j=0;j<=ny;j++)
{
for (int i=0;i<=nx;i++)
```

Figure 3.22. One representation of many invasive tumor cell growth scenarios within some tissue. Shade scale does not apply here but the program allows you to color the cells as you wish. In the electronic version this tissue sample is red.

```
{
cornerx[i][j]=0.1*abox+0.9*dx*float(i)+random*7.0*(0.5-(rand()%1000)/1000.0);
cornery[i][j]=0.1*bbox+(1.6*dy*float(j)+random*7.0*(0.5-(rand()%1000)/1000.0));
}
}

ifile=0;
for (int iiy=0;iiy<=ny-1;iiy++)
{
line1:
for(int iix=0;iix<=nx-1;iix++)
{
ix=int(float(nx)*(rand()%1000)/1000.0);
iy=int(float(ny)*(rand()%1000)/1000.0);
if(iplaced[ix][iy]==1||sqrt((cornerx[ix][iy]-abox/2.0)*(cornerx[ix][iy]-
abox/2.0)+cornery[ix][iy]*cornery[ix][iy])>0.35*abox)
{
goto line1;}
if(iplaced[ix][iy]==0&&sqrt((cornerx[ix][iy]-abox/2.0)*(cornerx[ix][iy]-
abox/2.0)+cornery[ix][iy]*cornery[ix][iy])<=0.35*abox)
{
iplaced[ix][iy]=1;
ifile=ifile+1;

sprintf(filename,"%d.pov",ifile);

ofstream myfile;
myfile.open(filename);

ifile=0;
for (int iiy=0;iiy<=ny-1;iiy++)
{
line1:
```

```
for(int iix=0;iix<=nx-1;iix++)
{
ix=int(float(nx)*(rand()%1000)/1000.0);
iy=int(float(ny)*(rand()%1000)/1000.0);
if(iplaced[ix][iy]==1||sqrt((cornerx[ix][iy]-abox/2.0)*(cornerx[ix][iy]-
abox/2.0)+cornery[ix][iy*cornery[ix][iy])>0.35*abox)
{
goto line1;
}
if(iplaced[ix][iy]==0&&sqrt((cornerx[ix][iy]-abox/2.0)*(cornerx[ix][iy]-
abox/2.0)+cornery[ix][iy*cornery[ix][iy])<=0.35*abox)
{
iplaced[ix][iy]=1;
ifile=ifile+1;

sprintf(filename,"%d.pov",ifile);

for (int j=0;j<=ny-1;j++)
{
for (int i=0;i<=nx-1;i++)
{
rad=0.03;
sph_rad=0.08;
red=2.5;
blue=2.0;
green=2.0;
if(iplaced[i][j]==1)
{
myfile<<"polygon{4 <"<<cornerx[i][j]<<","<<cornery[i][j]<<">"<<"<"
<<cornerx[i+1][j]<<","<<cornery[i+1][j]<<">"<<"<"<<cornerx[i+1][j+1]<<",
"<<cornery[i+1][j+1]<<">"<<"<"<<cornerx[i][j+1]<<","<<cornery[i][j+1]
<<">"<<endl;
myfile<<"texture {pigment {color rgb<"<<red<<","<<green<<","<<blue<<">
}"<<endl;
myfile<<"finish{specular 1}} "<<endl;
red=1.0;
blue=0.0;
green=0.0;
myfile<<"sphere { <"<<(cornerx[i][j]+cornerx[i+1][j])/2.0<<","<<(cornery[i]
[j]+cornery[i][j+1])/2.0<<",0.02>,"<<sph_rad<<endl;
myfile<<"texture {pigment {color rgb<"<<red<<","<<green<<","<<blue<<">
}"<<endl;
myfile<<"finish{specular 1}} "<<endl;
myfile<<"cylinder { <"<<cornerx[i][j]<<","<<cornery[i][j]<<","<<0.01<<
">,<"<<cornerx[i+1][j]<<","<<cornery[i+1][j]<<","<<0.01<<">,"<<endl;
myfile<<rad/2.0<<endl;
myfile<<"texture {pigment {color rgb<"<<red<<","<<green<<","<<blue<<">
}"<<endl;
myfile<<"finish{specular 1}} "<<endl;
myfile<<"cylinder { <"<<cornerx[i+1][j]<<","<<cornery[i+1][j]<<","<<0.
01<<">,<"<<cornerx[i+1][j+1]<<","<<cornery[i+1]
[j+1]<<","<<0.01<<">,"<<endl;
myfile<<rad/2.0<<endl;
myfile<<"texture {pigment {color rgb<"<<red<<","<<green<<","<<blue<<">
}"<<endl;
myfile<<"finish{specular 1}} "<<endl;
myfile<<"cylinder { <"<<cornerx[i][j+1]<<","<<cornery[i][j+1]<<","<<0.
01<<">,<"<<cornerx[i+1][j+1]<<","<<cornery[i+1]
[j+1]<<","<<0.01<<">,"<<endl;
myfile<<rad/2.0<<endl;
```

```
myfile<<"texture {pigment {color rgb<"<<red<<","<<green<<","<<blue<<">
}"<<endl;
myfile<<"finish{specular 1}} "<<endl;
myfile<<"cylinder { <"<<cornerx[i][j]<<","<<cornery[i][j]<<","<<0.01<<
">,<"<<cornerx[i][j+1]<<","<<cornery[i][j+1]<<","<<0.01<<">,"<<endl;
myfile<<rad/2.0<<endl;
myfile<<"texture {pigment {color rgb<"<<red<<","<<green<<","<<blue<<">
}"<<endl;
myfile<<"finish{specular 1}} "<<endl;
}
}
}
myfile.close();
}
}
```

PROBLEMS

3.1. The aspects of centering, spread, and trend all have physical origins in the fireworks shot shown in Figure 3.1. Identify and discuss as many of these as you can, but at least one physical origin for each aspect of the data set.

3.2. This chapter does not discuss modality of a data set, which addresses clustering, and peaks in distributions. Do a bit of outside research on modality and cumulants, and then discuss briefly in your own words.

3.3. Accuracy and precision are terms that compare centering and spread in a data set to target values, respectively. Illustrate data sets which have high accuracy and high precision, high accuracy and low precision, ... for a total of four illustrations. Include qualitative sketches of the data distributions. This one may be best suited for group discussion.

3.4. Let's get trendy. Illustrate data sets in which the data set is strongly correlated, weakly correlated and uncorrelated. In the correlated cases include one example of positive and one example of negative correlation for a total of 5 illustrations.

3.5. Looking at Equation 3.3 and considering its description, discuss what would happen to the calculated value if the absolute value signs in the formula were removed. Now write and run an algorithm to check your prediction and discuss.

3.6. Now deviating from accepted practice, write and execute a program that calculates the average deviation and standard deviation of a data set of 10,000 points where (a) all the points are coincident, (b) half the points have one value and the other half have a different value, and (c) half the points have one value and the other half have a different value but then two are removed and placed at the average value. Comment in any insight these special cases gives you.

3.7. Friction opposes the sliding of one surface on another but won't act unless the tendency to slide is present. Considering a block being pushed across the floor with kinetic friction acting such that it is decelerating, write and execute a program that calculates its motion until it stops. If you code this up somewhat like I do, it will require some conditionals when the block stops.

3.8.   Derive Equation 3.7. If you put the homework off until Sunday, then please derive carefully and watch out for other Sunday derivers as well!!

3.9.   Derive Equation 3.9 with friction in the pulley. First use a massless pulley and then go on to one with mass. How would you use a straight-line plot to measure g and the frictional torque from the pulley separately? Identifying slope and intercept will help.

3.10.   Explore the POV Raytracer and give examples of the things you found that you enjoyed the most or that interested you the most.

3.11.   Modify the sample visualization program so that something about the spheres changes differently from the way I had it set up. Run the program, render the POV files and then make a movie.

3.12.   Why did I make the statement "We will assume that the block is already moving so kinetic frictional torque is the only torque acting and then each wheel presents as a retarding friction f"? In other words, why is it useful to assume the block is already in motion and what complications arise if we don't make that assumption?

3.13.   Derive Equation 3.10

3.14.   Discuss how the friction terms compare between Equations 3.15 and 3.8.

3.15.   Derive Equation 3.12

3.16.   Derive Equation 3.15

3.17.   Draw a free body diagram for the cart in Section 3.8.1, where you explicitly show and the forces and torques acting on the system.

3.18.   Compare the friction terms for sliding block and cart; briefly discuss the significance of the comparison.

3.19.   Starting with Equation 3.17, determine the effective inertia and rotational frictional torque if the system were a single wheel of radius R. Then look at Equation 3.16 and determine effective mass and translational friction force. Reconcile the rotational and translational results.

3.20.   In the limiting case discussion in Section 3.8.2, why do these two different types of friction give the same overall dynamical results, and why do we get a different result if we remove sliding friction for the cart with wheels?

3.21.   Go journal hopping! Examine a few of the references for this chapter (or really anywhere) and find one or two ways to visualize data that I didn't discuss here and report out.

3.22.   Do the integral (Equation 3.23) to find the mass of a uniform sphere with radius 4 cm and density 3 kg/m^3 using both analytical and numerical techniques. Compare the results and discuss how close "close" needs to be for the numerical to be acceptable.

3.23. Code and run the program in Section 3.7 for various scenarios and parameter values. Discuss any interesting findings that jump put at you.

3.24. Using the program in the text, code and run the normal cell visualization with all the cells the same size.

3.25. There are other cell growth and atrophy conditions (with names) that I have not discussed here. Look up two of them and, starting with the code in the text, code and run visualizations for the conditions you identified.

3.26. The cell visualization programs show the step-by-step construction of various membranes. It's not a stretch to actually simulate aspects of real tissue growth. Research how some tissue of interest is built from cells and, starting with the code provided (or your own), simulate tissue growth. Discuss your results.

References

1. J.F. Epperson, *An Introduction to Numerical Methods and Analysis*, Second edition, John Wiley & Sons, Inc., Hoboken, New Jersey, October, 2013.

2. S.H. Strogatz, *Nonlinear Dynamics and Chaos: With Applications to Physics, Biology, Chemistry, and Engineering*, Westview Press, Boulder, CO, Published January 19, 2001.

3. M.W. Roth, C.M. Wilson, M.Z. Iqbal, Transient state groundwater flow in various geometries with many wells and impermeable barriers: Analytical methods, *Hydrol. Process.*, 17(3519), 2003.

4. H. Gould, J. Tobochnik, W. Christian, *An Introduction to Computer Simulation Methods: Applications to Physical Systems*, 3rd edition, Addison-Wesley Longman, Inc., Chicago, IL, 2006.

5. J. Stewart, *Calculus*, Cengage Learning, Blemont, CA, January 1, 2011.

6. G.B. Arfken, H.J. Weber, F.E. Harris, *Mathematical Methods for Physicists: A Comprehensive Guide*, 7th edition, Academic Press, Waltham, MA, January 17, 2012.

7. Div, Grad, Curl, and All that, *An Informal Text on Vector Calculus*, Harry Moritz Schey W.W. Norton, New York, 2005.

8. M.K. Balasubramanya, M.W. Roth, P. Tilton, B. Suchy, Molecular dynamics simulations of noble gas release from endohedral fullerene aggregates due to cage disintegration, *J. Comput. Theor. Nanosci.*, 5, 627–634, 2008.

9. P. Tilton, B. Suchy, M.K. Balasubramanya, M.W. Roth, Simulated dynamics of Ne@C60 clusters beyond dissociation, *Mol. Sim.*, 31(11), 951–958, 2007.

10. M.W. Roth, L. Firlej, B. Kuchta, M.J. Connolly, E. Maldonado, C. Wexler, Simulation and characterization of tetracosane on graphite: Molecular dynamics beyond the monolayer, *J. Phys. Chem. C*, 120, 984–994, 2016, doi: 10.1021/acs.jpcc.5b09677

11. E. Maldonado, M.W. Roth, Deterministic computer simulations of noble gas disks orbiting C60 fullerenes, *J. Comput. Theor. Nanosci.*, 5(11), 1–8, 2008.

12. R. He, F. Carta, M. Ashan, K. Bader, E. Maldonado, C. Delaney, T. Kidd et al., Formation and interaction of self-assembled pentacene structures on monolayer graphene, *Sci. Lett. J.*, 4, 199, 2015, http://www.cognizure.com/scilett.aspx?p=200638809

13. L. Zhao, R. He, A. Zabet-Khosousi, K.S. Kim, T. Schiros, M. Roth, P. Kim, G.W. Flynn, A. Pinczuk, A.N. Pasupathy, Dopant segregation in polycrystalline monolayer graphene, *Nano Lett.*, 15(2), 1428–1436, 2015, doi: 10.1021/nl504875x

14. E. Maldonado and M.W. Roth, A new method for simulating snowfall: Algorithm and application to snowdrift remediation with deflection fins, *Applied Mathematics and Computation*, submitted.

15. C.L. Pint and M.W. Roth, Simulated effects of odd-alkane impurities in a hexane monolayer on graphite, *Phys. Rev. B.*, 73, 115404, 2006.

16. M.W. Roth, B. Wandling, T. Kidd, P.M. Shand, A. Stollenwerk, Simulated structural and magnetic behavior of Mn-Ti intercalated dichalcogenide crystals. *J Phys Condens Matter.*, 28(18), 184001, May 11, 2016, doi: 10.1088/0953-8984/28/18/184001. Epub April 8, 2016.

17. E. Maldonado and M.W. Roth, Direct two—phase numerical simulation of snowdrift remediation using three—dimensional deflection fins, *Journal of Applied Fluid Mechanics*, 5(3), 71–78, 2012.

18. I. Ahrabi-Fard and M.W. Roth, Science: A source for creating new directions in volleyball, *American Volleyball Association Convention*. (December 13, 2001). http://faculty.chas.uni.edu/~rothm/Abstracts/vb_abs.pdf

Models of Everyday Things

CHAPTER 4

Things We See in the News: The Fun and the Dangerous

4.1 Modeling the Flight of Objects through Fluids: Using Science to Play a Better Game

Ideally, the science of computer simulations has the deeper purpose of helping people by adding unique insight to the behavior of systems not afforded in actual life. For example, imagine that an attorney contacts you and has a client with hearing loss because of a baseball hit to the head. Was it reasonable for the owners of the baseball complex to expect that balls would be traveling over their fence? Did the city construct the fence high enough? A computer simulation study could lend insight and help make the difference between winning or losing a lawsuit, preserving or spoiling a reputation, upholding or ruining lives. Or, a volleyball team practices their game at sea level and then plays at high altitude, underperforming dramatically. What did the difference in air density do and how could a coach make up for such a change in the playing environment by altering the practice environment? A won or lost season could result. It could even be that in cases such as an oil spill it would be important to track the motion of tarballs and predict their paths so as to know where to expect them. Simulating the path of such dangerous objects could help people prepare for cleanup and even save entire species of animals in fragile coastline environments. In all three cases above, computer simulations could lend tremendous insight because it is just not practical or sometimes even ethical to repeat what really happened. What's more, all three systems described above, and many more like them, can be described by the same model: objects (in many cases balls) passing through fluids. The general approach here is consistent throughout this text: formulating the model, obtaining governing equations and subsequently discretizing them. After discretization they can then be coded into the computer, the model validated, and then they can be used and applied in uncharted territory.

4.1.1 The Model: Rules of the Game as Equations

In modeling the flight of an object we have to know what alters its motion and so we have to determine the forces acting on it. If we can determine the net (or total) external force, then Newton's second law of motion tells us we can divide total force by mass to obtain acceleration, which gives us exactly what we are after: a tangible handle on changes in the object's motion. In mathematical terms, we start with Newton's second law of motion

$$\vec{F} = m\vec{a} \tag{4.1}$$

and obtain the object's acceleration:

$$\vec{a} = \frac{\vec{F}}{m}. \tag{4.2}$$

Equations 4.1 and 4.2 are vector equations, meaning that they automatically take all three directions into account at once.

The most influential force is that of gravity,

$$\vec{F}_G = -mg\hat{z} \tag{4.3}$$

Here $g = 9.81$ m/sec^2 is the acceleration of gravity at the surface of the earth, m is the mass of the object, and $\hat{z}$ is a unit vector in the vertical direction which, when coupled with the negative sign in front of Equation 4.3 indicates mathematically that gravity acts downward. Typical first-semester physics courses use only the above gravitational force to model projectile motion and hence provides students a solid understanding of two-dimensional kinematics. It should be noted that many expressions used in physics have limitations and are valid in certain situations and not others. Equation 4.3 applies when the variation in height of an object is small compared to the size of the earth or planet, and usually when it is confined to the surface of a planet. For rockets and spacecraft, we need a different model of gravity, which is dealt with elsewhere in this text. In reality, however, such a model does not take into account the fact that the fluid the object is passing through exerts both dynamic and static forces on the ball.

The most influential force from the surrounding fluid is drag resistance [1–3]. Drag resistance can be thought of as friction of an object with the air, because the ball uses energy to move the air around it but it does not get that energy back. In actuality, the energy taken from the ball heats up the air by stirring it and ultimately goes into making the universe expand which, for some of us may be a very useful and perspective—broadening thought on any given Monday morning. Where were we? Oh yes, the drag force increases with fluid density ρ, increases with projected area A (the area that the object's shadow would have) and increases with speed v. The model for drag is empirical, meaning that it is modeled to the physical phenomena rather than derived from

Table 4.1. Projected Areas for Various Common Objects

Object	Projected Area A
Sphere of radius R	πR^2
Cylinder of radius R and length L	$2RL$ sideways; πR^2 end on
Circular disc of radius R, axis rotated at angle θ with respect to its velocity	$\pi R^2 \cos \theta$
Rectangular prism of dimensions L, W and H	LW, LH or WH for three simple orientations

first principles. As a result, there are different powers of v used but because work and energy are important phenomena the power chosen is typically 2: [1–3]

$$\vec{F}_D = -\frac{1}{2} C_D \rho A v^2 \hat{v}. \tag{4.4}$$

Here ρ = 1.292 kg/m³ for air at 0°C and 1000 kg/m³ for fresh water. Moreover, A is the projected area of the object, which can be thought of as the area of its shadow when viewed as directly approaching. v is the speed of the ball in m/sec and $\hat{v}$ is a unit vector of the direction of the ball's velocity, which when coupled with the negative sign in the equation, indicates that this force always acts in opposition to the ball's motion.

Various projected areas are given in Table 4.1.

The drag coefficient C_D depends on the shape of the object, fluid surrounding it, and the speed of the object in the fluid. Various values for C_D in typical situations are shown in Table 4.2.

It should be noted that the rough sphere has more drag only for a certain speed range, but if the boundary layer trips, like for a golf ball traveling fast enough then drag actually reduces [4].

Table 4.2. Values of Drag Coefficients for Various Common Objects

Object	Drag Coefficient C_d
Most aircraft	0.01–0.09
Cars	0.2–0.4
Smooth Sphere	0.25
Bullet	0.4
Rough Sphere	0.57
Bicycle and cyclist	1.0–1.3
Flat plate perpendicular to flow	1.3
Skyscrapers	1.5–2.0
Brick	2.1

The next most important aerodynamic force is the Magnus force [1–5], which describes the lift that a ball can obtain because it is moving through a fluid and spinning:

$$\vec{F}_L = \left(\frac{1}{2} C_L \rho A v^2 \right) (\hat{\omega} \times \hat{v}). \tag{4.5}$$

Here $\vec{\omega}$ is the angular velocity of the ball's rotation in radians/sec and $\hat{\omega}$ is a unit vector in the direction of the angular velocity of the ball, which in our case may be thought of as the axis the ball is spinning about, pointing out of the pole of the ball rotating counterclockwise as viewed by an observer (Figure 4.1). The vector cross product $\hat{\omega} \times \hat{v}$ describes this lift force as being perpendicular to both the rotational axis of the

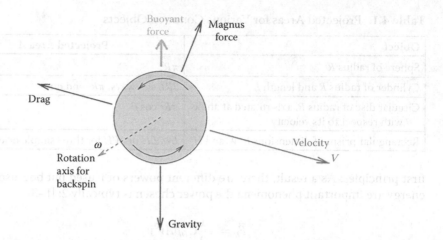

Figure 4.1. Diagram of all the forces acting on the baseball in our model. (Adapted from http://spiff.rit.edu/richmond/baseball/traj/traj.html.) The effect of air on spinning objects was noticed centuries ago and there is some very interesting reading I refer you to such as Isaac Newton, A letter of Mr. Isaac Newton, of the University of Cambridge, containing his new theory about light and color, *Philosophical Transactions of the Royal Society*, 7, 3075–3087, 1671–1672 [16], the work by Magnus himself, G. Magnus, Über die Abweichung der Geschosse, und: Über eine abfallende Erscheinung bei rotierenden Körpern, *Annalen der Physik*, 164(1), 1–29, 1853 [17], and Lord Rayleigh, On the irregular flight of a tennis ball, *Messenger of Mathematics*, 7, 14–16, 1877 [18].

ball and the ball's velocity. Furthermore, if the ball has a topspin, $F_{L, vertical} < 0$, the ball drops faster; if the ball has backspin $F_{L, vertical} > 0$, the ball gains lift. The reason for the Magnus force is that the air moves right along with the boundary of the rotating ball, so the ball "grabs" the air around it and rotates the flow pattern to make it asymmetric, hence resulting in asymmetric pressure differences and a vertical pressure differential. The lift coefficient C_L depends on the nature of airflow around the object and in general is not completely understood. For baseballs C_L is about 0.25 and for a rotating smooth cylinder C_L is 0.4. Whenever modeling an object moving through air one must determine C_D and C_L very carefully, and on a case-by-case basis.

The final force considered in this model is the buoyant force experienced by an object due to being immersed in a fluid of non-zero density (anything other than a vacuum). The expression for the buoyant force is

$$\vec{F}_B = \rho g V \hat{z}, \tag{4.6}$$

where V is the object's volume. It's not so influential for baseballs but central for balloons and tarballs. For the cases presented in this text, either the fluid is very light (air) or the objects do not exhibit large linear accelerations. When fluid mass or acceleration becomes important there is an additional force term—an inertial term, which is a shape-dependent factor times the mass of fluid displaced times the magnitude of the object's acceleration [6]. The specific model along with the issues it is designed to address will determine whether or not inertial fluid forces are important. In some cases, they would need to also be included in the model [6].

4.1.2 How the Simulation will Play Out: Putting the Model into Action

So we have all the forces acting on the ball, as shown in Figure 4.1, and hence the model. But how do we get a handle on the ball's motion? We need to get the equation(s) of motion next. Newton's second law of motion (*net vector force = mass × vector acceleration*) is employed in order to obtain the equation of motion (expression for acceleration) of the ball, which incorporates the previously discussed forces:

$$\vec{a} = -g\hat{z} - \frac{C_d \rho A v^2}{2m}\hat{v} + \frac{C_L \rho A v^2}{2m}\hat{\omega} \times \hat{v} + \frac{\rho}{\rho_b} g\hat{z}. \tag{4.7}$$

Here ρ_b is the average density of the ball, $\rho_b = m/\left(\frac{4}{3}\pi r^3\right)$. Now since Equation 4.7 is a vector equation it may be separated into three coupled scalar differential equations: one for the horizontal (x), lateral (y), and vertical (z) directions:

$$a_x = -\frac{C_d \rho A v v_x}{2m} - \frac{C_L \rho A v(\omega_y v_z - \omega_y v_z)}{2m\omega} \tag{4.8}$$

$$a_y = -\frac{C_d \rho A v v_y}{2m} + \frac{C_L \rho A v(\omega_z v_x - \omega_x v_z)}{2m\omega} \tag{4.9}$$

$$a_z = -\frac{C_d \rho A v v_z}{2m} + \frac{C_L \rho A v(\omega_y v_x - \omega_x v_y)}{2m\omega} + \left(\frac{\rho}{\rho_b} - 1\right)g. \tag{4.10}$$

Now we have the acceleration of the ball, but the computer still can't use it directly because we have to integrate it with respect to time—that is the acceleration tells us how the motion is changing, but we need to find the motion and position from it knowing the initial motion and position. Hence, the type of problem we are dealing with here is called an *initial value problem*.

4.1.3 Details for the Computer

If the computer could understand calculus, we would sit down with it and ask it to carry out the following operations

$$\vec{v}(t) = \vec{v}_0 + \int_0^t \vec{a}(t')dt' \tag{4.11}$$

$$\vec{r}(t) = \vec{r}_0 + \int_0^t \vec{v}(t')dt' \tag{4.12}$$

and we'd have the answer, namely the ball's position (x,y,z) as a function of time. Since the computer can't perform the integrals in Equations 4.11 and 4.12 directly, we'll have to help out more. The computer steps through time incrementally and so, no matter how small the time step, we need to discretize the equations of motion so they can be fed to the computer to loop through:

$$v_{ix}^{n+1} = v_{ix}^n + a_{ix}\Delta t \tag{4.13}$$

$$x_i^{n+1} = x_i^n + \left(\frac{v_{ix}^{n+1} + v_{ix}^n}{2}\right)\Delta t \tag{4.14}$$

$$v_{iy}^{n+1} = v_{iy}^n + a_{iy}\Delta t \tag{4.15}$$

$$y_i^{n+1} = y_i^n + \left(\frac{v_{iy}^{n+1} + v_{iy}^n}{2}\right)\Delta t \tag{4.16}$$

$$v_{iz}^{n+1} = v_{iz}^n + a_{iz}\Delta t \tag{4.17}$$

$$z_i^{n+1} = z_i^n + \left(\frac{v_{iz}^{n+1} + v_{iz}^n}{2}\right)\Delta t \tag{4.18}$$

So really, if you are reading this section and have to get right to the modeling, Equations 4.8 through 4.10 and 4.13 through 4.18 are the starting points. Here, the equations are employed at the nth incremental time step for the x, y, and z coordinates, separately. The time interval Δt is system-specific and needs to be carefully chosen. For systems with characteristic times ("relaxation times," or the time it takes the system to come to equilibrium after being disturbed), it's a good rule to pick the time step to be about 1/10 of the smallest characteristic time it has. That is to say, the time interval should safely capture the system's fastest vibration anywhere in it. For a baseball game, we have to pick our fights. We're not looking at internal vibrations of the bat or ball, and so we just have to try something reasonable. Standard intervals for sports simulations are on the order of 0.001 to 0.01 seconds. To know if the time step is too large, just double it and cut the total number of steps in half and see if the results are different. Keep doing the doubling and you will reach a point where results do change and you will know that the time step should not be chosen any larger. Keep going much past that and the simulation will become unstable and crash. Picking a good value of the time step is something of an art: it should be set small enough so that the results are trustable and the simulation does not "blow up," but large enough so that the system progresses in a reasonable time. Sure we could have amazing accuracy with a 1 fs time step for a baseball simulation but it would take too many steps to run practically. Note that Equations 4.13 through 4.18 have another subscript, i. The new subscript reflects that we could be modeling several particles and the model can address the motion of particle (i) separately. Some limits of many particle simulations are addressed in the homework for this section. It is also important to validate your simulations—to make sure they are working correctly. Two convenient validation cases are (*a*) the case where there are no effects from air present (well known), and (*b*) a baseball thrown without rotation [1].

4.1.4 Construction and Validation: Getting Ready for an Unknown Problem

Just like we can't play Beethoven's *Moonlight Sonata* before learning scales, we can't use and trust the results from an elaborate computer simulation before we have checked it in simple cases where we know the correct outcome. Such verification of one's simulation is called *validation* and should be done as the simulation is built incrementally from simple cases to the most elaborate.

In the case of a baseball, football, or other flying object the simplest case of motion involves no external forces—not even gravity. The motion of the ball should be a straight line at constant speed. Seemingly laughable, such a test case is invaluable in making sure the algorithm and integrator in the simulation run properly. Consider a volleyball. Its trajectory with no external forces is shown by curve 1 in Figure 4.2. It is important to note that the "calibration" here is that the ball is thrown so it goes a distance of 50 ft without any spin at sea level, and is thrown with the same initial velocity ($v_0 = 27$ m/sec at $\theta_0 = 0.811$ degrees off horizontal) for all cases shown in Figure 4.2, hence the upward slant of its path. In the next case, gravity was added curve 2 and the results compared with the well-known formula for the shape of a projectile in the absence of air resistance

$$y = x \tan \theta_0 - \frac{gx^2}{2(v_0 \cos \theta_0)^2}. \tag{4.19}$$

and agreed very favorably. Then, drag was added at sea level curve 3 and at mountain elevations curve 4 as described by changing air density. Finally, a backspin of 8 rev/sec was added at mountain elevations curve 5. So, the sequence of validation presented here gives a perspective—tells a story—as the computer model gets more complicated.

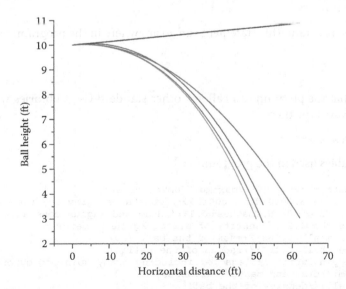

Figure 4.2. Plots of ball trajectories throughout the validation sequence for no forces (curve 1), adding gravity (curve 2), adding drag at sea level (curve 3), changing drag at sea level to drag at mountain elevations (curve 4) and adding back spin at mountain elevations (curve 5). In the electronic version curves 1–5 are colored black, blue, green, red and wine, respectively.

What's more, the ball is hit in the same way initially each time and so it is easy to compare the effects of the different force terms in the model. The effects of spin are very system-specific and in the cases of a baseball can make a difference of hundreds of feet in the range [1,2,5].

In the validation sequence here, the effects of buoyancy are not noticeable on the plots in Figure 4.2, which raises a very important question. How many significant figures should be kept in the model? How accurate is "accurate enough"? The ideal answer is that for all natural constants, one should keep as many significant figures as can be justified in the literature. Then, the model and the context of its results should determine digital accuracy. For example, in the validation above, we may keep 4 or 5 figures after the decimal point but in a real volleyball game we would need answers to only the nearest few centimeters. If we are after buoyancy effects, we'll need considerably more accuracy, and if we are investigating changes to the ball's path but changes in g due to altitude variation, then we may even need double precision arithmetic for the job. And, for the very theoretically-minded, there are some problems worth investigating that are just too fine for the calculations to address.

4.1.5 A Sample Program that Simulates Baseball Flight

Please see Baseball.cpp for the code related to the following example. First, include standard headers for C++ libraries as well as any special ones you will need.

```
#include<cstdlib>
#include<iostream>
#include<fstream>
#include<math.h>
#include <conio.h>
```

Now start the program (this statement can go anywhere in the program).

```
int main()
{
```

Make sure that the program can call any other standard C++ libraries without your having to always type them.

```
using namespace std;
```

Declare variables used in the program.

```
const int nstepmax=100000;//maximum number of steps
const double g=9.810000,dt=0.0001000;//constants (gravity; time step)
const double Cdrag=0.25,Cmagnus=0.25;//drag and Magnus constants
const double rho=1.29;//density of air in kg/cubic meter
const double mball=0.268;//mass of the ball (kg)
const double rball=0.107;//radius of the ball (m)
int idrag,ispin,iwrite;//flags for including drag, spin and output
double Cd,Cm;//drag and magnus variables
double rhoball;//density of the ball
double pi;//pi to be defined later as an arctangent
double Aball;//projected area of the ball
double t;//simulated time
double x,y,z;//cartesian coordinates of the ball
double v,vx,vy,vz;//speed and forward, lateral and vertical velocities
```

```
double vxold,vyold,vzold;//velocity placeholders
double ax,ay,az;//forward, lateral and vertical accelerations
double wx,wy,wz;//forward, lateral and vertical angular velocities
double theta;//launch angle off horizontal
double xstop;//forward (horizontal) stopping position of the ball
```

Allow writing data to files as output streams, and open an output file.

```
ofstream myfile;
myfile.open("baseball.res");
```

Start gathering data from the user of the program (aka "screen").

```
cout<<"Initial horizontal position from home plate (meters): "<<endl;
cin>>x;
cout<<"Initial lateral position (meters): "<<endl;
cin>>y;
cout<<"Initial height (meters): "<<endl;
cin>>z;
cout<<"Final horizontal position from home plate (meters): "<<endl;
cin>>xstop;
cout<<"Initial ball speed (meters/second): "<<endl;
cin>>v;
cout<<"Initial pitch angle above horizontal (degrees) : "<<endl;
cin>>theta;
cout<<"Enter 1 for drag or 0 for no drag: "<<endl;
cin>>idrag;
cout<<"Enter 1 for spin effects or 0 for none: "<<endl;
cin>>ispin;
```

If spin is included, find out about it.

```
if(ispin==1)
{
cout<<"x angular velocity (rev/sec): "<<endl;
cin>>wx;
cout<<"y angular velocity (rev/sec): "<<endl;
cin>>wy;
cout<<"z angular velocity (rev/sec): "<<endl;
cin>>wz;
}
```

Decide if you want the entire trajectory or not.

```
cout<<"Enter 1 to write trajectory to output file or 0 to skip: "<<endl;
cin>>iwrite;
```

Now initialize any calculated constants and carry out unit conversion.

```
pi=4.0*atan(1.0);

rhoball=3.0*mball/(4.0*pi*pow(rball,3));

Aball=pi*pow(rball,2);

vx=v*cos(theta*pi/180.0);
vy=0.0;
vz=v*sin(theta*pi/180.0);
```

```
wx=wx*2.0*pi;
wy=wy*2.0*pi;
wz=wz*2.0*pi;
```

Set the drag constant to zero but override with nonzero values if it is included in the simulation.

```
Cd=0.0;
if(idrag==1)
{
Cd=Cdrag;
}
```

Do the same for Magnus forces.

```
Cm=0.0;
if(ispin==1)
{
Cm=Cmagnus;
}
```

Now start the time loop...run at most for `nstepmax` steps.

```
for(int i=1;i<nstepmax;i++)
{
```

Calculate time.

```
t=float(i)*dt;
```

Now put the three velocity components in placeholders and calculate speed.

```
vxold=vx;
vyold=vy;
vzold=vz;
v=sqrt(pow(vx,2)+pow(vy,2)+pow(vz,2));
```

Calculate the three components of acceleration.

```
ax=-0.5*Cd*rho*Aball*v*vx+0.5*Cm*rho*Aball*(wy*vz-wz*vy);
ay=-0.5*Cd*rho*Aball*v*vy+0.5*Cm*rho*Aball*(wz*vx-wx*vz);
az=-0.5*Cd*rho*Aball*v*vz+0.5*Cm*rho*Aball*(wx*vy-wy*vx)+(rho/rhoball-1.00)*g;
```

Implement the first half of the integration algorithm: update velocities.

```
vx=vx+ax*dt;
vy=vy+ay*dt;
vz=vz+az*dt;
```

Now for the second half of the integration algorithm: update positions.

```
x=x+0.5*(vxold+vx)*dt;
y=y+0.5*(vyold+vy)*dt;
z=z+0.5*(vzold+vz)*dt;
```

We're still in the time loop...write a trajectory step out if we asked for that earlier.

```
if(iwrite==1)
{
myfile<<t<<" "<<x<<" "<<y<<" "<<z<<endl;
}
```

If the ball is where it should stop, write out some information and exit the time loop.

```
if(x>=xstop)
{
cout<<"Final time (seconds): "<<t<<endl;//write out important things
cout<<"Final position (x,y,z) in meters: "<<x<<" "<<y<<" "<<z<<endl;
cout<<"Final velocity (vx,vy,vz) in meters/sec: "<<vx<<" "<<vy<<"
"<<vz<<endl;
break;
}
```

The time loop closes with the next line.

```
}
```

Close and save the specified output file.

```
myfile.close();
```

Delay closing the screen to show output if on a MAC or PC.

```
getch();
```

End the program silently.

```
return(0);
}
```

4.2 Physics on the Field: Achieving More Efficient Football Tackles

In the previous example, we considered objects flying through the air. In our model, we dealt with how the earth and the air affected the ball's motion. We didn't really worry about when or how the ball was batted or what might happen to the bat. Certainly we didn't worry about when it landed. Now those are interesting things, and we could consider them some time, but they weren't part of the model at hand. Now consider a football game. We now have something new that can't be ignored, which is the fact that the interaction between the players is the central part of the game. In fact, most of the important physics happens during the tackling process because it is where the most important energy transfer takes place. On the football game, we're talking about energy transfer *between players*. There have been other cases, however, when considering energy transfer out of or into a system has proved very useful. Case in point: a discus throwers gearing up for the 1976 Olympics. During practice, they could throw the discus a certain distance but naturally they wanted to know if they could do better. Their body was approximated by an array of points, which was fed into a computer for analysis. It was found that there was substantial energy loss because due to friction between their feet and the ground. They modified their throw, based on the physical analysis and set a new world record. So back to football, even ignoring friction for the time being, we can still study what happens to energy during tackling. Through such investigations, we may be able to help players focus their energy in ways to more efficiently tackle and hence improve the game. For example, there could be situations where it is best if the tackle results in players' rotating, while in other cases it could be better to just budge them laterally. Also, it could be that sometimes it's better to grab onto somebody and move with them as you tackle them but other times in may be best just to contact them and not hold on. In the following sections, we're going to put together a simulation of the football tackle with the intent of understanding dynamics and energy transfer during the tackle.

4.2.1 The Model: Not Quite Like Play Diagrams in a Locker Room

As always, the first and perhaps most important step of designing a physical model is what aspects of the system are important to include: what will matter? And whenever I build a model, in the back of my mind is a little guy like in the cartoons who censors me when the model becomes too complicated. As you might imagine, it is very easy to include things that don't matter so much yet increase the difficulty enough, in either developing the model or running the resulting simulation, so that the purpose of doing the calculation in a reasonable time becomes defeated.

So, for the football problem we're talking about a tackle. First off, there are many things going on in a football game but, as for the fundamental physics of tackles, if we have studied one tackle we've studied them all, and it takes two to tackle. Thus, one simplification is that we can have two players: one tackling another. In addition, because the players are mainly confined to the field and don't spend too much time up in the air, it's not a bad approximation to have a two dimensional model...we don't need all three but certainly having just one is oversimplification. Since our model is two-dimensional, it justifies our being able to ignore gravity, which we could not do in the baseball problem. Last, important questions involve post-tackle rotation and whether or not the players stick together, So, what if we model the tackling player (tackler) and the tackled player (runner) as cylinders? That is not a bad approximation, but the problem we run into is we have to deal with cylinder collisions, even if they have length and no radius. I always encourage students to do the hard thing, but in the same breath, I would also say that it is important to think about how a much simpler model can give about the same payoff. Thinking about it, it's not so unreasonable to think that the tackle takes place at one location. So what if we were to model the tackled player as a cylinder and the tackler as a sphere? Such a model would indeed be able to address post-tackling rotation as well as whether the players stick together or not. It seems like a good start. Ultimately, the computer is going to run a program giving numbers so we have to have equations of some kind in our model that we can discretize and feed to the computer. And equations come from physics so that's our next question: what's the important physics in the model?

Because collisions are a central part of what we're looking at, one concept that we will need is momentum conservation. Now, you would typically run into momentum (aka linear momentum) conservation in a first year. Physics course...run into momentum conservation...I wasn't even trying to make a good joke! The momentum $\vec{p}$ of a single particle having mass m and velocity $\vec{v}$ is defined to be

$$\vec{p} = m\vec{v}. \tag{4.20}$$

The arrows over momentum and velocity indicate that they are vector quantities, and so they can apply in 1, 2, or 3 dimensions and have a direction (usually a plus or minus sign for each direction). Momentum conservation says that if there are no external forces on a system, its total momentum will be conserved, even when its constituents collide. Momentum conservation holds when the objects colliding bounce off each other (perfectly elastic) or stick together (perfectly inelastic). But because rotation of the cylinder is important in the problem, a quantity called angular momentum is also of issue. The angular momentum $\vec{L}$ of a rotating object is defined to be

$$\vec{L} = I\vec{\omega}. \tag{4.21}$$

Here, I is the object's rotational mass (aka moment of inertia); formulae for the moments of inertia of several common objects are shown in Table 4.3. As in the baseball problem, $\vec{\omega}$ is the objects angular velocity, which in this case can be thought of as its speed of rotation directed along the axis or axes it is rotating about. So why is angular momentum important in our football model? It matters because it is *conserved* in systems with no external torques (rotational forces), even through elastic or inelastic collisions.

So now we can put together the heart and soul of the physics in the problem. Treating the tackler as a point (sphere with zero size) and the tackled player as a cylinder we can obtain expressions for how the linear and rotational motion changes because of the collision. Now, the cylinder should be talked about a little bit more. For the collision dynamics, like the sphere it has no radius. If it did have a nonzero radius, the model of the collision would be much more involved than warranted in this problem. Don't worry...I've assigned that as a homework problem. But even though the cylinder has no radius for collision dynamics, it has to have a radius for figuring out its moment of inertia.

So now we can really get somewhere: consider the case of the elastic collisions where the players do not hold on to each other. If we consider the sphere colliding with a cylinder and we impose momentum conservation, angular momentum

Table 4.3. Moments of Inertia for Common Objects

Object	Moment of Inertia I
Sphere of radius R	$\frac{2}{5}MR^2$ about axis through center
Hoop	MR^2 about axis through center
Cylinder of radius R and length L	$\frac{1}{2}MR^2$ about long axis
	$\frac{1}{3}ML^2$ about an end
	$\frac{1}{12}ML^2$ about the middle
Spherical shell of radius R	$\frac{2}{3}MR^2$ about axis through center

conservation, and kinetic energy conservation—because of elasticity, we obtain the following three equations that tell us the two post-collision velocities, as well as post-collision angular velocity:

$$v_{1x} = v_{1x,old} + \frac{m_2 I_{new}}{m_1 m_2 (h-cm)^2 + (m_1+m_2)I_{new}} (\vec{v}_2 - \vec{v}_1) \cdot (\vec{r}_2 - \vec{r}_1) \cdot \frac{(x_2 - x_1)}{d^2} \qquad (4.22)$$

$$v_{2x} = v_{2x,old} + \frac{m_1 I_{new}}{m_1 m_2 (h-cm)^2 + (m_1+m_2)I_{new}} (\vec{v}_2 - \vec{v}_1) \cdot (\vec{r}_2 - \vec{r}_1) \frac{(x_2 - x_1)}{d^2} \qquad (4.23)$$

$$\omega_x = \frac{-2m_1 m\ (cm-h)}{m_1 m_2 (h-cm)^2 + (m_1+m_2)I_{new}} (\vec{v}_2 - \vec{v}_1) \cdot (\vec{r}_2 - \vec{r}_1) \frac{(y_2 - y_1)}{d^2}. \qquad (4.24)$$

If we do the same thing, but now consider the case when the players hold on to each other, we have a perfectly inelastic collision and the post-collision velocity and angular velocity are given by the following equations:

$$v_{1x} = v_{2x} = \frac{m_1 v_{1x,old} + m_2 v_{2x,old}}{m_1 + m} \qquad (4.25)$$

$$\omega_x = \frac{-m_2(v_{1y,old} - v_{2y,old})(h-cmnew)}{m_2(h-cm)^2 + \frac{m_1}{3h}[(h-cmnew)^3 + cmnew^3]}. \qquad (4.26)$$

And yes, you may have guessed that since I like you to know where things come from, but we don't have the room to show everything in detail, I strongly recommend that you derive Equations 4.22–4.26. Since the physics of how the tackle will change the players' motion is nailed down, we can now talk about what is going to happen in the program.

4.2.2 How the Program Will Work: Making it Look Like a Tackle

It makes sense for the runner to start running in a straight line. We could put curvature into their path, but it will be purely aesthetic, so a straight line it is for now. In fact, we can continue the runner in a straight line with speed v for distance $ltack$ until the tackle happens. Then the time until a tackle is just given by

$$ttack = \frac{ltack}{v_{10}}, \qquad (4.27)$$

The tackler, on the other hand, may have just noticed the runner changed direction and obviously needs to run and catch him/her. Therefore, as an approximation we could start tacklers moving in some direction angle θ_{10} off the vertical at some speed, and then they would change their motion in order to catch the runner. They would have to accelerate now, and the horizontal and vertical accelerations (acceleration vector) are given by

$$a_x = \frac{2[(xtack - x_{10}) - v_{10}\cos\theta_{10}ttack]}{ttack^2}, \tag{4.28}$$

and

$$a_y = \frac{2[(ytack - y_{10}) - v_{10}\sin\theta_{10}ttack]}{ttack^2}. \tag{4.29}$$

The program will start the players out and run until they get closer than a certain tolerance value, in which case the physics of the collision will be applied in order to get new velocities and angular velocity. Ignoring friction, the acceleration of the players are subsequently set to zero, and the program will run for a slightly longer time after $ttack$ (after the collision) so we can get a visual feeling for the post-collision motion.

4.2.3 Issues Relevant to the Computer

Equations 4.13 through 4.18 apply here also for stepping the system through time. Even though it is possible to calculate the forces on the players during the collision (yes, it's a homework problem), we model the players' velocities as instantaneously changing because of the collision and so the only acceleration we mandate is that of the tackler during the first part of the game. Because we do have player–player interactions here, we do have to insert a tolerance into the program, which is the way we have the computer tell us if the objects are touching or not. The value of the tolerance depends on how close you want to go, whether not you use double precision arithmetic or not in your code...I did in the code presented here just to illustrate the point. As mentioned in the previous example, the time step should be carefully chosen so that the program advances in a stable way. However, here there's another element to consider, that is, how much do you want the sphere and the cylinder to overlap during tackle? Such a consideration is very important because it is highly unlikely that in the course of the simulation there will be a point where the sphere is exactly touching the cylinder on a surface. Such an issue can usually be taken care of by setting a tolerance for overlap and then adjusting the time step to a tolerance is satisfied.

4.2.4 Development and Validation

So, as before it is going to be very important to build our computer program incrementally, checking in at each step along the way to make sure it works correctly. As was true for the baseball problem, probably the best starting point is to have two particles that are headed some direction at some speed, and moving in straight lines with no acceleration because there's no forces on them. This is a good check of the time the integration scheme offered in Equations 4.13–4.18. Once you're successful with this first step, you can then look at collisions or the initial phase of the program involving the path toward tackle? Since the paths taken toward tackle involve acceleration, it seems most reasonable to implement the collisions next. So, Equations 4.22–4.26 would be coded into the program, the players approaching each other on a straight line, and initial and final parameters of the collision checked. It seems reasonable to check initial and final velocities, initial and final rotation speeds of the cylinder, initial and final total momentum, and initial and final total angular momentum. The last two quantities are conserved through the collision and so should not change. Remember that there are two different kinds of collisions (elastic and inelastic), and because tackled players are presented as cylinders, there are a couple special points of tackle to look at such as the center of the cylinder and the ends. So then, once the collisions are implemented then you can work on the tackle approach. Equations 4.27 through 4.29 would then be coded into the program, the program could be run and the paths of the players checked. Once you are convinced that the players are moving correctly in the simulation, and that the tackles are happening correctly, then your code is validated and you're ready to do new things with it.

4.2.5 A Sample Program that Simulates a Football Tackle

Please review the provided program Football.cpp for code related to the following discussion. Start out by calling standard C++ libraries and any special ones you might need.

```
#include<cstdlib>
#include<iostream>
#include<fstream>
#include<math.h>
#include <conio.h>
```

Start the program (this statement can go anywhere in the program).

```
int main()
{
```

Make sure that the program can call any other standard C++ libraries without your having to always type them.

```
using namespace std;
```

Declare variables used in the program.

```
const int nstep=1000;//number of time steps in the program
int itype;//type of collision (1 for elastic; 2 for inelastic)
double dt;//time step
double t;//simulated time
double dist;//center-to-center separation of the players at any time
double wide;//width of the players' shoulders
double dvx;//difference in lateral velocity of the two players
double dvy;//difference in downfield velocity of the two players
double vij;//magnitude of velocity difference between the players
double vdotr;//dot product of velocity difference and relative displacement
of the players
double dx;//lateral relative displacent of the players
double dy;//downfield relative displacement of the players
double rat1;//ratio used in determining the reflected velocity of player 1
//after elastic collision
double rat2;//ratio used in determining the reflected velocity of player 2
//after elastic collision
double sum;//sum of the players' masses
```

```
double ltack;//distance from the runner (in the running direction) where //
the tackle takes place
double ttack;//time from the start of program until the tackle
double tackx;//lateral coordinate of tackle spot
double tacky;//downfield coordinate of tackle spot
double x0[2];//initial lateral coordinates of players
double y0[2];//initial downfield coordinates of players
double v0[2];//initial speeds of players
double theta0[2];//initial direction of players' velocities
double theta[2];//direction of players' velocities at any time
double m[2];//players' masses
double vx[2];//players' lateral velocities at any time
double vy[2];//players' downfield velocities at any time
double v[2];//players' speeds at any time
double x[2];//players' lateral positions at any time
double y[2];//players' downfield positions at any time
double vxold[2];//lateral velocity placeholder
double vyold[2];//downfield velocity dummy
double ax[2];//Lateral accelerations of players
double ay[2];//downfield accelerations of players
double hrun;//height of runner above ground
double htack;//height of tackle point above ground
double omegx;//lateral angular velocity of runner about center of mass //after
tackle
double omegy;//downfield angular velocity of runner about center of mass //
after tackle
double cmnew;//center of mass of runner after collision (doesn't change //
for elestic)
double icmnew;//moment of inertia of runner about the new center of mass //
after collison
double dummy;//dummy variable for input reading
double pxinit,pyinit;
double pxfin,pyfin;
double lxinit,lyinit;
double lxfin,lyfin;
```

Allow writing data to files as output streams, and open an output file.

```
ofstream myfile;
myfile.open("football_information.res");
ofstream myfile2;
myfile2.open("football_trajectories.res");
```

Get initial information from the user, otherwise known as "screen." Commented lines are suggestions for reasonable values.

```
cout<<"Enter the starting lateral coordinate of the runner (real in feet)";
cin>>dummy;
//dummy=0.0;
x0[0]=dummy;
cout<<"Enter the starting downfield coordinate of the runner (real in feet)";
cin>>dummy;
//dummy=14.0;
y0[0]=dummy;
cout<<"Enter the weight of runner (real in pounds)";
cin>>dummy;
//dummy=200.0;
m[0]=dummy;
cout<<"Enter the speed of runner (real in feet per second; may not be zero)";
```

```
cin>>dummy;
//dummy=2.0;
v0[0]=dummy;
cout<<"Enter the direction of the runner (real in degrees)";
cin>>dummy;
//dummy=90.0;
theta0[0]=dummy;
cout<<"Enter the height of the runner (real in feet)";
cin>>hrun;
//hrun=6.0;
cout<<"Enter the starting lateral coordinate of tackler (real in feet)";
cin>>dummy;
//dummy=4.0;
x0[1]=dummy;
cout<<"Enter the starting downfield coordinate of tackler real in feet)";
cin>>dummy;
//dummy=0.0;
y0[1]=dummy;
cout<<"Enter the weight of tackler (real in pounds)";
cin>>dummy;
//dummy=200.0;
m[1]=dummy;
cout<<"Enter the speed of tackler (real in feet per second)";
cin>>dummy;
//dummy=4.0;
v0[1]=dummy;
//cout<<"Enter the direction of tackler (real in degrees)";
//cin>>dummy;
dummy=88.0;
theta0[1]=dummy;
cout<<"How wide are the players' shoulders? (real in feet)";
cin>>wide;
//wide=0.1;
cout<<"Enter the point of tackle along direction of motion of runner (real
in feet)";
cin>>ltack;
//ltack=4.0;
cout<<"Enter the height of the tackle point (real in feet):";
cin>>htack;
cout<<"Enter the collision type (integer...1 for elastic; 2 for inelastic)";
cin>>itype;
```

Initialize any arrays and variables; make sure angles are in radians.

```
x[0]=x0[0];
y[0]=y0[0];
v[0]=v0[0];
x[1]=x0[1];
y[1]=y0[1];
v[1]=v0[1];
theta0[0]=theta0[0]*3.14159/180.0;
theta0[1]=theta0[1]*3.14159/180.0;
theta[0]=theta0[0];
theta[1]=theta0[1];
```

Calculate the velocity components from speeds and directions.

```
for(int j=0;j<=1;j++)
{
vx[j]=v[j]*cos(theta[j]);
```

```
vy[j]=v[j]*sin(theta[j]);
}
```

Calculate the time until tackle and write it out.

```
ttack=ltack/sqrt(pow(vx[0],2)+pow(vy[0],2));
cout<<"Time until tackle (sec) is: "<<ttack<< "\n";
myfile<<"Time until tackle (sec) was: "<<ttack<< "\n";
```

Calculate the time step and write it out.

```
dt=1.1*ttack/float(nstep);
cout<<"Time interval (sec) is: "<<dt<< "\n";
myfile<<"Time interval (sec) was: "<<dt<< "\n";
```

Calculate the position of the tackle and write it out.

```
tackx=x0[0]+ltack*cos(theta0[0]);
tacky=y0[0]+ltack*sin(theta0[0]);
```

Calculate the accelerations of the players consistent with the tackle time and position.

```
ax[0]=0.;
ay[0]=0.;
ax[1]=2.0*(tackx-x0[1]-v0[1]*cos(theta0[1])*ttack)/pow(ttack,2);
ay[1]=2.0*(tacky-y0[1]-v0[1]*sin(theta0[1])*ttack)/pow(ttack,2);
```

Start the time loop here.

```
for(int i=0;i<nstep;i++)
{
```

Calculate the time since the simulation started.

```
t=float(i)*dt;
```

Deal with elastic collisions first.

```
if(itype==1)
{
```

Elastic: Placehold the precollision velocities.

```
for(int j=0;j<=1;j++)
{
vxold[j]=vx[j];
vyold[j]=vy[j];
}
```

Elastic: Calculate displacements, distance, relative velocities, and dot product.

```
dx=x[1]-x[0];
dy=y[1]-y[0];
dist=sqrt(pow((x[1]-x[0]),2)+pow((y[1]-y[0]),2));
```

```
dvx=vxold[1]-vxold[0];
dvy=vyold[1]-vyold[0];
vij=sqrt(pow(dvx,2)+pow(dvy,2));
vdotr=dx*dvx+dy*dvy;
```

Elastic: If the "players" are touching and heading towards each other, color me a nerd and call it a collision.

```
if(dist<=wide&&vdotr<0.)
{
```

Elastic Collision: Zero out the tackler's acceleration...they caught each other.

```
ax[1]=0.;
ay[1]=0.;
```

Elastic Collision: Calculate the new moment of inertia and center of mass (running back). They are unchanged by the collision but still should be in the program.

```
icmnew=m[0]*pow(0.5*wide,2)/12.;
cmnew=hrun/2.;
```

Elastic Collision: Calculate ratios used to find post-collision velocities, see??

```
rat1=2.*m[1]*icmnew/(m[0]*m[1]*pow(htack-cmnew,2)+(m[0]+m[1])*icmnew);
rat2=2.*m[0]*icmnew/(m[0]*m[1]*pow(htack-cmnew,2)+(m[0]+m[1])*icmnew);
```

Elastic Collision: Calculate post-collision velocities; collision happens here.

```
vx[0]=vxold[0]+rat1*vdotr*dx/pow(dist,2);
vx[1]=vxold[1]-rat2*vdotr*dx/pow(dist,2);
vy[0]=vyold[0]+rat1*vdotr*dy/pow(dist,2);
vy[1]=vyold[1]-rat2*vdotr*dy/pow(dist,2);
```

Elastic Collision: Calculate post-collision angular velocity of the running back.

```
omegx=2.*m[0]*m[1]*(cmnew-htack)/(m[0]*m[1]*pow(htack-
cmnew,2)+(m[0]+m[1])*icmnew);
omegx=-omegx*vdotr*dy/(pow(dist,2));
omegy=2.*m[0]*m[1]*(cmnew-htack)/(m[0]*m[1]*pow(htack-
cmnew,2)+(m[0]+m[1])*icmnew);
omegy=omegy*vdotr*dx/(pow(dist,2));
```

Elastic Collision: Calculate pre- and post-collision (linear) momenta.

```
pxinit=m[0]*vxold[0]+m[1]*vxold[1];
pxfin=m[0]*vx[0]+m[1]*vx[1];
pyinit=m[0]*vyold[0]+m[1]*vyold[1];
pyfin=m[0]*vy[0]+m[1]*vy[1];
```

Elastic Collision: Calculate pre- and post-collision angular momenta.

```
lxinit=-(cmnew-htack)*m[1]*(vyold[1]-vyold[0]);
lxfin=icmnew*omegx-(cmnew-htack)*m[1]*(vy[1]-vy[0]);
lyinit=(cmnew-htack)*m[1]*(vxold[1]-vxold[0]);
lyfin=icmnew*omegy+(cmnew-htack)*m[1]*(vx[1]-vx[0]);
```

Elastic Collision: End collision.

```
}
```

Elastic: End elastic conditional.

```
}
```

Now consider inelastic collisions.

```
if(itype==2)
{
```

Inelastic: Placehold the pre-collision velocities.

```
for(int j=0;j<=1;j++)
{
vxold[j]=vx[j];
vyold[j]=vy[j];
}
```

Inelastic: Calculate displacements, distance, relative velocities, and dot product.

```
dx=x[1]-x[0];
dy=y[1]-y[0];
dist=sqrt(pow((x[1]-x[0]),2)+pow((y[1]-y[0]),2));
dvx=vxold[1]-vxold[0];
dvy=vyold[1]-vyold[0];
vij=sqrt(pow(dvx,2)+pow(dvy,2));
vdotr=dx*dvx+dy*dvy;
```

Inelastic: If the "players" are touching and heading towards each other, color me a nerd and call it a collision.

```
if(dist<=wide&&vdotr<0.)
{
```

Inelastic Collision: Zero out the tackler's acceleration...they caught each other.

```
ax[1]=0.;
ay[1]=0.;
```

Inelastic Collision: Calculate post-collision velocities; collision happens here.

```
sum=m[0]+m[1];
vx[0]=(m[0]*vxold[0]+m[1]*vxold[1])/sum;
vx[1]=vx[0];
vy[0]=(m[0]*vyold[0]+m[1]*vyold[1])/sum;
vy[1]=vy[0];
```

Inelastic Collision: Calculate the new moment of inertia and center of mass (both players). They stick together here.

```
cmnew=(m[0]*hrun/2.+m[1]*htack)/sum;
icmnew=m[1]*pow(htack-cmnew,2)+(m[0]/(3.*hrun))*(pow((hrun-
cmnew),3)+pow(cmnew,3));
```

Inelastic Collision: Calculate post-collision angular velocity.

```
omegx=((vyold[0]-vyold[1])*(htack-cmnew))*m[1];
omegx=-omegx/icmnew;
omegy=((vxold[1]-vxold[0])*(htack-cmnew))*m[1];
omegy=-omegy/icmnew;
```

Inelastic Collision: Calculate pre- and post-collision (linear) momenta.

```
pxinit=m[0]*vxold[0]+m[1]*vxold[1];
pxfin=m[0]*vx[0]+m[1]*vx[1];
pyinit=m[0]*vyold[0]+m[1]*vyold[1];
pyfin=m[0]*vy[0]+m[1]*vy[1];
```

Inelastic Collision: Calculate pre- and post-collision angular momenta.

```
lxinit=-(cmnew-htack)*m[1]*(vyold[1]-vyold[0]);
lxfin=icmnew*omegx-(cmnew-htack)*m[1]*(vy[1]-vy[0]);
lyinit=(cmnew-htack)*m[1]*(vxold[1]-vxold[0]);
lyfin=icmnew*omegy+(cmnew-htack)*m[1]*(vx[1]-vx[0]);
```

Inelastic Collision: End collision.

```
}
```

Inelastic Collision: End inelastic conditional.

```
}
```

We're still in the time loop, so we use a simple Euler algorithm to update the system.

```
for(int j=0;j<=1;j++)
{
x[j]=x[j]+vx[j]*dt+0.5*ax[j]*pow(dt,2);
y[j]=y[j]+vy[j]*dt+0.5*ay[j]*pow(dt,2);
vx[j]=vx[j]+ax[j]*dt;
vy[j]=vy[j]+ay[j]*dt;
}
```

Calculate the players' speeds and angles of travel.

```
for(int j=0;j<=1;j++)
{
v[j]=sqrt(pow(vx[j],2)+pow(vy[j],2));
theta[j]=atan2(vy[j],vx[j]);
}
```

At each time step, write out important information to the screen as well as the output files.

```
cout<<"\t"<<t<<"\t"<<x[0]<<"\t"<<y[0]<<"\t"<<x[1]<<"\t"<<y[1]<<"\t"<<omegx<<"\t"<<omegy<<"\n";
myfile2<<"\t"<<t<<"\t"<<x[0]<<"\t"<<y[0]<<"\t"<<x[1]<<"\t"<<y[1]<<"\t"<<omegx<<"\t"<<omegy<<"\n";
```

Time loop ends with the next line.

```
}
```

Close and save the specified trajectory file.

```
myfile2.close();
```

Now that the simulation is done, write selected information to the screen.

```
cout<<"Total pre-collision horizontal momentum (arbitrary
units):"<<pxinit<< "\n";
cout<<"Total post-collision horizontal momentum (arbitrary units):
"<<pxfin<< "\n";
cout<<"Total pre-collision vertical momentum (arbitrary units):
"<<pyinit<< "\n";
cout<<"Total post-collision vertical momentum (arbitrary units):
"<<pyfin<< "\n";
cout<<"Total pre-collision horizontal angular momentum (arbitrary
units):"<<lxinit<< "\n";
cout<<"Total post-collision horizontal angular momentum (arbitrary units):
"<<lxfin<< "\n";
cout<<"Total pre-collision vertical angular momentum (arbitrary units):
"<<lyinit<< "\n";
cout<<"Total post-collision vertical angular momentum (arbitrary units):
"<<lyfin<< "\n";
```

And write important summary information.

```
if(itype==1)
{
myfile<<"Collision type is elastic"<<"\n";
}
if(itype==2)
{
myfile<<"Collision type is inelastic"<<"\n";}
myfile<<"Total pre-collision horizontal momentum (arbitrary
units):"<<pxinit<< "\n";
myfile<<"Total post-collision horizontal momentum (arbitrary units):
"<<pxfin<< "\n";
myfile<<"Total pre-collision vertical momentum (arbitrary units):
"<<pyinit<< "\n";
myfile<<"Total post-collision vertical momentum (arbitrary units):
"<<pyfin<< "\n";
myfile<<"Total pre-collision horizontal angular momentum (arbitrary
units):"<<lxinit<< "\n";
myfile<<"Total post-collision horizontal angular momentum (arbitrary
units): "<<lxfin<< "\n";
myfile<<"Total pre-collision vertical angular momentum (arbitrary units):
"<<lyinit<< "\n";
myfile<<"Total post-collision vertical angular momentum (arbitrary units):
"<<lyfin<< "\n";
```

Close and save the specified output file.

```
myfile.close();
```

Delay closing the screen to show output if on a MAC or PC.

```
getch();
```

End the program silently.

```
return(0);
}
```

4.3 A Physics Nerd, a Cool Guy, and a Pool Table

The beauty of working with students who are not physics majors is that they have forced me to look at the world in a way that I'm not naturally used to. And gaining new perspective is always a good thing, especially when you're doing research. I was very pleased when a student of mine (the cool guy, Christian Leerberg, ...so you know who's left) in algebra-based physics class wanted to simulate a game of pool [7]. Fondly recalling that in high school and college I spent a significant amount of my time in pool halls and late night conversations, I was all ready to jump on the problem with him. And he was interested in chaos. He knew about it partially because of popularized notions of what chaos is and also because he read some science journals and had a good foundational understanding of the real deal. The word chaos has its roots in the Greek word for abyss and has the meaning of "chance in supreme." In that case it could describe my room growing up. Popular media often draws an equivalence of chaos and randomness, which is incorrect. Speaking from a physics perspective, a chaotic system is one where its outcomes are very sensitive to changes in initial conditions. Or another way to frame a definition of chaos is when two systems started out very closely together in their classical states (positions and velocities of its components) and diverge in behavior as time progresses. And because quantum mechanics and relativity are not necessarily needed to describe classical everyday systems, chaos is often called deterministic chaos. There are many real systems in physics which exhibit chaos, such as weather patterns, financial exchange, biological populations, blood flow, and many more.

So for pool...the need for a simulation was certainly there too. There's a lot of software out there that simulates pool games—many times very accurately and realistically—but they don't lend themselves nicely to scientific study. So it was really a fun project for a summer. We got the textbooks down, we thought about what we wanted the model to act like and do, and we built the program from scratch. Probably the most noticeable part came when he and I grabbed a balance, a meter stick, a stopwatch, and a couple calculators and went to our student union to measure various constants we needed for the program...all over lunch of course...and with curious onlookers.

4.3.1 Mathematical Game Rules

The model here is a good old classical one, which means that Newton's laws of motion describe how it steps through time—not quantum mechanics or relativity. Because we're using Newton's laws, figuring out the forces in the model is a central issue because these will determine how the motion of the objects change after the simulation starts up. So what's important? Because the balls exist mainly on a table, we can ignore gravity, or at least things falling because of gravity. As in the previous simulations, probably the best way to look at how to construct the model is to divide the interactions in the categories. In this case, balls can interact with each other, with the table, and with the air.

There are three types of ball-table interactions: the effect of friction on their linear motion, the effect of friction on the spin, and reflection of the boundaries. Because the effect on the linear motion involves loss of kinetic energy and the ball slowing down, we model the ball-table friction as kinetic friction: $\vec{f}_i = -\mu_k m_i g \hat{v}_i$, where m_i is the mass of ball i, g is the acceleration of gravity, and $\hat{v}_i$ is a unit vector in the direction of the velocity of ball i. We were able to measure the coefficient of friction experimentally and its value is assumed to be the same for all balls. The values of such important parameters are shown in Table 4.4. Secondly, the balls may be spinning but because of kinetic friction with the tabletop their spin will slow. The following describes the spin angular momentum the balls will have as a function of time, and the unit vector reflects the fact that the rotating about the z axis or vertical axis: $\vec{S}_i = S_{i0}e^{-\alpha t}\hat{z}$. The exponential function is inserted to describe how the spin damps with time due to rubbing friction with the table. The spin decay constant α is also recorded in Table 4.4.

Now, the spin of the ball does not affect its overall translational motion unless the spinning ball collides with another ball or a spinning ball reflects off the boundaries of the table. In the case of collision, the effect on velocity can be described with the following equations

Table 4.4. Important Parameters Used in the Simulation

Parameter	Symbol	Value
Cue ball mass	m_1	1.6×10^{-1} kg
Cue ball radius	r_1	2.79×10^{-2} m
Ball mass	m_2–m_{16}	1.7×10^{-1} kg
Ball radii	r_2–r_{16}	2.86×10^{-2} m
Coefficient of friction	μ_k	0.027
Horizontal table dimension	A	2.6 m
Vertical table dimension	B	1.3 m
Drag coefficient	C_D	0.5
Acceleration of gravity	G	9.8 m/sec^2
Time step	Δt	10^{-3} sec
Density of air	ρ	1.29 kg/m^3

Source: C. Leerberg, M.W. Roth, *Am. J. Undergrad. Res.*, 1(4), 27, 2003 [7].

Note: All measurements were made by the authors on a pool table in the UNI student union, and the dimensions of the table are not referenced explicitly in any of the equations.

$$v'_{ix} = v_{ix} + \beta \hat{v}_i \times \hat{S}_i \tag{4.30}$$

$$v'_y = -v_y \tag{4.31}$$

for a horizontal wall (table top or bottom) or alternatively by

$$v'_y = v_y + \beta \hat{v}_i \times \hat{S}_i \tag{4.32}$$

$$v'_x = -v_x \tag{4.33}$$

for a vertical wall (left or right boundaries). And if the ball hits a corner, all four equations are needed. Here, the constant β reflects to what degree the ball's spin couples with the edges of the table and, like α, can be measured experimentally by observing the behavior of balls with known spins. It seems reasonable that β (when included) should be taken to be the same for all balls.

Because the balls are confined to a table, the buoyant force is completely neglected in the model, and the only interaction between the ball and the air then is drag: $\vec{F}_{iD} = -\frac{1}{2}C_D\rho\pi r_i^2 v_i^2 \hat{v}_i$. Here ρ is the density of the air, r_i is the radius of ball i, C_D is the drag coefficient, v_i is the speed of the ball, and $\hat{v}_i$ is a unit vector in the direction of the ball's velocity. And as a reality check, negative sign out in front means that the driving force is always directed against the balls velocity...right on!

Now for the hard part...ball-ball interactions through collisions. It could be that one or more balls are touching at one time. The program determines the pairs of balls touching by figuring out which ones overlap with each other, and the algorithm will be described in more detail in the next section. Then the program uses the following relationship to figure out the new velocity from the old velocities and the properties of the balls.

$$\vec{v}_j{}' = \vec{v}_j - \left(\frac{(1+e)m_i}{m_i+m_j}\right)\frac{(\vec{v}_{ij}\cdot\vec{r}_{ij})(\vec{r}_j-\vec{r}_i)}{r_{ij}^2} - \gamma(S_i+S_j)\times\check{r}_{ij}. \tag{4.34}$$

Here $\vec{v}_{ij}$ is the velocity of ball (j) relative to that of (i), and e is the elasticity of the collision, which can be taken to range anywhere from $e = 0.01$ to $e = 1$. The parameter γ is an experimentally measured constant, which describes the strength of the spin–spin coupling.

4.3.2 How the Code Will Work

There are several options available for running the program: one could have a single cue ball that is hit with different speeds and different directions and the pocket it lands in is tracked. Alternatively, one could have all the balls on the table and the dynamics of an actual game can be simulated. It depends on what you're after, but here we will combine important elements of both setups.

So as with any dynamical system, it has to have an initial configuration, meaning positions and velocities of all 15 balls. Assuming we're modeling a standard game of pool, the cue ball will have an initial velocity and the system will step through time until it determines that there's a collision. In real life, a collision is intuitive but not for the computer. So we have to carefully think what it means to collide and communicate that to the computer through the program. Thinking of two identical balls colliding, we know that the one that is hit takes off with the same velocity as one that hit and one that hit it will stop. And if there's three balls in a row sitting still and one gets hit on the end, it's the ball on the extreme end that takes off. So there must be some material communication through the collision. In reality there must be some small deformation of the balls. To communicate such an idea to the computer we develop the notion of virtual collisions. In physics, something *virtual* is usually something that is intermediary or not tangible. So we assume the balls to be perfectly rigid and we can let them overlap a little bit then we use physical laws to implement the collision. The model's dynamics never really experiences a virtual collision; it is used only to implement the equations of momentum conservation and to propagate the collision information to all objects involved. So the program searches for pairs of balls who are heading towards each other and overlapping...and calls that a virtual collision. The system is not allowed to advance in time unless a configuration is reached where a virtual collision does not result in particle overlap; such a configuration is taken to be the real outcome of the collision. It is well known that uncertainty in the outcome of the simulation may arise when more than two balls collide and touch at once [7], which could also be the case. Since the algorithm searches for touching pairs, virtual collisions still proceed at different times when three or more balls collide at once. Such artificial delay introduces uncertainty in the outcome of the simulation and this effect must be alleviated. The post-collision velocities are therefore calculated as averages of the velocities obtained when the algorithm searches once in forward order and once in reverse order. The method used was validated by testing it in situations involving simple geometries where the results are known: one- and two-dimensional arrays of interacting and non-interacting masses undergoing collisions of varying elasticity—a mass colliding with two masses already touching for example.

4.3.3 Details for the Computer

Although the simulation of pool games is very similar to those of baseball and football already presented, it is still worthwhile to discuss specifics. After the forces on all N particles are calculated, the system is advanced in time utilizing a simple forward-difference scheme:

$$\vec{r}_i(t + \Delta t) = \vec{r}_i(t) + \vec{v}_i(t)\Delta t + \frac{1}{2}\frac{\vec{F}_i}{m_i}(\Delta t)^2$$

$$\vec{v}_i(t + \Delta t) = \vec{v}_i(t) + \frac{\vec{F}_i}{m_i}\Delta t.$$

The time step Δt in the simulation is chosen such that any physical overlap of two pool balls is small compared to either of their dimensions. Various efforts are made to rule out unimportant variables but still have the theory reflect what is encountered in the real world. For example, the simulation is run including all the types of interactions described earlier.

4.3.4 Development and Validation

Probably the best place to start coding is to get one ball moving with constant velocity on a table. It should travel in a straight line and although that sounds trivial, you'd be amazed at how challenging such a goal can be a times. Once you're successful with constant velocity motion, it may be useful to add in drag resistance and friction. Then, run the simulation and make sure it does what it needs to do...the ball slows down and stops. Next, you can implement the reflection boundary conditions, implementing the ball hitting the table edges. Once that works well, we need to implement the collision algorithm. And then, after all that works, you can throw in spam and make sure it doesn't affect the center mass motion of the balls but does affect the reflection at the table edges as well as the collisions. (See, you should never rely on a computer to spell check—it's spin, not spam.)

4.3.5 A Sample Program that Simulates the Behavior of Pool Balls

Please see the provided program Pool.cpp for code related to the following discussion. Start out by calling standard C++ libraries and any special ones you might need.

```
#include<cstdlib>
#include<iostream>
#include<fstream>
#include<math.h>
#include<conio.h>
```

Start the program (this statement can go anywhere in the program).

```
int main()
{
```

Make sure that the program can call any other standard C++ libraries without your having to always type them.

```
using namespace std;
```

Declare variables used in the program.

```
const int n=16,npkct=6;
const int nstep=3000;
const float t=0.005;
const float xq=2.0,yq=0.0,vxq=-2.0,vyq=0.0;
int ipkct[n];
int icoll,icmark,jcmark;
float x[n],y[n],z[n],vx[n],vy[n],vz[n];
float vxa[n],vya[n],vza[n],vxb[n],vyb[n],vzb[n];
float r[n],rm[n];
float fx[n],fy[n],fz[n];
float vxold[n],vyold[n],vzold[n];
float vxc[n],vyc[n];
float px[npkct],py[npkct];
float s[n],vrimx[n],vrimy[n];
float dx,dy,dz,rij,dvx,dvy,dvz,vdotr,rati,ratj,vij;
float mu,v,g,cd,rho,pi,cores;
float wleft,wright,wtop,wbot,ewal;
float rpkct,rdunk,vtol,alpha,beta;
float dummy;
```

Open the output file.

```
ofstream myfile;
myfile.open("pool.res");
```

Define important constants.

```
pi=4.*atan(1.);
cd=0.5;
rho=1.29;
g=9.8;
```

```
mu=0.027;
cores=0.95;
vtol=0.0;
alpha=0.001;
beta=0.1;
rpkct=0.05;
```

Locate the edges of the pool table.

```
wleft=0.0;
wbot=0.0;
wright=2.6;
wtop=1.3;
ewal=1.0;
```

Locate the pockets.

```
px[0]=wleft;
py[0]=wbot;
px[1]=wleft;
py[1]=wtop;
px[2]=0.5*(wleft+wright);
py[2]=wtop;
px[3]=wright;
py[3]=wtop;
px[4]=wright;
py[4]=wbot;
px[5]=0.5*(wleft+wright);
py[5]=wbot;
```

Initialize other arrays.

```
for(int i=0;i<=n-1;i++)
{
vxc[i]=0.;
vyc[i]=0.;
x[i]=0.;
y[i]=0.;
z[i]=0.;
vx[i]=0.;
vy[i]=0.;
vz[i]=0.;
r[i]=0.0286;
rm[i]=0.17;
ipkct[i]=-1;
s[i]=0.;
vrimx[i]=0.;
vrimy[i]=0.;
}
rm[15]=0.16;
r[15]=0.0279;
s[15]=0.;
```

Initial Conditions; Rack 'em up!

```
x[0]=0.2*wright+4.*0.0287*sqrt(3.);
x[1]=0.2*wright+3.*0.0287*sqrt(3.);
x[2]=0.2*wright+3.*0.0287*sqrt(3.);
```

```
x[3]=0.2*wright+2.*0.0287*sqrt(3.);
x[4]=0.2*wright+2.*0.0287*sqrt(3.);
x[5]=0.2*wright+2.*0.0287*sqrt(3.);
x[6]=0.2*wright+0.0287*sqrt(3.);
x[7]=0.2*wright+0.0287*sqrt(3.);
x[8]=0.2*wright+0.0287*sqrt(3.);
x[9]=0.2*wright+0.0287*sqrt(3.);
x[10]=0.2*wright;
x[11]=0.2*wright;
x[12]=0.2*wright;
x[13]=0.2*wright;
x[14]=0.2*wright;
y[0]=0.5*wtop+0.;
y[1]=0.5*wtop+0.0287;
y[2]=0.5*wtop-0.0287;
y[3]=0.5*wtop+2.*0.0287;
y[4]=0.5*wtop+0.;
y[5]=0.5*wtop-2.*0.0287;
y[6]=0.5*wtop+3.*0.0287;
y[7]=0.5*wtop+0.0287;
y[8]=0.5*wtop-0.0287;
y[9]=0.5*wtop-3.*0.0287;
y[10]=0.5*wtop+4.*0.0287;
y[11]=0.5*wtop+2.*0.0287;
y[12]=0.5*wtop+0.;
y[13]=0.5*wtop-2.*0.0287;
y[14]=0.5*wtop-4.*0.0287;
x[15]=x[0]+3.*0.0287;
x[15]=xq;
y[15]=0.5*wtop+yq;
```

Have only the cue ball moving.

```
vx[15]=vxq;
vy[15]=vyq;
```

Write out where the pockets are.

```
for(int i=0;i<=npkct-1;i++)
{
myfile<<px[i]<<" "<<py[i]<<endl;
}
myfile<<"   "<<endl;

icmark=-1;
jcmark=-1;
```

Start the time loop.

```
for(int itime=0;itime<=nstep-1;itime++)
{
```

Search through the balls and calculate the force acting on the ones not in pockets.

```
for(int i=0;i<=n-1;i++)
{
if(ipkct[i]==-1)
{
```

```
v=sqrt(vx[i]*vx[i]+vy[i]*vy[i]+vz[i]*vz[i]);
if(v!=0.)
{
fx[i]=-mu*rm[i]*g*(vx[i]/v)-0.5*cd*rho*pi*r[i]*r[i]*v*vx[i];
fy[i]=-mu*rm[i]*g*(vy[i]/v)-0.5*cd*rho*pi*r[i]*r[i]*v*vy[i];
fz[i]=0.;
}
if(v==0.)
{
fx[i]=0.;
fy[i]=0.;
fz[i]=0.;
}
}
}
```

Search through the balls and update the positions of the ones not in pockets.

```
for(int i=0;i<=n-1;i++)
{
if(ipkct[i]==-1)
{
dummy=x[i]+vx[i]*t+0.5*(fx[i]/rm[i])*t*t;
x[i]=dummy;
dummy=y[i]+vy[i]*t+0.5*(fy[i]/rm[i])*t*t;
y[i]=dummy;
dummy=z[i]+vz[i]*t+0.5*(fz[i]/rm[i])*t*t;
z[i]=dummy;
}
```

Still searching: if a ball is close enough to a pocket, dunk it!

```
for(int j=0;j<=npkct-1;j++)
{
rdunk=sqrt(pow(x[i]-px[j],2)+pow(y[i]-py[j],2));
if(rdunk<rpkct)
{
ipkct[i]=j;
}
}
```

Stop searching with the next line.

```
}
```

Put the velocities of the balls in a second array.

```
for(int i=0;i<=n-1;i++)
{
vxc[i]=vx[i];
vyc[i]=vy[i];
}
```

Update the velocities of the balls with Newton's second law.

```
for(int i=0;i<=n-1;i++)
{
```

Apply Newton's second law.

```
dummy=vx[i]+(fx[i]/rm[i])*t;
vx[i]=dummy;
if(vx[i]<0.&&vxc[i]>0.)
{
vx[i]=0.;
}
if(vx[i]>0.&&vxc[i]<0.)
{
vx[i]=0.;}
dummy=vy[i]+(fy[i]/rm[i])*t;
vy[i]=dummy;
if(vy[i]<0.&&vyc[i]>0.)
{
vy[i]=0.;
}
if(vy[i]>0.&&vyc[i]<0.)
{
vy[i]=0.;
}
dummy=vz[i]+(fz[i]/rm[i])*t;
vz[i]=dummy;
}
```

Update the velocities of the balls due to wall reflection.

```
for(int i=0;i<=n-1;i++)
{
if(x[i]-r[i]<=wleft)
{
vx[i]=-ewal*vx[i];
vy[i]=ewal*vy[i];
vrimy[i]=-r[i]*s[i];
dummy=vy[i]+beta*(-vrimy[i]);
vy[i]=dummy;
}
if(x[i]+r[i]>=wright)
{
myfile<<i<<" "<<x[i]+r[i]<<" "<<wright<<endl;
vx[i]=-ewal*vx[i];
vy[i]=ewal*vy[i];
vrimy[i]=r[i]*s[i];
dummy=vy[i]+beta*(-vrimy[i]);
vy[i]=dummy;
}
if(y[i]-r[i]<=wbot)
{
vx[i]=ewal*vx[i];
vy[i]=-ewal*vy[i];
vrimx[i]=r[i]*s[i];
dummy=vx[i]+beta*(-vrimx[i]);
vx[i]=dummy;
}
if(y[i]+r[i]>=wtop)
{
vx[i]=ewal*vx[i];
vy[i]=-ewal*vy[i];
vrimx[i]=-r[i]*s[i];
dummy=vx[i]+beta*(-vrimx[i]);
```

```
vx[i]=dummy;
}
}
```

Adjust the spin.

```
for(int i=0;i<=n-1;i++)
{
dummy=s[i]-alpha*s[i];
s[i]=dummy;
}
```

Evaluate Collisions in Forward Order (a).

```
for(int i=0;i<=n-1;i++)
{
vxa[i]=vx[i];
vya[i]=vy[i];
vza[i]=vz[i];
}
line1000:
for(int i=0;i<=n-1;i++)
{
vxold[i]=vxa[i];
vyold[i]=vya[i];
vzold[i]=vza[i];
}
icoll=0;
for(int i=0;i<=n-1;i++)
{
for(int j=i;j<=n-1;j++)
{
if(i!=j&&ipkct[i]==-1&&ipkct[j]==-1)
{
dx=x[j]-x[i];
dy=y[j]-y[i];
dz=z[j]-z[i];
rij=sqrt(dx*dx+dy*dy+dz*dz);
dvx=vxold[j]-vxold[i];
dvy=vyold[j]-vyold[i];
dvz=vzold[j]-vzold[i];
vij=sqrt(dvx*dvx+dvy*dvy+dvz*dvz);
vdotr=dx*dvx+dy*dvy+dz*dvz;
if(rij<r[i]+r[j])
{
if(vdotr<-vtol*rij*vij)
{
icoll=1;
icmark=i;
jcmark=j;
}
}
}
}
}
if(icoll==1)
{
for(int i=0;i<=n-1;i++)
{
```

```
for(int j=i;j<=n-1;j++)
{
if(i==icmark&&j==jcmark)
{
dx=x[j]-x[i];
dy=y[j]-y[i];
dz=z[j]-z[i];
rij=sqrt(dx*dx+dy*dy+dz*dz);
dvx=vxold[j]-vxold[i];
dvy=vyold[j]-vyold[i];
dvz=vzold[j]-vzold[i];
vij=sqrt(dvx*dvx+dvy*dvy+dvz*dvz);
vdotr=dx*dvx+dy*dvy+dz*dvz;
rati=(1.+cores)*rm[j]/(rm[i]+rm[j]);
ratj=(1.+cores)*rm[i]/(rm[i]+rm[j]);
vxa[i]=vxold[i]+rati*vdotr*dx/(rij*rij);
vxa[j]=vxold[j]-ratj*vdotr*dx/(rij*rij);
vya[i]=vyold[i]+rati*vdotr*dy/(rij*rij);
vya[j]=vyold[j]-ratj*vdotr*dy/(rij*rij);
vza[i]=vzold[i]+rati*vdotr*dz/(rij*rij);
vza[j]=vzold[j]-ratj*vdotr*dz/(rij*rij);
}
}
}
goto line1000;
}
```

Evaluate Collisions in Reverse Order (b).

```
for(int i=n-1;i>=0;i--)
{
vxb[i]=vx[i];
vyb[i]=vy[i];
vzb[i]=vz[i];
}
line1100:
for(int i=n-1;i>=0;i--)
{
vxold[i]=vxb[i];
vyold[i]=vyb[i];
vzold[i]=vzb[i];
}
icoll=0;
for(int i=n-1;i>=0;i--)
{
for(int j=n-1;j>=i;j--)
{
if(i!=j&&ipkct[i]==-1&&ipkct[j]==-1)
{
dx=x[j]-x[i];
dy=y[j]-y[i];
dz=z[j]-z[i];
rij=sqrt(dx*dx+dy*dy+dz*dz);
dvx=vxold[j]-vxold[i];
dvy=vyold[j]-vyold[i];
dvz=vzold[j]-vzold[i];
vij=sqrt(dvx*dvx+dvy*dvy+dvz*dvz);
vdotr=dx*dvx+dy*dvy+dz*dvz;
if(rij<r[i]+r[j])
{
```

```
if(vdotr<-vtol*rij*vij)
{
icoll=1;
icmark=i;
jcmark=j;
}
}
}
}
}
if(icoll==1)
{
for(int i=n-1;i>=0;i--)
{
for(int j=n-1;j>=i;j--)
{
if(i==icmark&&j==jcmark)
{
dx=x[j]-x[i];
dy=y[j]-y[i];
dz=z[j]-z[i];
rij=sqrt(dx*dx+dy*dy+dz*dz);
dvx=vxold[j]-vxold[i];
dvy=vyold[j]-vyold[i];
dvz=vzold[j]-vzold[i];
vij=sqrt(dvx*dvx+dvy*dvy+dvz*dvz);
vdotr=dx*dvx+dy*dvy+dz*dvz;
rati=(1.+cores)*rm[j]/(rm[i]+rm[j]);
ratj=(1.+cores)*rm[i]/(rm[i]+rm[j]);
vxb[i]=vxold[i]+rati*vdotr*dx/(rij*rij);
vxb[j]=vxold[j]-ratj*vdotr*dx/(rij*rij);
vyb[i]=vyold[i]+rati*vdotr*dy/(rij*rij);
vyb[j]=vyold[j]-ratj*vdotr*dy/(rij*rij);
vzb[i]=vzold[i]+rati*vdotr*dz/(rij*rij);
vzb[j]=vzold[j]-ratj*vdotr*dz/(rij*rij);
}
}
}
goto line1100;
}
```

Average Forward and Reverse Velocities.

```
for(int i=0;i<=n-1;i++)
{
vx[i]=0.5*(vxa[i]+vxb[i]);
vy[i]=0.5*(vya[i]+vyb[i]);
vz[i]=0.5*(vza[i]+vzb[i]);
}
```

Consider what happens when the balls touch and are spinning.

```
for(int i=0;i<=n-1;i++)
{
for(int j=i;j<=n-1;j++)
{
if(i!=j&&ipkct[i]==-1&&ipkct[j]==-1)
{
dx=x[j]-x[i];
dy=y[j]-y[i];
```

```
dz=z[j]-z[i];
rij=sqrt(dx*dx+dy*dy+dz*dz);
vij=sqrt(dvx*dvx+dvy*dvy+dvz*dvz);
vdotr=dx*dvx+dy*dvy+dz*dvz;
vrimx[i]=-r[i]*(dy/rij*s[i]);
vrimy[i]=-r[i]*(-dx/rij*s[i]);
vrimx[j]=r[j]*(dy/rij*s[j]);
vrimy[j]=r[j]*(-dx/rij*s[j]);
if(rij<=r[i]+r[j])
{
dummy=vx[i]+beta*(-vrimx[i]+vrimx[j]);
vx[i]=dummy;
dummy=vx[j]+beta*(-vrimx[j]+vrimx[i]);
vx[j]=dummy;
dummy=vy[i]+beta*(-vrimy[i]+vrimy[j]);
vy[i]=dummy;
dummy=vy[j]+beta*(-vrimy[j]+vrimy[i]);
vy[j]=dummy;
}
}
}
}
```

Write out important information to the screen and output file.

```
cout<<itime<<endl;
for(int i=0;i<=n-1;i++)
{
if(ipkct[i]==-1)
{
myfile<<float(itime)*t<<" "<<x[i]<<" "<<y[i]<<" "<<endl;
}
}
myfile<<" "<<endl;
```

Time loop ends with the line below.

```
}
```

Close and save the specified output file.

```
myfile.close();
```

Delay closing the screen to show output if on a MAC or PC.

```
getch();
```

End the program silently.

```
return(0);
}
```

4.4 Understanding Things of Danger in Hindsight and Foresight

4.4.1 Reconstructing Auto Collisions

There are many instances where it is useful to calculate the speeds of cars before a collision to demonstrate whether or not the weather conditions could be blamed for an accident as opposed to operator error or carelessness. The physics itself is very straightforward and involves momentum transfer through the collision. The reconstruction becomes an art when we need to determine things like road condition, tire age, passenger location, and such things. The program below is a numerical model of two cars colliding with varying elasticity (coefficient of restitution), and also includes static and kinetic friction between the car and road, and therefore you can calculate the extent of skid marks and get an idea of the real precollision speeds. If you carefully examine the code in the football tackle program regarding rotation, you can use the results to modify the sample code provided below so as to incorporate angular momentum transfer as well and, through that address rotational aspects of a collision. I have included results for a few different scenarios that may help you to validate against if needed.

Please review the provided program AutoCollisions.cpp and resulting example output that are related to the following discussion. The usual C++ headers and library calls

```
//A Simple C++ Example Program

#include<cstdlib>
#include<iostream>
#include<fstream>
#include<math.h>
#include<iomanip>
#include<stdio.h>

using namespace std;

//------Declare Variables------
```

Start the program and declare global constants. The variables are fairly straightforward and tskid is the amount ot time after the collision before skidding occurs. Before that, rolling friction is in effect and the code can be easily modified to include skidding before the collision as well.

```
int main()
{
const int nsteps=10000000,nevery=100;
const float tskid=2.0;
const float tol=0.001;
const float g=9.8;
const float length1=4.0,length2=4.5;
const float m1=1000.0,m2=100.0;
const float dt=0.0001;
const float muk=0.1,mus=0.6;
const float cores=1.0;
const float x10=50.0,x20=100.0;
const float vx10=20.0,vx20=4.0;
float x1,x2,vx1,vx2,ax1,ax2;
float time,tcoll;
float dummy1,dummy2,rat1,rat2;
```

Here is some I/O.

```
ofstream myfile;
myfile.open("collision_coords.txt");
```

Now initialize and start the time loop.

```
x1=x10;
vx1=vx10;
ax1=0.0;
x2=x20;
vx2=vx20;
ax2=0.0;
myfile<<"Pre-Collision positions and Velocities"<<endl;
for (int k=0;k<=nsteps;k++)
{
time=float(k)*dt;
```

Check for collision here and skip down if there is one.

```
if((x2-x1)<=(0.5*(length1+length2)))
{
tcoll=time;
goto line1000;
}
```

Motion integration (time update code).

```
x1=x1+vx1*dt+0.5*ax1*dt*dt;
vx1=vx1+ax1*dt;
x2=x2+vx2*dt+0.5*ax2*dt*dt;
vx2=vx2+ax2*dt;
```

More I/O.

```
if(float(k/nevery)==float(k)/float(nevery))
{
myfile<<time<<" "<<x1<<" "<<x2<<" "<<vx1<<" "<<vx2<<endl;

}

}
line1000:
```

More I/O.

```
myfile<<" "<<endl;
myfile<<"Collision time: "<<tcoll<<" seconds"<<endl;
myfile<<" "<<endl;
myfile<<"Post-Collision positions and velocities"<<endl;
```

Here we change the velocities through the collision.

```
dummy1=vx1;
dummy2=vx2;
```

```
rat1=(1.+cores)*m2/(m1+m2);
rat2=(1.+cores)*m1/(m1+m2);
vx1=dummy1+rat1*(dummy2-dummy1);
vx2=dummy2-rat2*(dummy2-dummy1);

dummy1=ax1;
dummy2=ax2;
for (int k=1;k<=nsteps;k++)
{
time=tcoll+float(k)*dt;
```

Here, we take skidding as well as rolling friction into account.

```
if(time<tcoll+tskid)
{
ax1=dummy1-muk*g*vx1/fabs(vx1);
ax2=dummy2-muk*g*vx2/fabs(vx2);
}
if(time>=tcoll+tskid)
{
ax1=dummy1-mus*g*vx1/fabs(vx1);
ax2=dummy2-mus*g*vx2/fabs(vx2);
}
x1=x1+vx1*dt+0.5*ax1*dt*dt;
vx1=vx1+ax1*dt;
x2=x2+vx2*dt+0.5*ax2*dt*dt;
vx2=vx2+ax2*dt;
```

Wrap it up when the velocities are small enough that they do not matter.

```
if(fabs(vx1)<=tol||fabs(vx2)<=tol)
{
goto line2000;
}
```

I/O every so many steps.

```
if(float(k/nevery)==float(k)/float(nevery))
{
if(time<tcoll+tskid)
{
myfile<<time<<" "<<x1<<" "<<x2<<" "<<vx1<<" "<<vx2<<" coasting"<<endl;
}
if(time>=tcoll+tskid)
{
myfile<<time<<" "<<x1<<" "<<x2<<" "<<vx1<<" "<<vx2<<" skidding"<<endl;}
}
}
line2000:
return(0);
}
```

4.4.2 Protecting Better against Bullet Impact

Whether in applications involving crime, combat, or law enforcement, the issue of injury due to bullet insult is a central one. To counteract injury and potential tragedies, the use of Kevlar vests and helmets is mandatory. The major complaint about bullet proof vests and helmets is the material's weight and the rigidity, both of which decrease range of motion. A student of mine (thank you Eddie!) was interested in simulating bullet impact on ceramic-composite protective vests focusing on new fiber materials that would parallel the bullet—stopping effectiveness of Kevlar but allow the decrease of weight and increase of maneuverability. We used MPM simulations (See Chapter 7) to search for the optimal thickness and structure of human-made silk thread by simulating bullet impacts on ceramic composite body armor. The composition of the bullet is also an issue since NATO allows lead AK-47 bullets but has banned steel ones. Our research is still in progress and we hope that the simulations can provide adequate data for leading engineers and experimentalists towards producing new armor effective against lead or steel bullets. Related reading can be found elsewhere [8–12].

Just a few specifics: our simulations began with a copper-clad AK-47 bullet heading towards the ceramic composite vest with an initial speed of 710 m/sec, as shown in Figure 4.3. The bullet is composed of a steel or lead core and a copper cladding, containing a total of 600,000 particles. Because the time between start and collision with the vest would allow only a fraction of a revolution for normally fired bullets, we chose to neglect rotation. The vest has a ceramic component as well as a fiber mesh (made of Kevlar or some other material), containing a total of 600,000 particles, yielding a 1.2×10^6 particle simulation. The vest is anchored to the computational grid so it will not drift when struck by the bullet. The system is advanced with a time step $\Delta t = 4 \times 10^{-9}$ sec and the nature of impact, as well as various physical quantities, are monitored throughout the simulation. Snapshots were created at various intervals, and various physical quantities are monitored through time; an example of such a time sequence for an impact is shown in Figure 4.4. The simulation results are very sensitive to parameters such as material breaking points and elastic moduli, as well as vest dimensions and particle number. We obtain blunt wounds, full penetration, ricochets

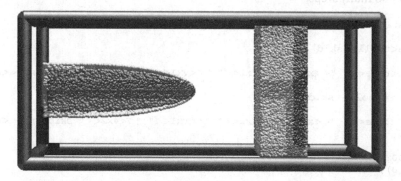

Figure 4.3. Initial configuration of the bullet/vest system (cutaway view showing bullet and vest structure). Different shadings in grayscale are helpful in identifying the following components that are shown in color in the book's electronic version. Purple particles show the core of the bullet, yellow is the outer copper shell, white indicates the ceramic component of the vest, red shows the spider web component and green particles anchor the vest to the computational box (blue frame).

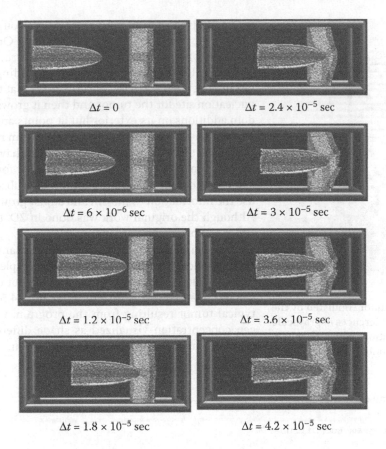

$\Delta t = 0$ $\Delta t = 2.4 \times 10^{-5}$ sec

$\Delta t = 6 \times 10^{-6}$ sec $\Delta t = 3 \times 10^{-5}$ sec

$\Delta t = 1.2 \times 10^{-5}$ sec $\Delta t = 3.6 \times 10^{-5}$ sec

$\Delta t = 1.8 \times 10^{-5}$ sec $\Delta t = 4.2 \times 10^{-5}$ sec

Figure 4.4. Evolution of a typical bullet impact simulation with time. Shading in grayscale helps identification of various system components shown in color in the book's electronic version as mentioned in Figure 4.3.

and full protection from injury using various parameters, and we're still in the process of combing through all of it to see a consistent picture that makes sense. There are other types of bullet impact simulations reported and it can be fruitful to check out what has been done! [8–12]

4.5 Diseases

4.5.1 Eden Model of Tumor Growth

Understanding tumor growth through modeling and simulation has been of interest in the research community for decades. To say "there are many models out there" is really an understatement, since there are medical research groups that focus on computer-based analysis, simulation, visualization, and treatment of tumors and a wide range of other medical issues. As such, I chose to cover a basic and well-known model in the

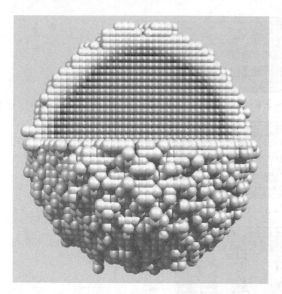

Figure 4.5. Snapshot of a tumor simulated by the sample program. Shading differences in grayscale delineate oxygen concentration, which is shown in color in the book's electronic version.

field and give you a starting point where you can take things and run with them if you chose to. One of the pioneering models was formulated by Murray Eden in 1961 [13]; I strongly encourage more reading in this fascinating area! The basic idea [13,14] is that there is a nucleation site for the tumor and then it grows by random additions on its exterior but at points adjacent to already existing ones. The growth algorithm results in the perimeter of the tumor being a fractal (having non-integer dimension). One can also account for oxygen and nutrition diffusion in the tumor and subsequently use the information to inform and adjust growth rates. Although the original work was done in 2D, the algorithm can be extended to 3D as well.

Below is a description of Eden3D.f, a sample program provided to you, to use that is a simple simulation of tumor growth; it is written in 3D but certainly can be collapsed to 2D. In addition, Figure 4.5 shows a typical tumor resulting from the program, with oxygen concentration visualized as shade differences in grayscale and color in the electronic version.

First name the program.

```
program EdenDiffuse3D
```

Now declare global constants.

```
parameter(nsteps=40000)
parameter(Nx=40,Ny=40,Nz=40,sphere_size=0.06)
parameter (npart_max=64000)
parameter(growth_time=0.0005,ngrowth=2000)
parameter(Cmax=100.0)
parameter(D=1.5e-02)
```

Declare program variables.

```
character*19 fileanim
integer i,j,k,kk,is,ix,iy,ig
integer iocc(npart_max),jocc(npart_max)
integer kocc(npart_max)
integer npart,ipart
integer irand,nevery,ifr
integer iwhere,jwhere,kwhere
real C(Nx,Ny,Nz),conc
real roll
real dx,dy,dz
real blue,green,red
real updatex,update,updatez
real dummy
```

Do some I/O

```
open(6,file='tumor.txt')
```

And perform needed variable initializations...

```
npart=0
dx=1.05*sphere_size
dy=1.05*sphere_size
dz=1.05*sphere_size
nevery=nsteps/200
```

...including making every site empty by setting the occupancy to zero.

```
      do 100 i=1,npart_max
      iocc(i)=0
      jocc(i)=0
      kocc(i)=0
  100 continue

      fileanim='TopView-0000000.pov'
      ifr=6
      npart=1
      do 110 i=1,Nx
      do 110 j=1,Ny
      do 110 k=1,Nz
      C(i,j,k)=Cmax
  110 continue
```

Place the first particle in the middle of the computational box by assigning nonzero values to the occupancy and set the oxygen concentration there to zero.

```
iocc(npart)=Nx/2
jocc(npart)=Ny/2
kocc(npart)=Nz/2
C(iocc(npart),jocc(npart),kocc(npart))=0.0
```

Now start the mail loop through the number of steps.

```
do 1000 is=1,nsteps
```

Pick a random site and look up where it is based on its occupancy.

```
      ipart=1+nint(rand()*float(npart-1))
      iwhere=iocc(ipart)
      jwhere=jocc(ipart)
      kwhere=kocc(ipart)
 3389 continue
```

Now use random numbers to select in which direction a new site will be picked adjacent to the one we already picked above.

```
      roll=rand()
      if(roll.lt.0.33) then
  211 continue
      irand=nint(2.0*(rand()-0.5))
      if(irand.eq.0) then
```

```
         goto 211
         endif
         iwhere=iwhere+irand
         else if (roll.ge.0.33.and.roll.le.0.67) then
212 continue
         irand=nint(2.0*(rand()-0.5))
         if(irand.eq.0) then
         goto 212
         endif
         jwhere=jwhere+irand
         else if (roll.gt.0.67) then
213 continue
         irand=nint(2.0*(rand()-0.5))
         if(irand.eq.0) then
         goto 213
         endif
         kwhere=kwhere+irand
         endif
```

Here's a safety statement so that if our random numbers didn't result in the successful selection of an adjacent site to the one we picked, start again!

```
         if(iocc(i).eq.iwhere.and.jocc(i).eq.jwhere
     1.and.kocc(i).eq.kwhere) then
         goto 3389
         endif
3378 continue
```

Unless we are at the maximum number of particles, make sure that there is maximum oxygen concentration except for at the sites, where it is zero.

```
         if(npart.eq.npart_max) then
         goto 7000
         endif
         npart=npart+1
         iocc(npart)=iwhere
         jocc(npart)=jwhere
         kocc(npart)=kwhere
         do 710 i=1,Nx
         do 710 j=1,Ny
         do 710 k=1,Nz
         C(i,j,k)=Cmax
710 continue
         do 9993 i=1,npart
         C(iocc(i),jocc(i),kocc(i))=0.0
9993 continue
```

Now look at oxygen diffusion in the tumor.

```
C#####Now Oxygen Diffusion################################
         do 8000 ig=1,ngrowth
         do 8200 kk=1,npart
         i=iocc(kk)
         j=jocc(kk)
         k=kocc(kk)
         updatex=(C(i+1,j,k)-2.0*C(i,j,k)+C(i-1,j,k))/dx**2
         updatey=(C(i,j+1,k)-2.0*C(i,j,k)+C(i,j-1,k))/dy**2
         updatez=(C(i,j,k+1)-2.0*C(i,j,k)+C(i,j,k-1))/dz**2
```

```
          dummy=C(i,j,k)+D*(updatex+updatey+updatez)*growth_time
          C(i,j,k)=dummy
 8200 continue
 8000 continue
C############################################################
```

Then let's visualize!

```
c#####Let's do some visualization#############################
          if (float(is/nevery).eq.float(is)/float(nevery)) then
          ifr=ifr+1
          write(fileanim(9:15),1001) ifr
 1001 format(i7.7)
          open(unit=ifr+6,file=fileanim)
          write(ifr+6,*) '#version 3.6;'
          write(ifr+6,*) '#include "colors.inc"'
          write(ifr+6,*) 'global_settings'
          write(ifr+6,*) '{'
          write(ifr+6,*) 'assumed_gamma 1.0'
          write(ifr+6,*) '}'
          write(ifr+6,*) 'camera'
          write(ifr+6,*) '{'
          write(ifr+6,*) 'location  <',float(Nx)*dx/2.0,',',float
     1(Ny)*dy/2.0,',10>'
          write(ifr+6,*) 'direction 1.5*z'
          write(ifr+6,*) 'right      4/3*x'
          write(ifr+6,*) 'look_at   <',float(Nx)*dx/2.0,',',float
     1(Ny)*dy/2.0,',0>'
          write(ifr+6,*) '}'
          write(ifr+6,*) 'background {
     1color red 0.7 green 0.7 blue 0.7 }'
          write(ifr+6,*) 'light_source'
          write(ifr+6,*) '{'
          write(ifr+6,*) '0*x'
          write(ifr+6,*) 'color red'
          write(ifr+6,*) '1.0  green 1.0  blue 1.0'
          write(ifr+6,*) 'translate <-30, 30, 30>'
          write(ifr+6,*) '}'
          write(ifr+6,*) 'light_source'
          write(ifr+6,*) '{'
          write(ifr+6,*) '0*x'
          write(ifr+6,*) 'color red'
          write(ifr+6,*) '0.5  green 0.5  blue 0.5'
          write(ifr+6,*) 'translate <30, 30, 30>'
          write(ifr+6,*) '}'
          write(ifr+6,*) 'light_source'
          write(ifr+6,*) '{'
          write(ifr+6,*) '0*x'
          write(ifr+6,*) 'color red'
          write(ifr+6,*) '0.1  green 0.1  blue 0.1'
          write(ifr+6,*) 'translate <30, -30, 30>'
          write(ifr+6,*) '}'
          DO 6673 i=1,npart
          if(kocc(i).lt.Nz/2.or.
     1(kocc(i).ge.Nz/2.and.jocc(i).lt.Ny/2))then
          WRITE(ifr+6,*) 'sphere { <',
     1float(iocc(i))*dx,',',float(jocc(i))*dy,',',
     2float(kocc(i))*dz,'>,'
          WRITE(ifr+6,*) sphere_size
          conc=C(iocc(i),jocc(i),kocc(i))
```

```
      if(conc.le.Cmax/4.0) then
      red=1.0-4.0*conc/Cmax
      green=0.0
      blue=1.0
      endif
      if(conc.gt.Cmax/4.0.and.conc.le.2.0*Cmax/4.0) then
      red=0.0
      green=4.0*(conc-Cmax/4.0)/Cmax
      blue=1.0-green
      endif
      if(conc.gt.2.0*Cmax/4.0.and.conc.le.3.0*Cmax/4.0)then
      red=4.0*(conc-2.0*Cmax/4.0)/Cmax
      green=1.0
      blue=0.0
      endif
      if(conc.gt.3.0*Cmax/4.0.and.conc.le.4.0*Cmax/4.0)then
      red=1.0+4.0*(conc-3.0*Cmax/4.0)/Cmax
      green=1.0
      blue=0.0
      endif
      if(conc.gt.4.0*Cmax/4.0) then
      red=2.0-4.0*(conc-4.0*Cmax/4.0)/Cmax
      green=red-1.0
blue=0.0
      endif
      WRITE(ifr+6,*) 'texture {pigment
     1{color rgb<',red,',',green,',',blue,'>}'
      WRITE(ifr+6,*) 'finish{specular 1}} }'
      endif
 6673 CONTINUE
      close(ifr+6)
      endif
c################################################################
 1000 continue

 7000 continue

      return
      end
```

4.5.2 Will the Real Epidemic Please Stand Up?

It can be exciting to jump into an emerging field of research because you can promote and influence the construction of our understanding of that area of human comprehension. Please scan Reference 15 and you'll see that computer modeling and simulation of epidemics certainly fits the bill for such a field. Not only are researchers determining how an epidemic is defined but (and not surprisingly) there are dramatically different outcomes for a given set of parameters and models. Think of a simple model that you can start with and code it up. Area of $N \times M$ squares, each one representing a diseased point. The disease lasts a certain time and within that time there is another (likely different) time interval in which it can disease a neighboring point with the same properties. How will you handle boundary conditions? Will the time intervals of diseasing be random or ordered in space? Random or ordered in time?

Try an epidemic simulation up and read Reference 15 again. Which one of the current models is yours most like? You'll have to do a bit of digging there and remember that Rome wasn't built in a day: go step-by step, validate and save successive versions of your code.

PROBLEMS

4.1. Suppose you work for an architectural firm and they are going out of business. They are offered a contract to build a ball stadium which might have the opportunity to have outdoor or indoor games. The question raised is whether or not the effects from differences in the air matter to the balls' flight. The firm is staking their future on your estimate and so you must investigate the effects of buoyancy for them to make absolutely certain that your answer is as complete as possible. Extend the validation in part (B) so that buoyancy effects are noticeable, and show this on a plot. What changes need to be made? In what other situations could such an effect matter?

4.2. You have some old friends from high school. They are real physics nerds and would like to know what changes are incurred in a volleyball's flight due only to changes in gravity by changing altitude by 0.5 mile. Using the formula $g = 9.81(1 - 2h/R)$, where h is the altitude above sea level and R is the radius of the Earth, use the validation case in part (B) to investigate the small effects in the ball's flight by such small changes in g. What changes need to be made? In what situations could such a theoretical problem matter?

4.3. Suppose you are hired to be on a committee to determine whether or not it is feasible to play certain sports in extreme outdoor weather conditions. Briefly discuss three computer models you might construct, what physics they may contain and what unique benefits the simulations bring to the table.

4.4. For a wedding stunt, a young couple would like to either throw a ball or maybe golf in a pool under water. They are wondering, though, if there are spherical objects possible of traveling far enough under water to make it worthwhile, before any budget is spent. Briefly discuss three specific models you would construct to investigate the feasibility of their ideas. Creativity is encouraged here. Be sure to address what unique benefits the simulation brings to the table.

4.5. Show that the units in all the terms of Equation 4.7 match.

4.6. Using Equation 4.7 as a starting point, show that Equations 4.8 through 4.10 result. You may have to review the algebra of vector cross products.

4.7. The choice of an appropriate time step can make or break a simulation. Consider the validation case where a ball is thrown with gravity as the only acting force. Investigate the importance of the choice of time step by making it too large or too small. Please be clear on how you determined what "too large" and "too small" means for you here. For your system and resources, what are the limiting time steps you obtained?

4.8. The model discussed here can handle more than one particle but they can't interact. Think of and briefly discuss two models where such a non-interaction assumption is reasonable. If the particles were to interact, what specific changes would have to be made to the model?

4.9. The method of integration presented here is Newton's method. There are others, though, which may or may not be better in certain cases. Using a particular model or validation case, implement the 4th order Runge–Kutta algorithm and compare the results to Newton's Method.

4.10.   Your city (Colorado Springs, CO) is on a budget and has hired you to help them determine the size of a new baseball field they are constructing. Consider a baseball batted with a speed of 80 miles per hour at an angle of 45° with respect to the horizontal. Construct a computer simulation of its path through air, considering topspin and backspin. Per your results for the minimum and maximum ranges of the ball, what dimensions would you advise them that the field should have?

4.11.   Suppose you did the work described in problem 8 but instead you used data for Sacramento, CA. What are the percent errors in your minimum and maximum ranges?

4.12.   Air density can depend on temperature, humidity, altitude, and local barometric pressure. Construct a computer simulation of the trajectory of a volleyball through air, and incorporate an expression for the air density calculable from any combination of the variables mentioned above.

4.13.   A fire is threatening the region of forest near where you live. Suppose you have been asked to track the smoke from a fire where there are northeasterly prevailing winds of 20 miles per hour. Considering the smoke to be buoyant spherical particles, construct a model that could track and/or predict the trajectory of the smoke. Note that there are some variables you just won't know precisely here so please justify your choices.

4.14.   You have a very prodigious nephew. You are out at the zoo one day and his helium balloon sadly floats away. The child would like to know where it went. Construct a computer simulation to track the balloon assuming 10 mile per hour southwesterly prevailing winds. Be sure to justify assumptions for any of the parameters you use in the simulation.

4.15.   You are out at a baseball game with your fiancé and you enter into a disagreement about the game. You used the example program to simulate the trajectory of the baseball but your better half points out that it would be more realistic to have the ball stop when it hits (a) the ground, or (b) a vertical wall or fence. Implement changes in the example program so the ball stops at the more realistic places suggested in (a) and (b).

4.16.   Modify the football tackling example program so as to include static and kinetic friction between the runner's feet and the ground.

4.17.   Consider your favorite football team. (a) Look up or make your best guess about the weather conditions on game days, and translate that information to your best guesses about static and kinetic friction coefficients. (b) Make the modifications suggested in Problem 4.16 and briefly discuss your results.

4.18.   Consider two aspects of tackling: tackle height and tackle type (elastic/inelastic). Use the example program to calculate the fractional change in translational kinetic energy. Discuss extreme cases (highest; lowest) and comment where each type of tackle may be the most beneficial.

4.19.   Consider two aspects of tackling: tackle height and tackle type (elastic/inelastic). Use the example program to calculate the fractional change in rotational kinetic energy. Discuss extreme cases (highest; lowest) and comment where each type of tackle may be the most beneficial.

4.20. Modify the football tackling program so the tackler is a sphere of nonzero radius. Run the simulation and compare results with the point sphere case.

4.21. Modify the football tackling program so the runner is a cylinder of nonzero radius. Run the simulation and compare results with the point sphere case.

4.22. Modify the football tackling example so as to calculate the forces on the players during the simulation. When player–player interactions are in effect, make sure that Newton's third law of motion is satisfied.

4.23. Modify the pool game simulation so the table is not rectangular any more but rather a curved shape of your choice. Run the simulation and make sure the dynamics are correct.

4.24. The program presented here for pool incorporates mathematics needed to model dampening spin but doesn't actually do it. Measure or estimate the dampening coefficient for spin and implement it in the program.

4.25. Consider the pool game simulation with the kinetic coefficient of friction set to $\mu = 0.05$, similar to a smooth ski on ice. Run the simulation and show that even such a small value for friction will easily bring the system to a halt by depleting its kinetic energy.

4.26. Modify the pool game example so as to calculate the forces on the balls during the simulation. When ball-ball interactions are in effect, make sure that Newton's third law of motion is satisfied.

4.27. Modify the car collision program so as to explicitly include precollision skidding as well. Validate your program and generate results for a few different situations.

4.28. Find or make up as realistic a scenario as you can for a car accident. Using results you generate with the program provided along with your best guesses about the situation, form your best understanding about how the accident could have unfolded and discuss.

4.29. Using a bullet or projectile weapon of your choice, code the initial condition for a bullet impact simulation into a Material Point Method simulation (see Chapter 7). Run and discuss your results.

4.30. Modify the 3D Eden tumor growth model so that the oxygen information feeds back into the growth rate. Run and discuss your results.

4.31. Modify the 3D Eden tumor growth model so that nutrition information feeds back into the growth rate. Run and discuss your results.

4.32. Collapse the 3D tumor growth simulation to 2D. Run, visualize, and compare your results to the 3D case.

4.33. The tumor growth model program provided here would be more realistic if it weren't constrained to the "cells" located at sites on a cubic lattice. Modify the program so as to remove the cubic lattice constraint. Run and compare your results with the sample code.

4.34. As I suggested in Section 4.5.2, formulate and code a very simple epidemic model and discuss what parameters you included and why you think they are important. Do contact time, stages of the disease, and demographics play a part? Can you easily promote or suppress an epidemic by adjusting parameters?

References

1. R.G. Watts, T. Bahill, *Keep Your Eye on the Ball*, W. H. Freeman and Company, San Francisco, CA, 2000.

2. M.W. Roth, I. Ahrabi-Fard, Integration of volleyball practice and competition in diverse atmospheric settings using computer simulations of passing jump serves, *Int. J. Volleyball Res.*, 5(1), 18, 2002. and M.W. Roth, I. Ahrabi-Fard, Passing a serve in different air densities, *Coaching Volleyball*, pp. 16–19, October/November, 2001. abstract

3. H. Gould, J. Tobochnik, W. Christian, *An Introduction to Computer Simulation Methods: Applications to Physical Systems*, 3rd edition, Addison-Wesley Longman, Inc., Chicago, IL, 2006.

4. L.A. Bloomfield, *How Things Work: The Physics of Everyday Life*, 5th edition, Wiley and Sons, Hoboken, NJ, 2013.

5. R.G Watts, R. Ferrer, The lateral force on a spinning sphere: Aerodynamics of a curveball, *American Journal of Physics*, 55(1), 40–44, 1987. http://aapt.scitation.org/doi/10.1119/1.14969

6. F. Odar, W.S. Hamilton, Forces on a sphere accelerating in a viscous fluid, *Journal of Fluid Mechanics*, 18(2), 302–314, 1964.

7. C. Leerberg, M.W. Roth, Computer modeling of pool games: Sensitivity of outcomes to initial conditions, *Am. J. Undergrad. Res.*, 1(4), 27, 2003.

8. J.A. Zukas, T. Nicholas, H.F. Swift, L.B. Greszczuk, D.R. Curran, *Impact Dynamics*, John Wiley and Sons, Inc., New York, ISBN 0-471-08677-0, 1982.

9. D.S. Preece, V.S. Berg, Bullet impact on steel and Kevlar/steel armor: Computer modeling and experimental data, *Published in the Proceedings of the ASME Pressure Vessels and Piping Conference - Symposium on Structures under Extreme Loading*, San Diego, CA, 6 pp, July 25–29, 2004.

10. R. Barauskas, A. Abraitiene, A. Vilkauskas, Simulation of a Ballistic impact of a deformable bullet upon a multilayer fabric package, *Comp. Ballist II*, 40, 41–51. doi 10.2495/CBAL050051, 2005.

11. http://www.mate.tue.nl/mate/pdfs/8206.pdf

12. V. Narayanamurthy, C. Lakshmana Rao, B.N. Rao, Numerical simulation of ballistic impact on armour plate with a simple plasticity model, *Defence Science Journal*, 64(1), 55–61, January 2014, doi:10.14429/dsj.64.4521

13. M. Eden, A two-dimensional growth process, *Proceedings of Fourth Berkeley Symposium on Mathematics, Statistics, and Probability*, Berkeley: University of California Press. 4, pp. 223–239, 1961.

14. T. Lahiri, R. Kumar, D.V. Rai, Simulating Various Types of Tumor Growth using Eden Model and its Modified Form, *International Journal of New Innovations in Engineering and Technology*, 3(2), 12–17, 2015.

15. C. Orbann, L. Sattenspiel, E. Miller, J. Dimka, Defining epidemics in computer simulation models: How do definitions influence conclusions?, *Epidemics*, 19, 24–32, 2017.

16. Isaac Newton, A letter of Mr. Isaac Newton, of the University of Cambridge, containing his new theory about light and color, *Philosophical Transactions of the Royal Society*, 7, 3075–3087, 1671–1672.

17. G. Magnus, Über die Abweichung der Geschosse, und: Über eine abfallende Erscheinung bei rotierenden Körpern, *Annalen der Physik*, 164(1), 1–29, 1853.

18. Lord Rayleigh, On the irregular flight of a tennis ball, *Messenger of Mathematics*, 7, 14–16, 1877.

CHAPTER 5

The Many Faces of Music

5.1 Introductory Thoughts

Music is a beautiful human creation that expresses love, makes social statements, and entertains, among many other things. The physical sources of music—vibrations and resonance—are exhibited by a wide variety of physical systems and expressed in different ways. In this chapter, we will dive into techniques and algorithms for simulating just some of the many systems where waves, oscillations, and resonance are important.

5.2 A Finite Difference Simulation of a Guitar String

Let's dive into modeling a guitar string starting with a simple program representing the vibration of a string clamped at both ends. As you go through this, take special note of the boundary conditions because it's a homework problem to implement free boundary conditions (unclamping one of the ends). In addition, in the homework I ask you to modify the initial conditions for the velocity of the string. I will be presenting finite difference as well as more sophisticated analytical approaches and each one has benefits and drawbacks. It is up to you to reflect on the problem you are trying to solve, what options you want in your simulations and what information you are trying to extract from the results.

First, we have the usual frontpiece to the C++ program, calling the relevant libraries. Please see the provided program GuitarString.cpp for related code.

```
#include<cstdlib>
#include<iostream>
#include<math.h>
#include<fstream>
using namespace std;
```

Now we begin the program.

```
int main()
{
```

First, it's good to declare the variables in the program.

```
const int nsteps=100000,nevery=500;
const int nx=1000;
const float L=2.0;
const float dt=0.0001;
const float vel=10.0;
const float rad=0.5;
float red,green,blue;
float x,xp1,dx;
float yk[nx],ykp1[nx],ykm1[nx],ydummy;
int activity[nx];
int ifile;
char filename[7];
```

Now we begin stepping the simulation through time. The one-dimensional wave equation for a guitar string on [0,L] with (clamped) boundary as well as initial conditions is

$$\begin{cases} v^2 \dfrac{\partial^2 u(x,t)}{\partial x^2} = \dfrac{\partial^2 u(x,t)}{\partial t^2} \\[2mm] u(x,0) = f(x) \\[2mm] \dfrac{\partial u}{\partial t}(x,0) = g(x) \\[2mm] u(0,t) = 0 \\[2mm] u(l,t) = 0 \end{cases} \qquad (5.1)$$

See [1] for a derivation. Now consider dividing the region into nx spaces. Considering u_i^m to be the string's displacement at the mth time step and the ith position along the string (Figure 5.1), and using central differences for the second derivatives we have

$$\begin{cases} v^2 \dfrac{u_{i+1}^m - 2u_i^m + u_{i-1}^m}{\Delta x^2} = \dfrac{u_i^{m+1} - 2u_i^m + u_i^{m-1}}{\Delta t^2} \\[2mm] u_i^0 = f(x_i) \\[2mm] \dfrac{u_i^1 - u_i^0}{\Delta t} = g(x_i) \\[2mm] u_0^m = 0 \\[2mm] u_{nx}^m = 0 \end{cases} \qquad (5.2)$$

If you want more visuals on the setup, the 1D problem here is a special case of the 2D Cartesian problem in the next section.

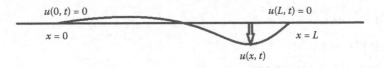

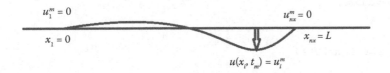

Figure 5.1. Illustration of the continuous guitar string boundary value problem (top) and the discretized problem (bottom).

Here we perform needed initializations.

```
dx=L/float(nx-1);

ifile=0;
```

First the initial position (shape) of the string.

```
for(int i=0;i<nx;i++)

{
x=float(i)*dx;
if(x<=L/3.0)
{
yk[i]=0.1*x;
}
if(x>L/3.0)
{
yk[i]=0.1*L/3.0-(0.1/3.0)*L*(x-L/3.0)/(2.0*L/3.0);
}
ykp1[i]=0.0;
if(x<=L/3.0)
{
ykm1[i]=0.1*x;
}
if(x>L/3.0)
{
ykm1[i]=0.1*L/3.0-(0.1/3.0)*L*(x-L/3.0)/(2.0*L/3.0);
}
activity[i]=1;
}
```

Now the clamped boundary conditions:

```
yk[0]=0.0;
activity[0]=0;
yk[nx-1]=0.0;
activity[nx-1]=0;
```

Note: zero-velocity initial conditions are implicit here but there is a homework problem where you are asked to modify the code and account for the string initially moving. You might want to even try new integration schemes!

```
for(int k=0; k<nsteps; k++)
for(int i=0;i<nx;i++)
{
if(activity[i]!=0)
{
ydummy=(vel*vel*dt*dt/(dx*dx))
*(yk[i+1]-2.0*yk[i]+yk[i-1])
+2.0*yk[i]-ykm1[i];
ykp1[i]=ydummy;
}
}


for(int i=0;i<nx;i++)

{
if(activity[i]!=0)
{
ykm1[i]=yk[i];
yk[i]=ykp1[i];
}
}
```

Write out animation files every so often.

```
if(k%nevery==0)
{
ifile=ifile+1;
sprintf(filename,"%d.pov",ifile);

ofstream myfile;
myfile.open(filename);

myfile<<"#version 3.6"<<endl;
myfile<<"#include "<<"\""<<"colors.inc"<<"\""<<endl;
myfile<<"global_settings"<<endl;
myfile<<"{"<<endl;
myfile<<"assumed_gamma 1.0"<<endl;
myfile<<"}"<<endl;
myfile<<"camera"<<endl;
myfile<<"{"<<endl;
myfile<<"location"<<"<"<<L/2.0<<","<<0.0<<","<<2.5<<">"<<endl;
myfile<<"direction 1.5*z"<<endl;
myfile<<"right 4/3*z"<<endl;
myfile<<"look_at <"<<L/2.0<<","<<0.0<<","<<0.0<<">"<<endl;
myfile<<"}"<<endl;
myfile<<"background { color red 0.5 green 0.5 blue 0.5 }"<<endl;
myfile<<"light_source"<<endl;
myfile<<"{"<<endl;
myfile<<"0*x"<<endl;
myfile<< "color red"<<endl;
myfile<<"1.0  green 1.0  blue 1.0"<<endl;
myfile<<"translate <-30, 30, -30>"<<endl;
myfile<<"}"<<endl;
myfile<<"light_source"<<endl;
myfile<<"{"<<endl;
myfile<<"0*x"<<endl;
myfile<<"color red"<<endl;
myfile<< "0.5  green 0.5  blue 0.5"<<endl;
myfile<<"translate <30, 30, -30>"<<endl;
myfile<<"}"<<endl;
```

```
for(int i=0;i<nx-1;i++)
{
red=1.0;
green=0.0;
blue=0.0;
x=float(i)*dx;
xp1=float(i+1)*dx;
myfile<<"cylinder
{
<"<<x<<","<<yk[i]<<","<<0.0<<">,<"<<xp1<<","<<yk[i+1]<<","<<0.0<<"
>,"<<endl;
myfile<<rad/40.0<<endl;
myfile<<"texture {pigment {color rgb<0,0,1>}"<<endl;
myfile<<"finish{specular 1}} }"<<endl;
}
}
}

return(0);
}
```

5.3 A Little Mathematical Overhead that Provides a Wealth of Understanding

There is another way to solve the wave equation that requires more mathematical over-
head but that gives a lot more physical insight into the system's behavior. The partial differential equation (and, in entirety the boundary value problem in Equation 5.1) can be
analytically solved. Reference 1 outlines the solution method in detail, and it is of the form

$$u(x,t) = \sum_{n=1}^{\infty} \sin\left(\frac{n\pi x}{L}\right) A_n \cos\left[\left(\frac{n\pi}{L}\right)^2 vt\right] + B_n \sin\left[\left(\frac{n\pi}{L}\right)^2 vt\right]. \tag{5.3}$$

Now, the preceding equation might not mean very much at first glance but it is very
significant. How significant is it? Why, it's so significant that... OK I'll knock it off with
the jokes. The reason it is significant is that it describes how much of each harmonic is
present in the sound made by the string. A harmonic is fairly easy to set up and observe—
it's a set of special frequencies (resonant frequencies) the system vibrates at such that
standing waves are set up—waves that hold their shape but just change their amplitude.
In fact the nth harmonic—the $(n-1)^{st}$ overtone occurs when the wave has the shape

$$u(x) = A\sin\left(\frac{n\pi x}{L}\right), \tag{5.4}$$

where A is an arbitrary constant [2].

So if all this is the case, then the constants A_n and B_n can be used to describe how
much of the nth harmonic is present in the initial sound and therefore the sound for all
time (or at least until dampening drowns it out):

$$\begin{cases} A_n = \int_0^L f(x)\sin\left(\frac{n\pi x}{L}\right) dx \\ \\ B_n = \frac{2L}{(n\pi)^2} \int_0^L g(x)\cos\left(\frac{n\pi x}{L}\right) dx \end{cases} \tag{5.5}$$

We can now take the theory one more step to get our hands on some really cool applications. The treatments of sound we have discussed so far involve the shape of the string and the sound it produces.

Now consider what sound really is—a compressional wave. If we look at just the sound (waveform) produced by any instrument with a vibrating string such as a piano we get plots like those in Figure 5.2 [3]. Specifically, the waveform tracks the sound amplitude as a function of time so we are looking at

$$y = f(t)$$

For a pure tone we would have a pure sine wave but you can see that the curves are more complex than that. Amazingly, we can decompose the plots in Figure 5.2 into how much of each frequency ω makes up the sound [3]. This process is known as taking a *Fourier Transform* of the sound wave and the resulting function is the frequency density spectrum, or Fourier spectrum $g(\omega)$:

$$g(\omega) = \int_{-\infty}^{\infty} f(t)e^{-i\omega t}dt \qquad (5.6)$$

as also shown in Figure 5.2. The true beauty of the Fourier transform treatment just outlined is that each musical instrument and each human voice has its unique Fourier spectrum—a concept that is widely used in music synthesis and you can see the wide variations across musical instruments by looking at the plots shown in Figure 5.3.

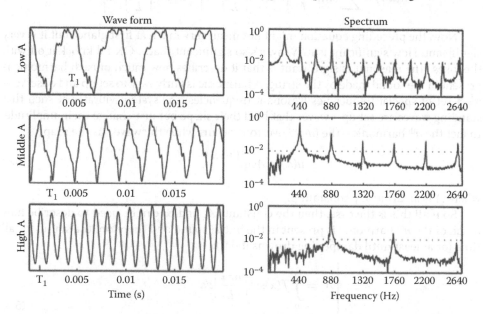

Figure 5.2. Wave forms and frequency spectra for three octaves of the A note on a piano (Adapted from M.R. Petersen (mark.petersen@colorado.edu), Applied Mathematics Department, University of Colorado, Boulder, CO.) [3].

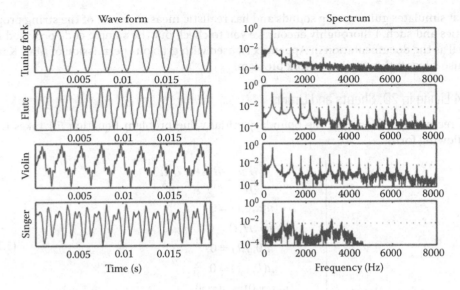

Figure 5.3. Wave forms and frequency spectra for the middle A note produced by various sound sources (Adapted from M.R. Petersen (mark.petersen@colorado.edu), Applied Mathematics Department, University of Colorado, Boulder, CO.) [3].

Pianos have strings with uniform mass density, so they are even along their length and yield a frequency spectrum not too terribly far away from pure tones. If we change the mass density of the string we can obtain a very different sound. Such an effect is seen with the *hyperpiano* [4,5] which is constructed with nonuniform strings. I encourage you to listen to it [5]; it's rather exotic indeed. The musical structure of the hyperpiano is shown in Figure 5.4 [4]; it comes complete with hyperoctaves, hyperthirds, hyperchords... as well.

One semester I was teaching a *Modeling and Simulation of Physical Systems* class and a student turned in a nice project that involves an integrated C++/Python code

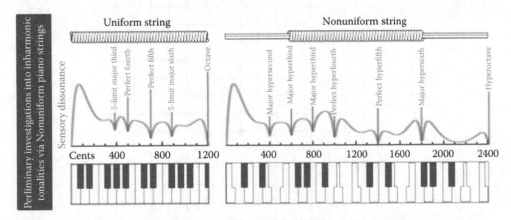

Figure 5.4. Comparison of the piano and hyperpiano. The horizontal axis is frequency in Hz and the vertical axis gauges dissonance, so dips in the curve indicate notes that don't clash (Adapted from K. Hobby, W.A. Sethares, *Applied Acoustics*, 114, 317–327, December 15, 2016.) [4].

that simulates guitar string sounds and has realistic measurements of the string properties and such. I thoroughly encourage you to check out the sample code provided as well as the documentation in Appendix A, used with gracious permission of Matt Karl. I also suggest this in a homework problem!

5.4 Living in 2D: Sheets and Drums

In rectangular geometry, the boundary/initial value problem Equation 5.1 takes the following form:

$$
\begin{cases}
v^2 \nabla^2 u = v^2 \left[\dfrac{\partial^2 u}{\partial x^2} + \dfrac{\partial^2 u}{\partial y^2} \right] = \dfrac{\partial^2 u}{\partial t^2} \\[2mm]
u(x,0,t) = 0 \\
u(x,H,t) = 0 \\
u(0,y,t) = 0 \\
u(L,y,t) = 0 \\
u(x,y,0) = \alpha(x,y) \\
\dfrac{\partial u}{\partial t}(x,y,0) = \beta(x,y)
\end{cases}
\tag{5.7}
$$

The above initial boundary value problem applies for a uniform rectangular sheet that is clamped around the edges and—you guessed it—it's mentioned in the homework to formulate the problem with any number of free edges. Sticking with the order of approach we had for the 1D problem previously discussed we can use a grid such as in Figure 5.5 and indexing shown in Figure 5.6 and write the above equation in finite difference form:

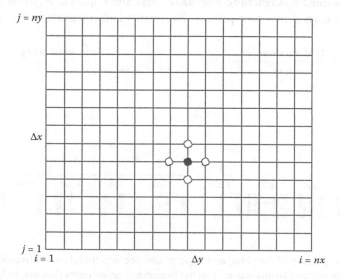

Figure 5.5. 2D Cartesian finite difference grid used, for example in the solution of Equation 5.8.

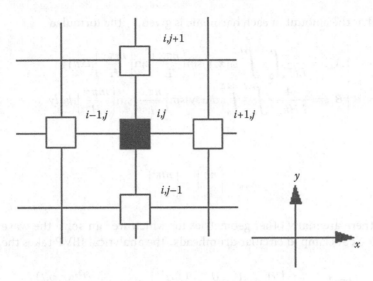

Figure 5.6. Detail of 2D Cartesian finite-difference indexing.

$$
\begin{cases}
v^2 \dfrac{u^m_{i+1,j} - 2u^m_{i,j} + u^m_{i-1,j}}{\Delta x^2} + \dfrac{u^m_{i+1,j} - 2u^m_{i,j} + u^m_{i-1,j}}{\Delta y^2} = \dfrac{u^{m+1}_{i,j} - 2u^m_{i,j} + u^{m-1}_{i,j}}{\Delta t^2} \\[2ex]
u^m_{i,0} = 0 \\[1ex]
u^m_{i,ny} = 0 \\[1ex]
u^m_{0,j} = 0 \\[1ex]
u^m_{nx,j} = 0 \\[1ex]
u^0_{i,j} = \alpha(x_i, y_j) \\[1ex]
\dfrac{u^{m+1}_{i,j} - u^m_{i,j}}{\Delta t} = \beta(x_i, y_j)
\end{cases}
\tag{5.8}
$$

For thorough treatments of constructing finite difference calculations I recommend references such as Carnahan, Luther, and Wilkes [6]. Just like we did in the 1D treatment we can recast Equation 5.8 to look like an update to the displacement:

$$
u^{m+1}_{i,j} = [\text{function of neighboring } u^m_{i,j}] * (\Delta t)^2
$$

As homework, I ask you to consider the 1D problem and construct the 2D finite difference updating calculation from that.

Similar to the 1D case, you can check references, such as Reference 1 to verify that the analytical solution is of the form

$$
u(x,y,t) = \sum_{n,m=1}^{\infty} \sin\left(\frac{n\pi x}{L}\right) \sin\left(\frac{m\pi y}{H}\right) A_{nm} \cos[k_{mn} v t] + B_{nm} \sin[k_{mn} v t]
\tag{5.9}
$$

And that the amount of each harmonic is given by the formulae

$$
\begin{cases}
A_{nm} = \dfrac{4}{LH} \displaystyle\int_0^H \int_0^L \alpha(x,y)\sin\left(\dfrac{n\pi x}{L}\right)\sin\left(\dfrac{m\pi y}{H}\right)dxdy \\[4mm]
B_{nm} = \dfrac{4}{LHk_{mn}} \displaystyle\int_0^H \int_0^L \beta(x,y)\sin\left(\dfrac{n\pi x}{L}\right)\sin\left(\dfrac{m\pi y}{H}\right)dxdy
\end{cases}
\tag{5.10}
$$

where

$$
k_{nm} = \left(\frac{n\pi}{L}\right)^2 + \left(\frac{m\pi}{H}\right)^2
$$

Now there are many other geometries for which we can solve the wave equation [7]... let's look at clamped circular drumheads. The analytical IBVP takes the form

$$
\begin{cases}
v^2\nabla^2 u = v^2 = \left(\dfrac{\partial^2}{\partial r^2} + \dfrac{1}{r}\dfrac{\partial}{\partial r} + \dfrac{1}{r^2}\dfrac{\partial^2}{\partial \phi^2}\right)u(r,\phi,t) = \dfrac{\partial^2 u(r,\phi,t)}{\partial t^2} \\[3mm]
u(a,\phi,t) = 0 \\[2mm]
u(r,\phi,0) = f(r,\phi) \\[2mm]
\dfrac{\partial u}{\partial t}(r,\phi,0) = g(r,\phi)
\end{cases}
\tag{5.11}
$$

and the discretized form looks like

$$
\begin{cases}
v^2 \dfrac{u_{i+1,j}^m - 2u_{i,j}^m + u_{i-1,j}^m}{\Delta r^2} + \dfrac{u_{i,j+1}^m - u_{i,j-1}^m}{\Delta r} = \dfrac{u_{i,j}^{m+1} - 2u_{i,j}^m + u_{i,j}^{m-1}}{\Delta t^2} \\[3mm]
u_{nr,j}^m = 0 \\[2mm]
u_{i,j}^0 = f(r_i,\theta_j) \\[2mm]
\dfrac{u_{i,j}^{m+1} - u_{i,j}^m}{\Delta t} = g(r_i,\theta_j)
\end{cases}
\tag{5.12}
$$

A visual of the finite difference indexing is shown in Figures 5.7 and 5.8. The solution is of the form

$$
u(r,\phi,t) = \sum_{m=0}^{\infty}\sum_{n=1}^{\infty} J_m(k_{mn}r)\begin{Bmatrix} A_{mn}\cos(m\phi)\cos(k_{mn}vt) + B_{mn}\sin(m\phi)\cos(k_{mn}vt) + \\ C_{mn}\cos(m\phi)\sin(k_{mn}vt) + D_{mn}\sin(m\phi)\sin(k_{mn}vt) \end{Bmatrix}
\tag{5.13}
$$

and again tells us how much of the sound is comprised of the harmonics formed by the Bessel functions radially and the sines and cosines azimuthally. Comparing Equation 5.13 to Equation 5.9 we can see that the Bessel functions are analogous oscillatory functions to sines and cosines but with the zeroes unevenly spaced. To appreciate the beauty and symmetry of circular drum harmonics please check out treatments such as in Reference

8, and for a detailed treatment of the problem I encourage you to look at texts such as Reference 1.

It is convenient that the Fourier transform of a drum sound proceeds exactly like it does for any other system making noise because the output we are analyzing is sound amplitude as a function of time, which they all produce.

5.5 Sometimes You Win and Sometimes You Lose: Advantages and Disadvantages of Each Method

Yeah, I'm thinking the same thing… Good time Charlie's got a craving for computer simulations! No, actually we need to talk about the two different approaches to solution of the problem. Do you pick a finite difference or the analytical approach? In actuality the answer is one that comes up so often in computational physics and computer modeling: season to flavor. For example, the analytical method works for a limited set of boundary conditions and so if you need to apply unique or irregularly-shaped boundary conditions, the finite difference scheme is your friend. The analytical method of solution, however, is exact and automatically gives a wealth of knowledge about and insight into the system's symmetry, resonance behavior, and sound spectrum. In the end you will have to consider the scope of the problem you are trying to solve, how the solutions needs to be expressed and what elements if the physics and engineering matter.

5.6 When Resonance Isn't Your Friend: The Tacoma Narrows Bridge

Every object has a set of resonant frequencies—frequencies it naturally tends to vibrate at, and if we can put energy into such systems at resonant frequencies then we can promote the modes greatly and even cause an object to break apart. This is the idea behind a singer's voice breaking a glass and an ultrasound machine destroying kidney stones. On July 1, 1940, the Tacoma Narrows Bridge spanning the Tacoma Narrows strait of Puget Sound between Tacoma and the Kitsap Peninsula opened to the public [9]. It quickly earned the nickname "Galloping Gertie" because a phenomenon called dynamic aeroelasticity resulted in low speed winds causing fluttering in the structure, in turn driving oscillations in the structure at resonant frequencies. The bridge

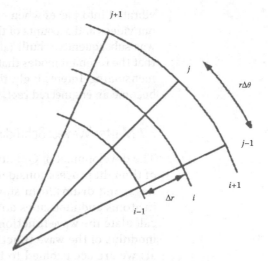

Figure 5.7. Detail of the 2D plane polar finite difference scheme such as used in numerical solution of the spherical drum head problem.

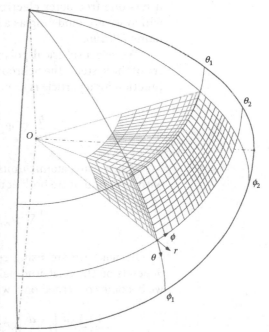

Figure 5.8. Detail of the 3D spherical polar finite difference setup, often used where one is unable to simplify a problem and reduce its dimensions by exploiting its symmetry (Adapted from W. Zhang, Y. Shen, L. Zhao, *J. Int.*, 188, 1359–1381, 2012.) [13].

vibrated into pieces when experiencing 40 mph winds and I encourage you to check out videos and accounts of the collapse [10]. Yep, there's a car on that thing. The bridge was subsequently rebuilt (albeit the construction was delayed by World War II) such that the resonant modes that destroyed it are suppressed through dampening and redimensioning. Interestingly, the part of the span that fell into the water was left there to become an engineered reef.

5.7 Matter Waves: Schrödinger's Equation

The development of Quantum Mechanics in the 1920's [11] represented a new type of thought process considering matter as being a wave. Just like how guitar strings, flags, and drums form standing wave patterns at resonant frequencies, electrons in atoms and molecules do also. Using Schrödinger's Equation we can, in principle calculate the wave function for the electrons in a given system and then the square modulus of the wave function gives the electron density and the s, p d and f orbitals we are accustomed to in chemistry are nothing other than standing resonant electron density waves. The problem below deals with calculating energy levels in a magnesium atom confined in a C60 Fullerene using the Schrodinger wave equation [12]. The atom is a magnesium$^+$ ion which we are treating like hydrogen because it has one free outer electron, and the atom is being confined in a C60 Fullerene which we are modeling as a smeared out shell of charge much like jelly—hence the name *Jellium*.

We select spherical-polar geometry, which is best reflective of the natural symmetry of the system. The Schrödinger Equation describing the quantum mechanical wave function for a particle is then

$$-\frac{\hbar}{2m}\nabla^2\Psi_{n\ell m}(\vec{r})+V(\vec{r})\Psi_{n\ell m}(\vec{r})=E_{n\ell m}\Psi_{n\ell m}(\vec{r}). \tag{5.14}$$

When using atomic units Equation 5.14 becomes much more convenient to work with as $\hbar$ and m are both equal to unity. We then obtain

$$-\frac{1}{2}\nabla^2\Psi_{n\ell m}(\vec{r})+V(\vec{r})\Psi_{n\ell m}(\vec{r})=E_{n\ell m}\Psi_{n\ell m}(\vec{r}). \tag{5.15}$$

Because we are using spherical geometry, the part of the wave function that depends on the colattitudinal as well as azimuthal angles is immediately solvable and we become concerned only with the radial part of the wave function [12]:

$$-\frac{1}{2}\left\{\frac{1}{r^2}\frac{d}{dr}\left(r^2\frac{d}{dr}\right)\right\}R_{n\ell}(r)+\frac{\ell(\ell+1)}{r^2}R_{n\ell}(r)+V(r)R_{n\ell}(r)=E_{n\ell}R_{n\ell}(r). \tag{5.16}$$

The reduction of Equation 5.15 to the one-dimensional form (Equation 5.16) significantly reduces the computational time and memory demands.

So now it starts with the program headers and variable declarations; please review EndoMg.cpp (provided) for code related to the following discussion.

```
#include<cstdlib>
#include<iostream>
#include<fstream>
#include<math.h>

int main()
{

using namespace std;

const int n=50,nd=51;
const int ifreq=1,itmax=50000;
const int stripd=1,sweep=0;
const double eps1=1.0e-07,eps2=1.0e-03,eps3=1.0e-04,eps4=1.0e-01;

int l;
int tag1,tag2;
int ibegin[nd],ifinish[nd];
int nm1,jlow,jhigh,ilow,ihigh,klow,khigh;
int iter;
int ip1,im1,jp1,jm1;

double amat[n][n];
double r[n];
double a,b[n],c;
double alpha,beta,gama;
double deltr,rad;
double bmat[nd][nd];
double v[nd],x[nd][nd];
double rlngth,sum,sumsq,subsum;
double dummy;
```

Now I/O and variable initialization.

```
ofstream myfile;
myfile.open("EndoMg.res");

for(int i=0;i<=n-1;i++)
{
b[i]=0.0;
r[i]=0.0;
for(int j=0;j<=n-1;j++)
{
amat[i][j]=0.0;
}
}

for(int i=0;i<=nd-1;i++)
{
for(int j=0;j<=nd-1;j++)
{
x[i][j]=0.0;
bmat[i][j]=0.0;
}
}

for(int i=0;i<=nd-1;i++)
{
v[i]=0.0;
```

```
ibegin[i]=0;
ifinish[i]=0;
}
```

The electron interacts with two things, and each interaction provides a potential that we will stuff into Schrödinger's Equation to solve for the wave function. First, there is the interactions with the Mg nucleus $V_{core}(r)$, and second is that with the jellium-shell, $V_{shell}(r)$. Then, in Equation 6.16 we have $V(r) = V_{core}(r) + V_{shell}(r)$. The valence electron–core interaction has the form [12]

$$V_{core}(r) = \frac{-\gamma}{r} + \frac{\alpha}{r}e^{-\beta r}. \tag{5.17}$$

The electron–core interaction has terms which describe point charge interactions ("Coulombic") as well as exponential screening due to the extended electron distribution in the core. The potential parameters α, β, and γ for Mg^{2+} as well as H for comparison are given in Table 5.1 [12]. The second term in the potential, from the confining jellium cage, looks like [12]

$$V_{shell}(r) = \begin{cases} 0, & r < R_s - \Delta r \\ U_0, & R_s - \Delta r \leq r \leq R_s + \Delta r, \\ 0, & r > R_s + \Delta r \end{cases} \tag{5.18}$$

where the parameters U_0, R_s and Δr are also given in Table 5.1.

And here are more initializations, including the parameters, radius of the cage, orbital angular momentum quantum number l, and spatial divisions:

```
alpha=20.657;
beta=2.55;
gama=2.0;
```

Table 5.1. Potential Parameters for the Hydrogen and Magnesium Valence Electron Potentials as well as Those for the Jellium Shell

Parameter	Value
α	0 (H) 20.657 (Mg)
β	0 (H) 2.550 (Mg)
γ	1.000 (H) 2.000 (Mg)
U_0	−0.302
$R_s(Å)$	3.04
$\Delta r(Å)$	1.00

Source: W. Even et al., *Int. J. Mol. Sci.*, 5, 333–346, 2004 [12].
Note: For hydrogen, β may be any real number. (all values are in atomic units unless otherwise indicated.)

```
l=0;
rad=10.0;

rad=rad/0.529177;
deltr=rad/float(n+1);
a=1.0/pow(deltr,2);
c=1.0/pow(deltr,2);

myfile<<"Divisions= "<<n<<endl;
myfile<<"Quantum number l: "<<l<<" Cage radius= "<<rad*0.529177<<"
Angstroms."<<endl;
```

Now Equation 5.16 must be discretized in order to pursue a numerical solution. The finite-difference method selected is similar to those used to solve the time-independent Schrödinger Equation in Cartesian coordinates and for time-dependent eigenvalue problems having cylindrical symmetry [6]. Including several computational details outlined in Reference 12 we arrive at the following boundary value problem:

$$\frac{d^2 u_{n\ell}(r)}{dr^2} - 2V(r)u_{n\ell}(r) - 2\frac{\ell(\ell+1)}{r^2}u_{nl}(r) = -2E_{n\ell}u_{n\ell}(r);$$

$$u_{n\ell}(0) = u_{n\ell}(R) = 0.$$

(5.19)

Space is divided into finite one-dimensional elements of width Δr. Subsequent discretization of the derivative in Equation 5.19 results in

$$\frac{u_{i+1} - 2u_i + u_{i-1}}{\Delta r^2} - 2V(r_i)u_i - 2\frac{\ell(\ell+1)}{r_i^2}u_i = -2E_{nl}u_i,$$

(5.20)

for $i = 1, 2, 3, \ldots, N$, with $u_i = u_{n\ell}(r_i)$.

We now have N simultaneous eigenvalue equations for each radial element index (i) and these equations may be cast in the following form:

$$\begin{bmatrix} B_1 & A & 0 & 0 & \cdots & 0 \\ A & B_2 & A & 0 & \cdots & 0 \\ 0 & A & B_3 & A & \cdots & 0 \\ 0 & 0 & A & B_4 & \cdots & 0 \\ \vdots & \vdots & \vdots & \vdots & \cdots & \vdots \\ 0 & 0 & 0 & 0 & \cdots & B_N \end{bmatrix} \begin{bmatrix} u_1 \\ u_2 \\ u_3 \\ u_4 \\ \vdots \\ u_N \end{bmatrix} = E_{nl} \begin{bmatrix} u_1 \\ u_2 \\ u_3 \\ u_4 \\ \vdots \\ u_N \end{bmatrix},$$

(5.21)

where $A = 1/\Delta r^2$ and $B_i = -2/\Delta r^2 - 2V(r_i)$.

```
for(int i=1;i<=n;i++)
{
r[i-1]=deltr*float(i);
```

```
b[i-1]=-2.0/pow(deltr,2)
+(2.*gama/r[i-1])-(2.*alpha*(1./r[i-1])*exp(-beta*r[i-1]))
-(float(1)*float(1+1)/pow(r[i-1],2));
}
```

The Eigenvalue problem in Equation 5.21 is now solved using a routine matrix library, or one can also code up and use Rutishauser's Method (below, from Reference 6), the latter of which provides more experience and affords more insight into the symmetry of such problems. Both the routine library solution as well as Rutishauser's give the energy of the system (eigenvalues) and its wave function (eigenvectors), but the standard library does a bit better with the eigenvectors.

Here we construct the required matrices:

```
amat[0][0]=b[0];
amat[0][1]=a;

for (int i=2;i<=n-1;i++)
{
amat[i-1][i-2]=c;
amat[i-1][i-1]=b[i-1];
amat[i-1][i]=a;
}

amat[n-1][n-2]=c;
amat[n-1][n-1]=b[n-1];

for (int i=1;i<=n;i++)
{
for (int j=1;j<=n;j++)
{
dummy=amat[i-1][j-1]*pow(deltr,2);
amat[i-1][j-1]=dummy;
}
}
```

Now compute energy eigenvalues using Rutishauser's Method:

```
l=0;

for(int i=0;i<=nd-1;i++)
{
v[i]=0.0;
ibegin[i]=0;
ifinish[i]=0;
}

nm1=n-1;

tag1=0;
tag2=0;

if(stripd==0)
{
goto line3;
}
ibegin[1]=1;
ibegin[n]=nm1;
```

```
ifinish[1]=2;
ifinish[n]=n;
if(n<=2)
{
goto line5;
}

for(int j=2;j<=nm1;j++)
{
ibegin[j]=j-1;
ifinish[j]=j+1;
}
goto line5;

line3:;
for(int j=1;j<=n;j++)
{
ibegin[j]=1;
ifinish[j]=n;
}
line5:;

for(int i=1;i<=n;i++)
{
jlow=ibegin[i];
jhigh=ifinish[i];
for(int j=jlow;j<=jhigh;j++)
{
bmat[i][j]=amat[i][j];
}
}

for(int iter=1;iter<=itmax;iter++)
{

for(int j=1;j<=n;j++)
{
ilow=ibegin[j];
for(int i=ilow;i<=j;i++)
{
sum=0.0;
im1=i-1;
klow=ibegin[i];
if(klow>im1)
{
goto line8;
}
for(int k=klow;k<=im1;k++)
{
sum=sum+bmat[i][k]*bmat[k][j];
}
line8:;
bmat[i][j]=bmat[i][j]-sum;
}
jp1=j+1;
ihigh=ifinish[j];
if(jp1>ihigh)
{
goto line15;
}
```

```
for(int i=jp1;i<=ihigh;i++)
{
sum=0.0;
klow=ibegin[i];
jm1=j-1;
if(klow>jm1)
{
goto line10;
}
for(int k=klow;k<=jm1;k++)
{
sum=sum+bmat[i][k]*bmat[k][j];
}
line10:;
bmat[i][j]=(bmat[i][j]-sum)/bmat[j][j];
}
}

line15:
for(int i=1;i<=n;i++)
{
jlow=ibegin[i];
im1=i-1;
if(jlow>im1)
{
goto line21;
}

for(int j=jlow;j<=im1;j++)
{
bmat[i][j]=bmat[i][i]*bmat[i][j];
ip1=i+1;
khigh=ifinish[i];
if(ip1>khigh)
{
goto line20;
}
for(int k=ip1;k<=khigh;k++)
{
bmat[i][j]=bmat[i][j]+bmat[i][k]*bmat[k][j];
}
line20:;
}
line21:;
jhigh=ifinish[i];
for(int j=i;j<=jhigh;j++)
{
jp1=j+1;
khigh=ifinish[j];
if(jp1>khigh)
{
goto line23;
}
for(int k=jp1;k<=khigh;k++)
{
bmat[i][j]=bmat[i][j]+bmat[i][k]*bmat[k][j];
}
line23:;
}
}
```

```
for(int i=1;i<=n;i++)
{
jlow=ibegin[i];
jhigh=ifinish[i];
for(int j=jlow;j<=jhigh;j++)
{
if(abs(bmat[i][j])<1.0E-10)
{
 bmat[i][j]=0.0;
}
}
}

l=l+1;
subsum=0.0;
for(int i=2;i<=n;i++)
{
subsum=subsum+abs(bmat[i][i-1]);
}
if(!(l==ifreq&&subsum<eps4&&sweep==1))
{
goto line42;
}

for(int j=1;j<=nm1;j++)
{
for(int i=1;i<=n;i++)
{
if(abs(bmat[j][j]-bmat[i][i])<eps2&&j!=i)
{
goto line37;
}
}
jp1=j+1;
for(int it=jp1;it<=n;it++)
{
int i=n+jp1-it;
v[i]=bmat[i][j];
ip1=i+1;
if(i==n)
{
goto line32;
}
for(int k=ip1;k<=n;k++)
{
v[i]=v[i]+bmat[i][k]*v[k];
}
line32:;
v[i]=v[i]/(bmat[j][j]-bmat[i][i]);
}
for(int it=jp1;it<=n;it++)
{
int i=n+jp1-it;
x[i][j]=x[i][j]+v[i];
im1=i-1;
if(jp1>im1)
{
goto line34;
}
for(int k=jp1;k<=im1;k++)
{
```

```
x[i][j]=x[i][j]+x[i][k]*v[k];
}
line34:;
}

for(int i=1;i<=n;i++)
{
for(int k=jp1;k<=n;k++)
{
bmat[i][j]=bmat[i][j]+bmat[i][k]*v[k];
}
}
for(int i=jp1;i<=n;i++)
{
for(int k=1;k<=n;k++)
{
bmat[i][k]=bmat[i][k]-v[i]*bmat[j][k];
}
}
line37:;
}

if(stripd==0)
{
goto line41;
}
for(int j=1;j<=n;j++)
{
ibegin[j]=1;
ifinish[j]=n;
}
line41:;
line42:;

if(!(l==ifreq||iter==itmax||subsum<eps1))
{
goto line50;
}
l=0;
line50:;
if(subsum<eps1)
{
goto line52;
}

}

line52:;
for(int i=1;i<=n;i++)
{
x[i][i]=0.0;
}

//End Rutishauser's Method
//Now write out desired information
for(int i=1;i<=n;i++)
{
//if(bmat[i][i]>0.0) Commented out so write eigenvalues of both signs
//{
```

```
myfile<<" "<<endl;
myfile<<"Energy eigenvalue (eV):"<<endl;
myfile<<bmat[i][i]*25.2114/(-2.0*pow(deltr,2))<<endl;
//}
}
//
myfile.close();
return(0);
}
```

There are situations where you might need to have the atom set off-center or wind up with a situation for another calculation where you can't exploit symmetry in order to simplify the problem, in which case you might need a 2D [12] or even 3D finite difference scheme.

PROBLEMS

5.1. Use the guitar string simulation program to model the behavior of the string for various initial displacements and velocities. Create a few animations and discuss the validity of special cases. Predict what initial conditions should give rise to standing waves and verify through simulation. In as many cases as you can, find animations of real guitar strings to compare your results to. Hint: For nonzero initial velocities you will need to add code that deals with the guitar string initially moving.

5.2. Working with the spatial derivatives (slope of the string) at the endpoints, modify the guitar string simulation program so that it has free boundary conditions on one end. Predict the behavior for a few special cases including standing waves and verify through simulation and animation.

5.3. Real guitar strings attenuate and lose energy with time as they vibrate. What are physical sources of this attenuation? Using observations of real systems, code in exponential attenuation, run a few simulations of guitar strings in with various initial conditions and discuss your results.

5.4. Starting with the guitar string simulation program, code in Fourier transforms. Run the code for a few cases and compare/contrast the results with frequency spectra of real instruments.

5.5. Write down the boundary value problem for the vibrating rectangular membrane with any number of free edges you choose. Discuss qualitatively how the resonant frequencies will differ from those for the system with all clamped edges.

5.6. Find and write down the 2D updating equation for the displacement arising from the discussion directly after Equation 5.8.

5.7. Code the 2D circular drum head problem in any language you choose. Validate and discuss results for a few runs.

5.8. Code the 3D wave equation (vibrating cavity). Validate the program and discuss.

5.9. You will have to do some outside digging for this one— code the 3D cylindrical or spherical wave equation (vibrating cavity). Validate and discuss results for a few runs.

5.10. Run the endohedral Magnesium sample program. To validate, adjust the potential parameters so that you are modeling the energy levels of a hydrogen atom. Discuss your results and in particular the robustness of the program.

5.11. Compare the two methods of solving for the motion of a guitar string by coding/running the second method as well as running the first. Discuss applications where each case would seem more beneficial to use.

5.12. Pick a standard eigenvalue/eigenvector solving library and compare it to the results Rutishauser's Method gives for the same problem. Discuss one benefit and one drawback for each method.

5.13. I have provided the integrated C++/Python code with little direction. (a) Starting with the sample code provided, get it running and obtain results for nylon as well as metal strings. Discuss the differences in your results. (b) Note that the sound is something like that of an organ...how could you make the sound more realistic? See how far you can get.

References

1. R. Haberman, *Applied Partial Differential Equations with Fourier Series and Boundary Value Problems*, 5th Edition., P. Higher (Ed.), Upper Saddle River, NJ, 2012.

2. https://www.youtube.com/watch?v=cnH2ltfW48U

3. M.R. Petersen (mark.petersen@colorado.edu), Applied Mathematics Department, University of Colorado, Boulder, CO 80309-0526.

4. K. Hobby, W.A. Sethares, *Applied Acoustics*, 114, 317–327, December 15, 2016.

5. https://www.youtube.com/watch?v=fXXv6Ls1u-M

6. B. Carnahan, H.A. Luther, J.O. Wilkes, *Applied Numerical Methods*, John Wiley and Sons, Inc., New York/London/Sydney/Toronto, 1969.

7. G. Arfken, H. Weber, F.E. Harris, *Mathematical Methods for Physicists 7th Edition A Comprehensive Guide* ISBN: 9780123846549 eBook ISBN: 9780123846556 Imprint: Academic Press Waltham, MA, Published Date: January 17, 2012.

8. http://falstad.com/circosc/

9. https://www.aps.org/publications/apsnews/201611/physicshistory.cfm

10. https://www.youtube.com/watch?v=j-zczJXSxnw

11. D.J. Griffiths, *Introduction to Quantum Mechanics*, Cambridge University Press, Cambridge CB2 8BS, United Kingdom, 2017.

12. W. Even, J. Smith, M.W. Roth, H.A. Schuessler, Calculated electronic behavior and spectrum of Mg+@C60 using a simple jellium-shell model, *Int. J. Mol. Sci.*, 5, 333–346, 2004.

13. W. Zhang, Y. Shen, L. Zhao, Three-dimensional anisotropic seismic wave modelling in spherical coordinates by a collocated-grid finite-difference method, *Geophys. J. Int.*, 188, 1359–1381, 2012.

CHAPTER 6

///

Going with the Flow

6.1 Introductory Thoughts

There are many times when it is important to understand how fluids can transport things both harmful and beneficial. Of course, there are many times also when it's great to have a cup of coffee and look out the window thinking about physics, but we should stay on task here. Right off the bat, I want to discuss the differences between fluids and liquids so when you talk to other people about this they don't think you're wet behind the ears. A fluid is any substance that flows on an observable timescale given a pressure, density, or energy gradient. So, put in everyday terms, a fluid is something we can watch flow. Air, helium, water, syrup, and gasoline are examples of fluids. Liquids, however, are special types of fluids. Liquids are very difficult to compress and are dense, whereas gases can be compressed and are much lighter.

Thinking microscopically it's useful to imagine a fluid as being comprised of many small noninteracting particles that bounce (really, scatter—they never come in contact; that's relevant for nuclear processes) off one another when they get close enough. For gases there is enough space between the particles so that compression is possible, but in the case of liquids the particles are so close already that compressing them causes them to be close enough to feel a strong repulsion. Our simple model is not a bad approximation for conceptualizing the relationship between gases, fluids, and liquids and later in this book will see that we can dramatically improve our model with just a little bit of work. Thinking macroscopically, fluids take the shape of their container but a liquid maintains its volume while a gas expands to fill its container. Thinking from an engineering perspective, fluids don't support sheer forces like solids do. There's many ways to think about fluids and liquids, and it's a homework problem to identify a material that is categorized in a way you wouldn't normally expect and to discuss the details.

There are many instances when fluids move and carry things with them as they flow. Most of the cases involve pollutants, such as wind transporting smoke from oil well fires, forced air ventilation transporting aerosol contaminants at the workplace, saline contamination leeching into a freshwater river, or radioactive materials being carried by the ocean currents after a tsunami. There are beneficial cases too, such as air fresheners, filtration systems, or ocean currents carrying plankton to marine life for sustenance. It seems to me that fluid flow transporting things breaks down into two different categories. In some cases, the fluid is just moving around barriers and boundaries. Even if there aren't boundaries we can still have boundary conditions: we then say that "free boundary conditions" are in effect. In other cases, the fluid is actually moving through something and, as you might imagine, behaves much differently as opposed to just flowing in the presence of boundaries. This chapter introduces the models required to simulate fluid flow in both cases, as well as techniques for modeling material transport through them. And interestingly, since the heat equation and flow through porous medium equations are isomorphic (of the same form and constants/variables can be compared directly), heat can be thought of as something that flows and transports energy. Yep, we cover that in this chapter as well!

6.2 How Fluids Move around Boundaries

6.2.1 Understanding the Effect of Boundaries on Fluids

Let's take this discussion in chunks. First, it's important to talk about why boundaries even matter to fluids and affect their motion. After we do that, we will discuss the second chunk: the actual mathematical model required to describe fluid motion with boundary conditions. The model requires calculus, and I have always believed that physics is one of the best teaching tools for calculus that is around. Then, after we have outlined the model, we will bite off the third chunk and talk about how to tell it all to the computer so it can calculate what we need. In this section, we are also assuming a perfectly incompressible fluid and one of the problems I pose later is to consider the changes that arise from a fluid being compressible.

So, for the first part of the conversation: Why do boundaries matter to fluids? Consider a deck of playing cards sitting on a level table as shown in Figure 6.1. Now, press down gently on the top with your hand and move the top card in a particular direction. You are likely to get the results shown in Figure 6.1 where the top card traveled the greatest distance, while going down the deck the cards moved progressively less with the bottom card not moving at all. Of course, my example is idealized and in reality you would not get perfect motion of the deck as shown in Figure 6.1, but you certainly should see the described trend in motion from top to bottom. Ideally, you would notice that the bottom card moved with the table and the top card moved with your hand, but both cards moved with the boundaries they were in contact with. And what's more, throughout the deck there was varying degrees of motion. What causes the communication in motion between the top in the bottom of the deck? Many times

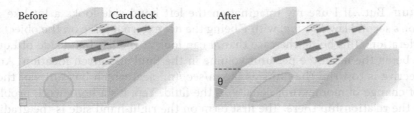

Figure 6.1. Deck of cards sliding, illustrating the effects of viscosity in fluids.

in physics it's useful to take away what's present and then see what doesn't happen. Imagine there were no friction between the cards. Friction is the force that opposes the tendency for two surfaces in contact to slide. It is one of the two forces surfaces in contact generate; the other is the normal force and is perpendicular to the surface and that can be used to simply adjust the strength of friction. So, in the absence of friction between the cards, only the card touching the hand would have moved—if we assume that contact is frictionless—and then the deck would not have responded at all to the hand's motion. So that's for a deck of cards, but we can actually do a "trick" now and extend our thoughts to the microscopic regime. If the deck of cards behaves like it did because the cards interact with each other, then at a microscopic level, there must be a type of friction that results from the atoms' and molecules' interacting with each other as well as with the boundary. And indeed, there is, and it arises from the fluid having *viscosity*.

6.2.2 The Mathematical Model of Fluid Flow with Boundary Conditions

6.2.2.1 The General Equations
Now, let's dive into the mathematical model used for fluid flow. It is good for you to keep in mind that I am presenting the mathematical framework for our computer simulation, but not the *derivation* of the model or math review. For a great review on vector differential and integral calculus [1] as well as the mathematics of fluid flow I refer you to References 2–6. Fluid flow is typically modeled by a set of partial differential equations known as the Navier–Stokes equations for incompressible fluid flow:

$$\rho \frac{d\vec{v}}{dt} = -\vec{\nabla}P - \mu\nabla^2\vec{v} - \rho(\vec{v}\cdot\vec{\nabla})\vec{v} \tag{6.1}$$

We can utilize the fact that derivatives are rates of change to obtain a verbal description of what is happening in Equation 6.1. To become familiar with the variables at hand, ρ is the fluid density (1.3 kg/m^3 for air), $\vec{v}$ is the velocity with Cartesian components $(v_x, v_y, v_z) = (u, v, w)$ at any point in the fluid, P is the pressure at any point, and μ is the fluid's kinematic viscosity (1.23 × 10^{-4} St for air). The left hand side is the fluid density multiplied by its acceleration (the first derivative, or rate of change, of velocity). Thinking in terms of units, we have something like kg/m^3 × m/s^2, which is a kg/(m^{2}s^2). Got nothing immediately? Yeah, I didn't either when I was first learning this stuff. But... if I use my imagination the left hand side looks a lot like *ma* in Newton's second law $F = ma$, with F being the net external force on the object under consideration. This being the case than, I can look at the left hand side of Equation 6.1 as being the net force per unit volume in the fluid at a given location. Another perspective is to write the units as (kg m/sec)/(m³ sec) which has units of the time rate of change of momentum density in the fluid. Yep, it's a homework problem to verify the relationship there. The first term on the right-hand side is the gradient of the pressure in the fluid at any point. The *gradient* operator [1] involves spatial derivatives, or spatial rates of change of pressure but they are partial derivatives, meaning that you're looking at a rate of change with respect to only one variable out of many present. If you imagine the pressure being a surface, with high values being hills and low ones being valleys, the gradient makes its living by pointing in the steepest direction uphill on that surface. With a minus sign in front, then the combination points downhill. The middle term on the right-hand side has units of Equation 6.1 deals with the momentum of the system and is a mathematical relationship stating that any acceleration in the fluid comes from three sources: pressure gradients, momentum diffusion through viscosity, and momentum diffusion through convection. The middle term in the right-hand side of Equation 6.1 is a diffusion term... meaning that the velocity is being diffused throughout the system because of the presence of viscosity. This second term would describe how the deck of cards moved in the example discussed earlier. The last term on the right-hand side describes a process called advection, where momentum is transferred throughout the fluid due to its own velocity field—a process which arises from collisions (scattering, actually). And it's a homework problem to make sense out of the units on those terms. Because the middle term in the right hand side of Equation 6.1 is called the *dispersion* term, the entire relationship is often called the *advection-dispersion equation*. So, now we can finally describe what is happening in Equation 6.1 verbally. Changes in the momentum of a fluid at any point arise from a combination of pressure gradients at that point, the motion of neighboring fluid or boundaries at that point and finally any moving fluid that is pushing it there.

Although necessary, Equation 6.1 alone is not sufficient to model the system adequately. The equation of continuity

$$\frac{\partial \rho}{\partial t} + \rho \vec{\nabla} \cdot \vec{v} = 0 \tag{6.2}$$

is a statement of mass conservation which balances the change in density to the net fluid influx or outflux at any point in the system. Together, Equations 6.1 and 6.2 describe the physical behavior of the fluid phase of the system.

6.2.2.2 What the Equations Look Like for a Rectangular System
The extent and symmetry of any model you are using is very important to think through. A three-dimensional (3D) model is used in this case because we know we will be looking at novel deflection fin shapes as well as side winds. Moreover, we will be modeling buildings whose symmetry is such that Cartesian coordinates will be the most convenient to use here. A 2D model simply can't capture the variation in

geometry we are interested in. Equations 6.1 and 6.2 then take the following explicit forms:

$$\rho\frac{du}{dt} = -\frac{dP}{dx} - \mu\left[\frac{d^2u}{dx^2} + \frac{d^2u}{dy^2} + \frac{d^2u}{dz^2}\right] - \rho u\frac{du}{dx} - \rho v\frac{du}{dy} - \rho w\frac{du}{dz} \tag{6.3}$$

$$\rho\frac{dv}{dt} = -\frac{dP}{dy} - \mu\left[\frac{d^2v}{dx^2} + \frac{d^2v}{dy^2} + \frac{d^2v}{dz^2}\right] - \rho u\frac{dv}{dx} - \rho v\frac{dv}{dy} - \rho w\frac{dv}{dz} \tag{6.4}$$

$$\rho\frac{dw}{dt} = -\frac{dP}{dz} - \mu\left[\frac{d^2w}{dx^2} + \frac{d^2w}{dy^2} + \frac{d^2w}{dz^2}\right] - \rho u\frac{dw}{dx} - \rho v\frac{dw}{dy} - \rho w\frac{dw}{dz} \tag{6.5}$$

$$\frac{\partial\rho}{\partial t} + \rho\left[\frac{du}{dx} + \frac{dv}{dy} + \frac{dw}{dz}\right] = 0 \tag{6.6}$$

Equations 6.3 through 6.5 and 6.6 have many possible solutions, which is in general true for partial differential equations. For a *unique* solution to be obtained, boundary conditions are needed for velocities, pressures, or their derivatives, which are shown in Table 6.1.

Upon inspecting the boundary conditions, I hope you see a pattern. The air is moving at the speed of the boundaries where it is in contact with them, so everywhere the velocity components are zero, there is a stationary boundary. Moreover, we are using

Table 6.1. Boundary Conditions for Phase I—the Navier–Stokes Equations

	Horizontal Velocity	Lateral Velocity	Vertical Velocity	Pressure
Ground	$u=0$	$v=0$	$w=0$	$\frac{dP}{dz}=0$
Ceiling	$u=0$	$v=0$	$w=0$	$\frac{dP}{dz}=0$
Inlet	$\frac{du}{dx}=0$	$\frac{dv}{dx}=0$	$\frac{dw}{dx}=0$	$P=P_i(y,z)$
Outlet	$\frac{du}{dx}=0$	$\frac{dv}{dx}=0$	$\frac{dv}{dx}=0$	$P=P_o(y,z)$
Front	$\frac{du}{dy}=0$	$\frac{dv}{dy}=0$	$\frac{dv}{dy}=0$	$P=P_f(x,z)$
Back	$\frac{du}{dy}=0$	$\frac{dv}{dy}=0$	$\frac{dv}{dy}=0$	$P=P_b(x,z)$
Interior barriers	$u=0$	$v=0$	$w=0$	$\nabla_n P=0$

Source: E. Maldonado, M.W. Roth, *Journal of Applied Fluid Mechanics*, 5(3), 71–78, 2012 [7].

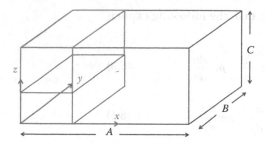

Figure 6.2. The computational cell of dimensions (A,B,C) used in the 3D snow simulations. The Cartesian coordinate system, building outline, and region where snow is created is shown. The figure in the electronic version is colored (E. Maldonado, M.W. Roth, Journal of Applied Fluid Mechanics, 5(3), 71–78, 2012.) [7].

pressure gradients to drive the airflow and so we specify pressures on various walls to achieve the wind speeds and directions we want. You will see that we also have places where we require the *derivatives of the velocities* to vanish. We require such a condition because we not only need information about either the velocity or its derivative on all surfaces but we are specifically saying that the velocities are uniform over the surface. It is a homework problem to discuss why the zero-derivative velocity conditions match the pressure boundary conditions. Finally, as seen in the lower right-hand corner of Table 6.1, we require the normal component of the gradient of the pressure to vanish on interior boundaries. That is to say that, close to the interior boundaries, the pressure isn't changing. If it were, that would mean that there would be flow into or out of the boundary, which doesn't make sense for a hard wall.

In addition to boundary conditions we need initial conditions, describing how the system starts out. To seek a solution without initial conditions would be like trying to find a map to your favorite restaurant and not knowing the starting place. Reasonable initial conditions $u(x,y,z,t=0)=u_0(x,y,z)$, $v(x,y,z,t=0)=v_0(x,y,z)$, $w(x,y,z,t=0)=w_0(x,y,z)$, and $P(x,y,z,t=0)=P_0(x,y,z)$ must be specified, which match the boundary conditions for the system being described.

In this work, the computational cell is chosen to be of dimensions $(A,B,C)=(40\text{ m},$ 20 m, 12 m)$ so that it can contain a building of reasonable size, as shown in Figure 6.2: [7] (10 m, 13.3 m, 3 m). Moreover, the deflection fin can be placed at any position within the computational cell and its lateral ends can be bent. The high pressure is chosen such that the maximum gusting air speed above the building is about 35 miles per hour. In our simulations we tried to do things that made sense, but you will notice something that does not: the no-slip boundary condition implemented on the top of the simulation! We are implementing that condition only as a computational utility, and the boundary is far enough away from the top of the building so as not to matter.

6.2.3 That's Fine for Us; Now Tell the Computer about the Problem

It would save us a lot of time if we were able to "speak" the equations to solve to the computer directly, but we can't do that. They have to be discretized and coded in, compiled, executed, debugged, improved, compiled, executed, …you get it. It's best (normally) to not reinvent the wheel, so we choose to follow what has been done in in previous work [7] involving the SIMPLE algorithm [8,9]. The computational cell shown in Figure 6.2 is divided into n_x spaces horizontally, n_y spaces laterally, and n_z spaces vertically such that distance increments are $\Delta x = A/(n_x-1)$, $\Delta y = B/(n_y-1)$, and $\Delta z = C/(n_z-1)$. As a matter of computational convenience, the grid points are labeled as being either active or inactive, and the inactive cells are used to enforce any constant-value or free boundary conditions. The initial conditions are implemented for u_{ijk}, v_{ijk}, w_{ijk}, and P_{ijk} for $i \subset [0,n_x-1]$, $j \subset [0,n_y-1]$, and $k \subset [0,n_z-1]$. Now, the velocity field is updated using a discretized form of Equations 6.3 through 6.5 over all active cells. And, you guessed it, it is a homework problem for you to do the discretization!:

$$
\begin{aligned}
u_{i,j,k}^{n+1} = u_{i,j,k}^{n} - \Bigg(&\frac{\mu}{\rho}\Bigg\{ \frac{u_{i+1,j,k}^{n}-2u_{i,j,k}^{n}+u_{i-1,j,k}^{n}}{(\Delta x)^2} + \frac{u_{i,j+1,k}^{n}-2u_{i,j,k}^{n}+u_{i,j-1,k}^{n}}{(\Delta y)^2} \\
&+ \frac{u_{i,j,k+1}^{n}-2u_{i,j,k}^{n}+u_{i,j,k-1}^{n}}{(\Delta z)^2}\Bigg\} + u_{i,j,k}^{n}\frac{u_{i+1,j,k}^{n}-u_{i-1,j,k}^{n}}{2\Delta x} + v_{i,j,k}^{n}\frac{u_{i,j+1,k}^{n}-u_{i,j-1,k}^{n}}{2\Delta y} \\
&+ w_{i,j,k}^{n}\frac{u_{i,j,k+1}^{n}-u_{i,j,k-1}^{n}}{2\Delta z} + \frac{1}{\rho}\Bigg[\frac{P_{i+1,j,k}^{n}-P_{i-1,j,k}^{n}}{2\Delta x}\Bigg] \Bigg)\Delta t
\end{aligned}
\tag{6.7}
$$

$$
\begin{aligned}
v_{i,j,k}^{n+1} = v_{i,j,k}^{n} - \Bigg(&\frac{\mu}{\rho}\Bigg\{ \frac{v_{i+1,j,k}^{n}-2v_{i,j,k}^{n}+v_{i-1,j,k}^{n}}{(\Delta x)^2} + \frac{v_{i,j+1,k}^{n}-2v_{i,j,k}^{n}+v_{i,j-1,k}^{n}}{(\Delta y)^2} + \frac{v_{i,j,k+1}^{n}-2v_{i,j,k}^{n}+v_{i,j,k-1}^{n}}{(\Delta z)^2}\Bigg\} \\
&+ u_{i,j,k}^{n}\frac{v_{i+1,j,k}^{n}-v_{i-1,j,k}^{n}}{2\Delta x} + v_{i,j,k}^{n}\frac{v_{i,j+1,k}^{n}-v_{i,j-1,k}^{n}}{2\Delta y} + w_{i,j,k}^{n}\frac{v_{i,j,k+1}^{n}-v_{i,j,k-1}^{n}}{2\Delta z} + \frac{1}{\rho}\Bigg[\frac{P_{i,j+1,k}^{n}-P_{i,j-1,k}^{n}}{2\Delta y}\Bigg] \Bigg)\Delta t
\end{aligned}
\tag{6.8}
$$

$$
\begin{aligned}
w_{i,j,k}^{n+1} = w_{i,j,k}^{n} - \Bigg(&\frac{\mu}{\rho}\Bigg\{ \frac{w_{i+1,j,k}^{n}-2w_{i,j,k}^{n}+w_{i-1,j,k}^{n}}{(\Delta x)^2} + \frac{w_{i,j+1,k}^{n}-2w_{i,j,k}^{n}+w_{i,j-1,k}^{n}}{(\Delta y)^2} \\
&+ \frac{w_{i,j,k+1}^{n}-2w_{i,j,k}^{n}+w_{i,j,k-1}^{n}}{(\Delta z)^2}\Bigg\} + u_{i,j,k}^{n}\frac{w_{i+1,j,k}^{n}-w_{i-1,j,k}^{n}}{2\Delta x} \\
&+ v_{i,j,k}^{n}\frac{w_{i,j+1,k}^{n}-w_{i,j-1,k}^{n}}{2\Delta y} + w_{i,j,k}^{n}\frac{w_{i,j,k+1}^{n}-w_{i,j,k-1}^{n}}{2\Delta z} + \frac{1}{\rho}\Bigg[\frac{P_{i,j+1,k}^{n}-P_{i,j,k-1}^{n}}{2\Delta z}\Bigg] \Bigg)\Delta t
\end{aligned}
\tag{6.9}
$$

The time step is chosen (as discussed earlier in this text) so that the simulation doesn't run too slowly and waste time or run too choppy and lack accuracy or convergence. The method is quite like those used in previous studies (review the successive approximation algorithm of Goldilocks and the three bears regarding selection of furniture and dinnerware size). We use $\Delta t = 10^{-4}$ s while constant-value as well as constant-derivative boundary conditions are implemented where required. Once through is not enough so the algorithm continuously repeats until results converge based upon the average fractional changes of the velocity components being less than a specified tolerance value. The updates in iteration may introduce some error in the velocity field, hence violating the continuity equation (Equation 6.6), so we *correct the pressure* by adding P' to the pressure P_0 everywhere such that

$$
\nabla^2 P' = -\vec{\nabla}\cdot\vec{v}.
\tag{6.10}
$$

In discretized form the pressure correction is calculated in an iterative loop as

$$
\begin{aligned}
P_{i,j,k}^{\prime n} = &\frac{P_{i+1,j,k}^{\prime n}(\Delta x)^2 + P_{i-1,j,k}^{\prime n}(\Delta x)^2 + P_{i,j+1,k}^{\prime n}(\Delta y)^2 + P_{i,j-1,k}^{\prime n}(\Delta y)^2 + P_{i,j,k+1}^{\prime n}(\Delta z)^2 + P_{i,j,k-1}^{\prime n}(\Delta z)^2}{2((\Delta x)^2 + (\Delta y)^2 + (\Delta z)^2)} \\
&- \frac{u_{i+1,j}^{n}-u_{i-1,j}^{n}}{2\Delta x} - \frac{v_{i,j+1}^{n}-v_{i,j-1}^{n}}{2\Delta y} - \frac{w_{i,j+1}^{n}-w_{i,j-1}^{n}}{2\Delta z}
\end{aligned}
\tag{6.11}
$$

The iterative loop also includes any constant value as well as constant derivative pressure conditions. When the pressure correction is converged based upon our criteria, we then add the correction to the pressure field

$$P_{i,j,k}^{n} = P_{0\,i,j,k}^{n} + P_{i,j,k}'^{n}.$$ (6.12)

And then the velocity field:

$$u_{i,j,k}^{n} = u_{i,j,k}^{n} - \frac{P_{i+1,j,k}^{n} - P_{i-1,j,k}^{n}}{2\Delta x}\Delta t$$ (6.13)

$$v_{i,j,k}^{n} = v_{i,j,k}^{n} - \frac{P_{i,j+1,k}^{n} - P_{i,j-1,k}^{n}}{2\Delta y}\Delta t$$ (6.14)

$$w_{i,j,k}^{n} = w_{i,j,k}^{n} - \frac{P_{i,j,k+1}^{n} - P_{i,j,k-1}^{n}}{2\Delta z}\Delta t$$ (6.15)

Then constant value, as well as constant derivative velocity boundary conditions, are again implemented to ensure proper calculation of the pressures and velocities.

6.3 A Sample Program that Calculates Wind Velocity in Cartesian Coordinates

Please review the provided code Snow_Wind_Velocity.cpp as we dive into this program. First, Include relevant C++ libraries.

```
#include<cstdlib>
#include<iostream>
#include<fstream>
#include<math.h>
#include<iomanip>
#include<stdio.h>
using namespace std;
```

Begin Program.

```
int main()
{
```

Declare variables and global constants.

```
const int nx=36,ny=36,nz=36,nsteps=1000;
const float pxmax=500.0,pymax=0.0;
const double dt=5.0e-04;
const double acell=40.0,bcell=20.0,ccell=12.0;
const double rho=1.24,nu=0.1;
char fileanim[10];
int ifr,ix1step,ix2step,iy1step,iy2step,iz1step,iz2step;
int ixfinb,ixfint,iyfinb,iyfint,izfinb,izfint;
double dpress;
double p[nx][ny][nz],pp[nx][ny][nz];
double u[nx][ny][nz],v[nx][ny][nz];
double dum1[nx][ny][nz],dum2[nx][ny][nz],dum3[nx][ny][nz];
double ract[nx][ny][nz];
```

```
double dx,dy,dz;
double a,b,c,d,e;
double d1x,d2x,d1y,d2y,d1z,d2z,cxp1,cxm1,cyp1,cym1,czm1,czp1,px,py,pz;
double d3x,d3y,d3z;
double diffx,diffy,diifz;
double uold[nx][ny][nz],vold[nx][ny][nz],usum,vsum,wsum,psum;
double wold[nx][ny][nz];
double up[nx][ny][nz],vp[nx][ny][nz],wp[nx][ny][nz];
double veltol,ptol;
double pdum;
double ppost,ppre;
```

Perform needed initializations.

```
dx=acell/float(nx-1);
dy=bcell/float(ny-1);
dz=ccell/float(nz-1);
a=2.0*dx;
b=2.0*dy;
f=2.0*dz;
c=1.0/(dx*dx);
d=1.0/(dy*dy);
g=1.0/(dz*dz);
e=2.0*(c+d+g);
veltol=(1.0e-05)*float(nx)*float(ny)*float(nz);
ptol=(1.0e-04)*float(nx)*float(ny)*float(nz);
```

Assign integer array indices to internal boundaries (fins and steps).

```
ix1step=0;
ix2step=nx/4;
iy1step=ny/6;
iy2step=5*ny/6;
iz1step=0;
iz2step=nz/2;
ixfinb=nx/3;
ixfint=nx/3;
iyfinb=ny/6;
iyfint=5*ny/6;
izfinb=nz/2-nz/4;
izfint=nz/2+nz/4;
ifr=0;
```

Implement initial conditions.

```
for(int i=0;i<=nx-1;i++)
{
for(int j=0;j<=ny-1;j++)
{
for(int k=0;k<=nz-1;k++)
{
p[i][j][k]=0.0*float(nx-1-i)/float(nx-1);
pp[i][j][k]=0.0;
u[i][j][k]=0.0;
v[i][j][k]=0.0;
w[i][j][k]=0.0;
up[i][j][k]=0.0;
vp[i][j][k]=0.0;
wp[i][j][k]=0.0;
```

```
uold[i][j][k]=0.0;
vold[i][j][k]=0.0;
wold[i][j][k]=0.0;
dum1[i][j][k]=0.0;
dum2[i][j][k]=0.0;
dum3[i][j][k]=0.0;
ract[i][j][k]=1.0;
}
}
}
```

Now start with constant boundary conditions.

Inlet wall boundary condition(s):

```
for(int j=0;j<=ny-1;j++)
{
for(int k=0;k<=nz-1;k++)
{
ract[0][j][k]=0.0;
p[0][j][k]=50.0;
}
}
```

Right wall boundary condition(s):

```
for(int j=0;j<=ny-1;j++)
{
for(int k=0;k<=nz-1;k++)
{
ract[nx-1][j][k]=0.0;
p[nx-1][j][k]=0.0;
}
}
```

Floor boundary condition(s):

```
for(int i=0;i<=nx-1;i++)
{
for(int j=0;j<=ny-1;j++)
{
ract[i][j][0]=0.0;
u[i][j][0]=0.0;
v[i][j][0]=0.0;
w[i][j][0]=0.0;
}
}
```

Ceiling boundary condition(s):

```
for(int i=0;i<=nx-1;i++)
{
for(int j=0;j<=ny-1;j++)
{
ract[i][j][nz-1]=0.0;
u[i][j][nz-1]=0.0;
v[i][j][nz-1]=0.0;
w[i][j][nz-1]=0.0;
}
}
```

Step boundary condition(s):

```
for(int i=ix1step;i<=ix2step;i++)
{
for(int j=iy1step;j<=iy2step;j++)
{
for(int k=iz1step;k<=iz2step;k++)
{
ract[i][j][k]=0.0;
u[i][j][k]=0.0;
v[i][j][k]=0.0;
w[i][j][k]=0.0;
}
}
}
```

Fin boundary condition(s):

```
for(int i=ixfinb;i<=ixfint;i++)
{
for(int j=iyfinb;j<=iyfint;j++)
{
for(int k=izfinb;k<=izfint;k++)
{
ract[i][j][k]=0.0;
u[i][j][k]=0.0;
v[i][j][k]=0.0;
w[i][j][k]=0.0;
}
}
}
```

Start time loop.

```
for(int istep=1;istep<=nsteps;istep++)
{
```

Calculation of velocities:

```
for(int i=0;i<=nx-1;i++)
{
for(int j=0;j<=ny-1;j++)
{
for(int k=0;k<=nz-1;k++)
{
if(ract[i][j][k]!=0.0)
{
uold[i][j][k]=u[i][j][k];
vold[i][j][k]=v[i][j][k];
wold[i][j][k]=w[i][j][k];
d1x=u[i][j][k]*(u[i+1][j][k]-u[i-1][j][k])/a;
d2x=v[i][j][k]*(u[i][j+1][k]-u[i][j-1][k])/b;
d3x=w[i][j][k]*(u[i][j][k+1]-u[i][j][k-1])/f;
d1y=u[i][j][k]*(v[i+1][j][k]-v[i-1][j][k])/a;
d2y=v[i][j][k]*(v[i][j+1][k]-v[i][j-1][k])/b;
d3y=w[i][j][k]*(v[i][j][k+1]-v[i][j][k-1])/f;
d1z=u[i][j][k]*(w[i+1][j][k]-w[i-1][j][k])/a;
d2z=v[i][j][k]*(w[i][j+1][k]-w[i][j-1][k])/b;
d3z=w[i][j][k]*(w[i][j][k+1]-w[i][j][k-1])/f;
```

```
px=(1.0/rho)*(p[i+1][j][k]-p[i-1][j][k])/a;
py=(1.0/rho)*(p[i][j+1][k]-p[i][j-1][k])/b;
pz=(1.0/rho)*(p[i][j][k+1]-p[i][j][k-1])/f;
diffx=nu*(c*(u[i+1][j][k]-2.0*u[i][j][k]+u[i-1][j][k]) +
d*(u[i][j+1][k]-2.0*u[i][j][k]+u[i][j-1][k])+
g*(u[i][j][k+1]-2.0*u[i][j][k]+u[i][j][k-1]));
diffy=nu*(c*(v[i+1][j][k]-2.0*v[i][j][k]+v[i-1][j][k]) +
d*(v[i][j+1][k]-2.0*v[i][j][k]+v[i][j-1][k])+
g*(v[i][j][k+1]-2.0*v[i][j][k]+v[i][j][k-1]));
diffz=nu*(c*(w[i+1][j][k]-2.0*w[i][j][k]+w[i-1][j][k]) +
d*(w[i][j+1][k]-2.0*w[i][j][k]+w[i][j-1][k])+
g*(w[i][j][k+1]-2.0*w[i][j][k]+w[i][j][k-1]));
dum1[i][j][k]=u[i][j][k]+(diffx-px-d1x-d2x-d3x)*dt;
dum2[i][j][k]=v[i][j][k]+(diffy-py-d1y-d2y-d3y)*dt;
dum3[i][j][k]=w[i][j][k]+(diffz-pz-d1z-d2z-d3z)*dt;
}
}
}
}
```

Update velocity arrays.

```
for(int i=0;i<=nx-1;i++)
{
for(int j=0;j<=ny-1;j++)
{
for(int k=0;k<=nz-1;k++)
{
if(ract[i][j][k]!=0.0)
{
u[i][j][k]=dum1[i][j][k];
v[i][j][k]=dum2[i][j][k];
w[i][j][k]=dum3[i][j][k];
}
}
}
}
```

Velocity derivative boundary conditions:

Inlet wall:

```
for(int j=0;j<=ny-1;j++)
{
for(int k=0;k<=nz-1;k++)
{
u[0][j][k]=u[1][j][k];
v[0][j][k]=v[1][j][k];
w[0][j][k]=w[1][j][k];
}
}
```

Outlet wall:

```
for(int j=0;j<=ny-1;j++)
{
for(int k=0;k<=nz-1;k++)
```

```
{
u[nx-1][j][k]=u[nx-2][j][k];
v[nx-1][j][k]=v[nx-2][j][k];
w[nx-1][j][k]=w[nx-2][j][k];
}
}
```

Back wall:

```
for(int i=0;i<=nx-1;i++)
{
for(int k=0;k<=nz-1;k++)
{
u[i][0][k]=u[i][1][k];
v[i][0][k]=v[i][1][k];
w[i][0][k]=w[i][1][k];
}
}
```

Front wall:

```
for(int i=0;i<=nz-1;i++)
{
for(int k=0;k<=nz-1;k++)
{
u[i][ny-1][k]=u[i][ny-2][k];
v[i][ny-1][k]=v[i][ny-2][k];
w[i][ny-1][k]=w[i][ny-2][k];
}
}
```

You could also have ceiling, floor step, and fin bc's here too.

Interior calculation for P'–iterative:

```
for(int i=0;i<=nx-1;i++)
{
for(int j=0;j<=ny-1;j++)
{
for(int k=0;k<=nz-1;k++)
{
pp[i][j][k]=0.0;
}
}
}

for(int kstep=1;kstep<=1000;kstep++)
{

for(int i=0;i<=nx-1;i++)
{
for(int j=0;j<=ny-1;j++)
{
for(int k=0;k<=nz-1;k++)
{

if(ract[i][j][k]!=0.0)
{
if(k==nz-2)
{
```

```
pp[i][j][k+1]=pp[i][j][k];
}
if(k==1)
{
pp[i][j][k-1]=pp[i][j][k];
}
```

Horizontal wall faces—zero derivative:

```
if(j>=iy1step&&j<=iy2step)
{
if(i>=ix1step&&i<=ix2step&&k==iz2step+1)
{
pp[i][j][k-1]=pp[i][j][k];
}
if(i==ix2step+1&&k<=iz2step)
{
pp[i-1][j][k]=pp[i][j][k];
}
}
```

Side wall zero derivatives:

```
if(i>=ix1step&&i<=ix2step)
{
if(k<=iz2step)
{
pp[i][iy1step-1][k]=pp[i][iy1step-2][k];
pp[i][iy2step+1][k]=pp[i][iy2step+2][k];
}
}

dpress=(u[i+1][j][k]-u[i-1][j][k])/a+
v[i][j+1][k]-v[i][j-1][k])/b+(w[i][j][k+1]-w[i][j][k-1])/f;
dpress=dpress;
dum1[i][j][k]=(pp[i+1][j][k]*c+pp[i-1][j][k]*c+
pp[i][j+1][k]*d+pp[i][j-1][k]*d+
pp[i][j][k+1]*g+pp[i][j][k-1]*g
-dpress+)/e;
}
}
}
}
psum=0.0;
for(int i=0;i<=nx-1;i++)
{
for(int j=0;j<=ny-1;j++)
{
for(int k=0;k<=nz-1;k++)
{
if(ract[i][j][k]!=0.0)
{
ppre=pp[i][j][k];
pp[i][j][k]=dum1[i][j][k];
ppost=pp[i][j][k];
psum=psum+fabs(ppost-ppre);
}
}
}
}
```

```
if(psum<=ptol)
{
break;
goto line1000;
}
}
line1000:
```

Correct velocities with P':

```
for(int i=0;i<=nx-1;i++)
{
for(int j=0;j<=ny-1;j++)
{
for(int k=0;k<=nz-1;k++)
{
if(ract[i][j][k]!=0.0)
{
dum1[i][j][k]=-(pp[i+1][j][k]-pp[i-1][j][k])*dt/dx;
dum2[i][j][k]=-(pp[i][j+1][k]-pp[i][j-1][k])*dt/dy;
dum3[i][j][k]=-(pp[i][j][k+1]-pp[i][j][k-1])*dt/dz;
up[i][j][k]=up[i][j][k]+dum1[i][j][k];
vp[i][j][k]=vp[i][j][k]+dum2[i][j][k];
wp[i][j][k]=wp[i][j][k]+dum3[i][j][k];
}
}
}
}
```

Now correct the pressure:

```
for(int i=0;i<=nx-1;i++)
{
for(int j=0;j<=ny-1;j++)
{
for(int k=0;k<=nz-1;k++)
{
pdum=p[i][j][k];
p[i][j][k]=pdum+pp[i][j][k];
}
}
}
```

Now for the pressure derivative boundary conditions.

Ceiling:

```
for(int i=0;i<=nx-1;i++)
{
for(int j=0;j<=ny-1;j++)
{
p[i][j][nz-1]=p[i][j][nz-2];
}
}
```

Floor:

```
for(int i=1;i<=nx-1;i++)
{
```

```
for(int j=0;j<=ny-1;j++)
{
p[i][j][0]=p[i][j][1];
}
}
```

Step:

```
for(int i=0;i<=ixstep;i++)
{
for(int j=iy1step;j<=iy2step;j++)
{
p[i][j][iz2step]=p[i][j][iz2step+1];
}
}
for(int k=0;k<=iz2step;k++)
{
for(int j=iy1step;j<=iy2step;j++)
{
p[ix2step][j][k]=p[ix2step+1][j][k];
}
}
```

Fin:

```
for(int i=ixfinb;i<=ixfint;i++)
{
for(int j=iyfinb;j<=iyfint;j++)
{
p[i][j][izfinb]=p[i][j][izfinb-1];
p[i][j][izfint]=p[i][j][izfint+1];
}
}

for(int j=iyfinb;j<=iyfint;j++)
{
for(int k=izfinb;k<=izfint;k++)
{
p[ixfinb][j][k]=p[ixfinb-1][j][k];
p[ixfint][j][k]=p[ixfint+1][j][k];
}
}

for(int i=ixfinb;i<=ixfint;i++)
{
for(int k=izfinb;k<=izfint;k++)
{
p[i][iyfinb][k]=p[i][iyfinb-1][k];
p[i][iyfint][k]=p[i][iyfint+1][k];
}
}
```

You could also have left inlet and outlet here.

Now update velocities.

```
for(int i=0;i<=nx-1;i++)
{
```

```
for(int j=0;j<=ny-1;j++)
{
for(int k=0;k<=nz-1;k++)
{
if(ract[i][j][k]!=0.0)
{
u[i][j][k]=u[i][j][k]+up[i][j][k];
v[i][j][k]=v[i][j][k]+vp[i][j][k];
w[i][j][k]=w[i][j][k]+wp[i][j][k];
}
}
}
}
```

Velocity derivative boundary conditions:

Inlet wall:

```
for(int j=0;j<=ny-1;j++)
{
for(int k=0;k<=nz-1;k++)
{
u[0][j][k]=u[1][j][k];
v[0][j][k]=v[1][j][k];
w[0][j][k]=w[1][j][k];
}
}
```

Outlet wall:

```
for(int j=0;j<=ny-1;j++)
{
for(int k=0;k<=nz-1;k++)
{
u[nx-1][j][k]=u[nx-2][j][k];
v[nx-1][j][k]=v[nx-2][j][k];
w[nx-1][j][k]=w[nx-2][j][k];
}
}
```

Back wall:

```
for(int i=0;i<=nx-1;i++)
{
for(int k=0;k<=nz-1;k++)
{
u[i][0][k]=u[i][1][k];
v[i][0][k]=v[i][1][k];
w[i][0][k]=w[i][1][k];
}
}
```

Front wall:

```
for(int i=0;i<=nz-1;i++)
{
for(int k=0;k<=nz-1;k++)
{
```

```
u[i][ny-1][k]=u[i][ny-2][k];
v[i][ny-1][k]=v[i][ny-2][k];
w[i][ny-1][k]=w[i][ny-2][k];
}
}
```

You could also have ceiling floor step fin boundary conditions here.

Calculate velocity tolerances.

```
usum=0.0;
vsum=0.0;
for(int i=0;i<=nx-1;i++)
{
for(int j=0;j<=ny-1;j++)
{
for(int k=0;k<=nz-1;k++)
{
if(ract[i][j][k]!=0.0)
{
usum=usum+fabs(u[i][j][k]-uold[i][j][k]);
vsum=vsum+fabs(v[i][j][k]-vold[i][j][k]);
wsum=vsum+fabs(w[i][j][k]-wold[i][j][k]);
}
}
}
}
```

Optional: Write out info on how convergence is coming along.

```
if(usum<=veltol&&vsum<=veltol&&wsum<=veltol)
{
//      break;
//      goto line2000;
}
//      cout<<istep<<" "<<usum/veltol<<" "<<vsum/veltol<<" "<<psum/
        ptol<<endl;
}

line2000:
```

Now write out boundary condition information and the velocity field.

```
ifr=ifr+1;
sprintf(fileanim,"%d.txt",ifr);
ofstream myfile;
myfile.open(fileanim);
myfile<<nx<<" "<<acell<<" "<<ny<<" "<<bcell<<" "<<nz<<ccell<<
" "<<ix1step<<" "<<ix2step<<
" "<<iy1step<<" "<<iy2step<<
" "<<iz1step<<" "<<iz2step<<
" "<<ixfint<<" "<<iyfinb<<
" "<<iyfint<<" "<<iyfinb<<
" "<<izfint<<" "<<izfinb<<<<endl;
for(int i=0;i<=nx-1;i++)
{
for(int j=0;j<=ny-1;j++)
```

```
{
for(int k=0;k<=nz-1;k++)
{
myfile<<float(i)*dx<<" "<<float(j)*dy<<" "<<float(k)*dz<<" "<<
u[i][j][k]*15.0<<" "<<v[i][j][k]*15.0<<" "<<" "<<w[i][j][k]*15.0<<
15.0*sqrt(u[i][j][k]*u[i][j][k]+v[i][j][k]*v[i][j][k]+w[i][j][k]*w[i][j]
[k])<<endl;
}
}
}
```

Wrap it up.

```
myfile.close();
return (0);
}
```

6.4 Snow in July

6.4.1 Theoretical and Mathematical Framework

So, in the previous section we were able to calculate the air velocity field $\vec{v}_{air}$. But since the air is carrying snow we have to figure out details of placing snowflakes in the simulation box, calculating forces on the flakes, and stepping things through time. First off, what is a snowflake to us? Ideally, it would be the beautifully symmetric crystals that nature offers but we will remove the lovely, intricate detail and model snowflakes as spheres of radius R. I know that the approximation has the same delicacy of painting one's nails with a roller, but it is what it needs to be.

Every process in nature is governed by a differential equation. In our case here, we need to understand what the force is on each flake and how it affects the snowflake's velocity. The differential equation governing the motion of each snowflake (i) is Newton's second law $ma = F$ but it looks like the following:

$$m\frac{d\vec{v}_i}{dt} = -mg\hat{z} + a(\vec{v}_{air} - \vec{v}_i) + \sum_{j=1}^{n} \theta\left(1 - \frac{r_{ij}}{2R}\right)(b\hat{r}_{ji} + c\hat{v}_{ji}). \tag{6.16}$$

And of course we need to understand what each term is modeling before we discretize and code it up as will be discussed in the next section. It is good to keep in mind that Equation 6.16 is a vector equation which deals with all three components of motion in a single expression. The left hand term is the velocity vector of a snowflake (i) in an orthogonal Cartesian coordinate system with the origin at the lower left corner of the building wall. Then the left hand side of Equation 6.16 is nothing other than *mass* × *acceleration*, or *ma*. The first term on the right hand side is a constant gravitational force in the vertical direction and the negative sign signifies that the force is downward. The second term is a viscous drag term that takes into account the force that the air exerts on each snowflake, which can be advancing, retarding or zero. The last term represents snowflake–snowflake interactions and is zero if the particles are not touching, as prescribed by the Heaviside step function θ. Within the sum, the $b\hat{r}_{ji}$ term expresses the normal force and is equal to a constant magnitude repulsive force directed along a line separating the flakes' centers if the snowflakes are touching. The

Table 6.2. Values for Various Constants for Direct Snowfall Simulations

Constant	Value
m	5×10^{-8} kg
g	6.8 m/s^2
a	6.8 N/kg
b	100 N
c	20 N
R	10 cm

Source: E. Maldonado, M.W. Roth, *Journal of Applied Fluid Mechanics*, 5(3), 71–78, 2012 [7].

$c\hat{v}_{ji}$ term is a friction term and is assumed to be constant in magnitude and directed opposite the relative velocity of two touching snowflakes. Here $\hat{r}_{ji}$ is the unit vector pointing to snowflake (i) from snowflake (j), $\hat{v}_{ij}$ is the relative velocity vector pointing to snowflake (j) from snowflake (i). The values of important constants are shown in Table 6.2 [7].

It should be noted that the size of the snowflakes are exaggerated so as to accelerate piling and accumulation but this difference from the size of real snowflakes does not alter the conclusions of the study. In addition, there are many tunable parameters in the simulation and values for such parameters in Table 6.2 are initial guesses that were tuned to reasonable behavior in the simulation.

The equation of motion (Equation 6.16) now needs to be discretized (so the computer can use it) and numerically integrated with respect to (stepped through) time utilizing Newton's method. The presence of boundaries has not appeared yet in the discussion of forces on the snowflakes. We can either calculate forces on the flakes and incorporate them into Equation 6.16 or we can implement boundary conditions directly on the motion of the flakes. The latter approach is much easier and more accurate. As time advances, then there are four types of boundary conditions which must be implemented: vertical surface, horizontal surface, free boundary, and drifting as shown in Table 6.3 [7].

6.4.2 Telling the Computer How to Solve the Problem

We chose to start the simulation by initially placing n spherical snowflakes at random positions within the region between the roof of the building and the upper boundary, and assigned them initial velocities of zero. To make the simulation as realistic as possible, the roof of the building loads with snow for the first quarter of the simulation, because wind blowing snow off the loaded roof contributes to drifting in its own

Table 6.3. Boundary Conditions for Direct Snowfall Simulations

Surface Type	Boundary Condition
Surfaces	
Walls and fin	Reversal of velocity component normal to the surface
Ground	Sticking: $\vec{v}_i = 0$
Free Boundaries	
Inlet, outlet, front, back	Return snowflake to random initial position
Drifting	
Settled snowflakes	If $v_{xi} > 0.001$ m/s then $v_{xi} = 0$.

Source: E. Maldonado, M.W. Roth, *Journal of Applied Fluid Mechanics*, 5(3), 71–78, 2012 [7].

unique way. When the remaining three quarters of the simulation begins, the air velocity (u,v,w) at the snowflake's position is determined using air velocity calculations described earlier. Next, the force on each snowflake is calculated using Equation 6.16 and the acceleration is obtained by dividing the force by the snowflake's mass.

$$a_{ix} = \frac{a}{m}(u - v_{ix}) + \sum_{j=1}^{n} \theta\left(1 - \frac{r_{ij}}{2R}\right)\left(\frac{b}{m}\hat{r}_{ji} \cdot \hat{x} + \frac{c}{m}\hat{v}_{ji} \cdot \hat{x}\right) \tag{6.17}$$

$$a_{iy} = \frac{a}{m}(v - v_{iy}) + \sum_{j=1}^{n} \theta\left(1 - \frac{r_{ij}}{2R}\right)\left(\frac{b}{m}\hat{r}_{ji} \cdot \hat{y} + \frac{c}{m}\hat{v}_{ji} \cdot \hat{y}\right) \tag{6.18}$$

$$a_{iz} = -g + \frac{a}{m}(v - v_{iz}) + \sum_{j=1}^{n} \theta\left(1 - \frac{r_{ij}}{2R}\right)\left(\frac{b}{m}\hat{r}_{ji} \cdot \hat{z} + \frac{c}{m}\hat{v}_{ji} \cdot \hat{z}\right) \tag{6.19}$$

Here, the summation is over all neighboring snowflakes and θ is the Heaviside theta (step) function, which takes into account that snowflake pairs interact with each other only for certain pair separations. Then the trajectories of the flakes are obtained by numerically integrating the acceleration with respect to time using Newton's method with a time step of 0.001 seconds:

$$v_{ix}^{n+1} = v_{ix}^{n} + a_{ix}\Delta t \tag{6.20}$$

$$x_i^{n+1} = x_i^{n} + \left(\frac{v_{ix}^{n+1} + v_{ix}^{n}}{2}\right)\Delta t \tag{6.21}$$

$$v_{iy}^{n+1} = v_{iy}^{n} + a_{iy}\Delta t \tag{6.22}$$

$$y_i^{n+1} = y_i^{n} + \left(\frac{v_{iy}^{n+1} + v_{iy}^{n}}{2}\right)\Delta t \tag{6.23}$$

$$v_{iz}^{n+1} = v_{iz}^{n} + a_{iz}\Delta t \tag{6.24}$$

$$z_i^{n+1} = z_i^{n} + \left(\frac{v_{iz}^{n+1} + v_{iz}^{n}}{2}\right)\Delta t \tag{6.25}$$

The system is advanced, and any necessary boundary conditions are enforced for each snowflake and then the time integration loop continues. After a certain number of steps (in many cases, one tenth of the total), another set of snowflakes are placed at initial random positions above the wall and the simulations continues until the total number of steps has been reached.

Figure 6.3 shows the development of a typical simulated storm and ending configuration after 180,000 time steps and 200,000 particles have fallen. Drifts have not accumulated yet but it is quite easy to see patterns. Figure 6.4 shows resulting snowfall patterns including side winds.

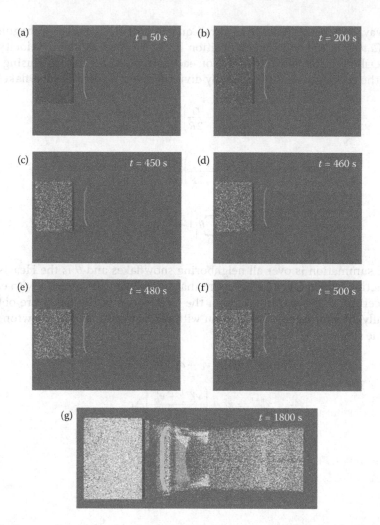

Figure 6.3. Progression of the simulated snowstorm, with snow falling vertically on the roof (a–c), the wind picking up and spreading the snow (d–f), and example final configuration with the deflection fin emphasized (g). Shading helps distinguish the building and fins in the gray-scale printing and the online version is in color (E. Maldonado, M.W. Roth, *Journal of Applied Fluid Mechanics*, 5(3), 71–78, 2012.) [7].

6.5 A Sample Program that Simulates a Snowstorm

For code related to the following discussion please review the provided program Snow_ Dynamics.cpp. Include relevant C++ libraries.

```
#include<cstdlib>
#include<iostream>
#include<fstream>
#include<math.h>
#include<iomanip>
```

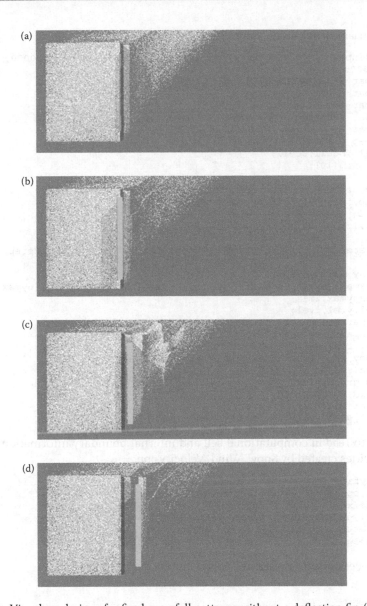

Figure 6.4. Visual renderings for final snowfall patterns without a deflection fin (a) and with selected fins (yellow) having typical effects of shadowing and structure in the drifting (b–d). Shading helps distinguish the building and fins in the grayscale printing and the online version is in color (E. Maldonado, M.W. Roth, *Journal of Applied Fluid Mechanics*, 5(3), 71–78, 2012.) [7].

```
#include<stdio.h>
using namespace std;
```

Begin Program.

```
int main()
{
```

Declare variables and assign values to global constants

```
const int npart=200000,nxmax=36,nymax=36,nzmax=36,nsteps=180000,
nevery=18000;
const float dt=1.0e-02;
const float sphere_size=0.06;
const float g=6.81,smass=5.0e-08;
const float fconst=10.0;
const float L=2.0,theta=5.0*float(ABCD)*(3.14)/180.0;
const float pi=3.14159265;
float acell,bcell,ccell;
char fileanim[10];
int n,ifr;
int nx,ny,nz,ix1step,ix2step,iy1step,iy2step,iz1step,iz2step;
int ixfinb,ixfint,iyfinb,iyfint,izfinb,izfint;
int seed=3;
int ix,iy,iz;
float
x1step,x2step,xfinb,xfint,y1step,y2step,yfinb,yfint,z1step,z2step,zfinb,
zfint;
float ax,ay,az,dx,dy,dz;
float u[nxmax][nymax][nzmax],v[nxmax][nymax][nzmax],w[nxmax][nymax][nzmax];
float xmax,ymax,zmax;
float xvel,yvel,zvel;
float x[npart],y[npart],z[npart];
float vx[npart],vy[npart],vz[npart];
float vrel,rij;
float dummy,xdum,ydum,zdum,sdum,udum,vdum,wdum;
float drag;
float red,green,blue;
float vxold,vyold,vzold;
float count,xleft,xright,ybot,ytop;
float uvel,vvel,wvel;
```

Get ready to read in computational cell and internal boundary information as well as wind velocities created by Snow_Wind_Velocity.cpp.

```
fstream myfile_in;
myfile_in.open("velocities.txt");
velocities.xt is 1.txt produced by Snow_Wind_Velocity.cpp
myfile_in>>nx>>acell>>ix1step>>ix2step>>ixfinb>>ixfint;
myfile_in>>ny>>bcell>>iy1step>>iy2step>>iyfinb>>iyfint;
myfile_in>>nz>>ccell>>iz1step>>iz2step>>izfinb>>izfint;
```

Perform necessary initializations.

```
ifr=6;
drag=smass*g;
srand(seed);
for(int i=0;i<=npart-1;i++)
{
x[i]=0.0;
y[i]=0.0;
z[i]=0.0;
vx[i]=0.0;
vy[i]=0.0;
vz[i]=0.0;
}
for(int i=0;i<=nxmax-1;i++)
{
for(int j=0;j<=nymax-1;j++)
```

```
{
for(int k=0;k<=nzmax-1;k++)
{
u[i][j][k]=0.0;
v[i][j][k]=0.0;
w[i][j][k]=0.0;
}
}
}

dx=acell/float(nx-1);
dy=bcell/float(ny-1);
dz=ccell/float(nz-1);
```

Include calculation of the locations of boundaries in the simulation.

```
x1step=acell*float(ix1step)/float(nx-1);
x2step=acell*float(ix2step)/float(nx-1);
xfinb=acell*float(ixfinb)/float(nx-1);
xfint=acell*float(ixfint)/float(nx-1);
y1step=bcell*float(iy1step)/float(ny-1);
y2step=bcell*float(iy2step)/float(ny-1);
yfinb=bcell*float(iyfinb+6)/float(ny-1);
yfint=bcell*float(iyfint-6)/float(ny-1);
z1step=ccell*float(iz1step)/float(nz-1);
z2step=ccell*float(iz2step)/float(nz-1);
zfinb=ccell*float(izfinb)/float(nz-1);
zfint=ccell*float(izfint)/float(nz-1);
xmax=acell;
ymax=bcell;
zmax=ccell;
for(int i=0;i<=nx-1;i++)
{
for(int j=0;j<=ny-1;j++)
{
for(int k=0;k<=nz-1;k++)
{
```

Now read in windspeed information and assign values to velocity arrays.

```
myfile_in >> xdum>>ydum>>zdum>>udum>>vdum>>wdum;
u[i][j][k]=udum;
v[i][j][k]=vdum;
w[i][j][k]=wdum;
}
}
}
myfile_in.close();n an appropriate initial random position
```

Start time loop and be sure to initialize the "name" (integer number) of each snow flake and place in.

```
n=-1;
for(int it=1;it<=nsteps;it++)
{
if(n<npart)
{
n=n+1;
x[n]=x1step+(x2step-x1step)*0.001*(rand()%1000);
y[n]=y1step+(y2step-y1step)*0.001*(rand()%1000);
```

```
z[n]=z2step+(zmax-z2step)*0.001*(rand()%1000);
vx[n]=0.0;
vy[n]=0.0;
vz[n]=0.0;
}
```

Enforce periodic boundary conditions.

```
for(int i=0;i<=n;i++)
{
vxold=vx[i];
vyold=vy[i];
vzold=vz[i];
ix=int(float(nx)*x[i]/xmax);
iy=int(float(ny)*y[i]/ymax);
iz=int(float(nz)*z[i]/zmax);
if(ix==nx)
{
ix=nx-1;
}
if(iy==ny)
{
iy=ny-1;
}
if(iz==nz)
{
iz=nz-1;
}
```

Factor in some periodic gusting.

```
uvel=u[ix][iy][iz]*0.5*(1.0+cos(20.0*pi*float(it)/float(nsteps)));
vvel=v[ix][iy][iz]*0.5*(1.0+cos(20.0*pi*float(it)/float(nsteps)));
wvel=w[ix][iy][iz]*0.5*(1.0+cos(20.0*pi*float(it)/float(nsteps)));
vrel=sqrt(pow(vx[i]-uvel,2)+pow(vy[i]-vvel,2)+pow(vz[i]-wvel,2));
```

Let the snow fall vertically for the first part of the simulation and experience drag for the later part.

```
if(it>nsteps/4)
{
ax=drag*(uvel-vx[i])/smass;
ay=drag*(vvel-vy[i])/smass;
az=(drag*(wvel-vz[i]))/smass - g;
}

if(it<=nsteps/4)
{
ax=0.0;
ay=0.0;
az=- g;
}
```

Here include attractive or repulsive forces between the flakes when they are in contact. This part must be tuned to preference!

```
for(int j=0;j<=n;j++)
{
if(j!=i)
{
```

```
rij=sqrt(pow(x[j]-x[i],2)+pow(y[j]-y[i],2)+pow(z[j]-z[i],2));
if(rij<=2.0*sphere_size)
{
ax=ax+fconst*(x[i]-x[j]);
ay=ay+fconst*(y[i]-y[j]);
az=az+fconst*(z[i]-z[j]);
}
if(rij<=3.0*sphere_size&&rij>2.0*sphere_size)
{
ax=ax-0.2*fconst*(x[i]-x[j]);
ay=ay-0.2*fconst*(y[i]-y[j]);
az=az-0.2*fconst*(z[i]-z[j]);
}
}
}
```

Update velocities (integrate them with respect to time).

```
dummy=vx[i]+ax*dt;
vx[i]=dummy;
if(x[i]>xfinb+dx&&fabs(vx[i])<1.0e-01)
{
vx[i]=0.0;
vy[i]=0.0;
}
dummy=vy[i]+ay*dt;
vy[i]=dummy;
dummy=x[i]+0.5*(vxold+vx[i])*dt;
x[i]=dummy;
dummy=y[i]+0.5*(vyold+vy[i])*dt;
y[i]=dummy;
dummy=vz[i]+az*dt;
vz[i]=dummy;
dummy=z[i]+0.5*(vzold+vz[i])*dt;
z[i]=dummy;
}
```

If a snowflake leaves the computational call, put another one back in!

```
for(int i=0;i<=n;i++)
{
if(x[i]>xmax||x[i]<0.0||
y[i]>ymax||y[i]<0.0||z[i]>zmax)
{
x[i]=x1step+(x2step-x1step)*0.001*(rand()%1000);
y[i]=y1step+(y2step-y1step)*0.001*(rand()%1000);
z[i]=z2step+(zmax-z2step)*0.001*(rand()%1000);
vx[i]=0.0;
vy[i]=0.0;
vz[i]=0.0;
}
}
```

Snowflakes interact with boundaries - in a variety of ways that can be tuned.

```
for(int i=0;i<=n;i++)
{
if(x[i]>=x1step&&x[i]<=x2step)
{
if(y[i]>=y1step&&y[i]<=y2step)
```

```
{
if(z[i]<=z2step)
{
z[i]=z2step;
vx[i]=0.0;
vy[i]=0.0;
vz[i]=0.0;
}
}
}


if(x[i]>x2step)
{
if(z[i]<=0.0)
{
z[i]=0.0;
vx[i]=vx[i]*0.85;
vy[i]=0.0;
vz[i]=fabs(vz[i])*0.1;
}
}


if(y[i]<y1step||y[i]>y2step)
{
if(z[i]<=0.0)
{
z[i]=0.0;
vx[i]=0.0;
vy[i]=0.0;
vz[i]=0.0;
}
}


if(x[i]>xfinb&&x[i]<xfinb+dx)
{
if(vx[i]>0.0)
{
if(y[i]>yfinb&&y[i]<yfint)
{
if(z[i]>zfinb&&z[i]<zfint)
{
vx[i]=-vx[i];
}
}
}
}
if(z[i]>zfinb&&z[i]<zfint)
{
if(y[i]<=yfinb&&y[i]>=yfinb-L*cos(theta))
{
if(fabs(xfinb+(y[i]-yfinb)*tan(theta)-x[i])<0.01)
{
vx[i]=vx[i]-2*(-vx[i]*cos(theta)-vy[i]*sin(theta))*(-cos(theta));
vy[i]=vy[i]-2*(-vx[i]*cos(theta)-vy[i]*sin(theta))*(-sin(theta));
}
}

if(y[i]<=yfint+L*cos(theta)&&y[i]>=yfint)
{
```

```
if(fabs(xfint+(yfint-y[i])*tan(theta)-x[i])<0.01)
{
vx[i]=vx[i]-2*(-vx[i]*cos(theta)+vy[i]*sin(theta))*(-cos(theta));
vy[i]=vy[i]-2*(-vx[i]*cos(theta)+vy[i]*sin(theta))*sin(theta);
}
}
}

}
```

Write out .pov visualization files with specified frequency.

```
if (float(it/nevery)==float(it)/float(nevery))
{
ifr=ifr+1;
sprintf(fileanim,"%d.pov",ifr);
ofstream myfile;
myfile.open(fileanim);
myfile<<"#version 3.6"<<endl;
myfile<<"#include "<<"\""<<"colors.inc"<<"\""<<endl;
myfile<<"global_settings"<<endl;
myfile<<"{"<<endl;
myfile<<"assumed_gamma 1.0"<<endl;
myfile<<"}"<<endl;
myfile<<"camera"<<endl;
myfile<<"{"<<endl;
myfile<<"location <"<<2<<","<<1<<","<<6<<">"<<endl;
myfile<<"direction 1.5*z"<<endl;
myfile<<"right     -4/3*x"<<endl;
myfile<<"sky z"<<endl;
myfile<<"look_at <"<<2<<","<<1<<","<<0<<">"<<endl;
myfile<<"}"<<endl;
myfile<<"background { color red 0.1 green 0.1 blue 0.1 }"<<endl;
myfile<<"light_source"<<endl;
myfile<<"{"<<endl;
myfile<<"0*x"<<endl;
myfile<<"color red"<<endl;
myfile<<"1.0  green 1.0  blue 1.0"<<endl;
myfile<<"translate <-30, 30, 30>"<<endl;
myfile<<"}"<<endl;
myfile<<"light_source"<<endl;
myfile<<"{"<<endl;
myfile<<"0*x"<<endl;
myfile<<"color red"<<endl;
myfile<<"0.5  green 0.5  blue 0.5"<<endl;
myfile<<"translate <30, 30, 30>"<<endl;
myfile<<"}"<<endl;
myfile<<"light_source"<<endl;
myfile<<"{"<<endl;
myfile<<"0*x"<<endl;
myfile<<"color red"<<endl;
myfile<<"0.1  green 0.1  blue 0.1"<<endl;
myfile<<"translate <30, -30, 30>"<<endl;
myfile<<"}"<<endl;
myfile<<"box{<"<<x1step/10.0<<" "<<y1step/10.0<<" "<<z1step/10.0<<">
<"<<endl;
myfile<<"x2step/10.0<<","<<y2step/10.0<<","<<z2step/10.0<<">"<<endl;
myfile<<"texture {pigment"<<endl;
myfile<<"{color rgb<0.5,0,0>}"<<endl;
myfile<<"finish{specular 0.0}} "<<endl;
```

```
myfile<<"box{<"<<xfinb/10.0<<","<<yfinb/10.0<<","<<zfinb/10.0<<">"<<endl;
myfile<<"<"<<xfint/10.0<<","<<yfint/10.0<<","<<zfint/10.0<<">"<<endl;
myfile<<"texture {pigment"<<endl;
myfile<<"{color rgb<0.8,0.8,0.0>}"<<endl;
myfile<<"finish{specular 0.0}} }"<<endl;

myfile<<"polygon{4,<"<<xfinb/10.0<<","<<yfinb/10.0<<","<<zfinb/10.0<<">,<"
<<xfinb/10.0<<","<<yfinb/10.0<<","<<zfint/10.0<<">,<"
<<(xfinb+L*sin(theta))/10.0<<","<<(yfinb-L*cos(theta))/10.0<<","<<zf
int/10.0<<">,<"
<<(xfinb+L*sin(theta))/10.0<<","<<(yfinb-L*cos(theta))/10.0<<","<<zf
inb/10.0<<"> pigment {color rgb<1,1,0>}}";

myfile<<"polygon{4,<"<<xfinb/10.0<<","<<yfint/10.0<<","<<zfinb/10.0<<">,<"
<<xfinb/10.0<<","<<yfint/10.0<<","<<zfint/10.0<<">,<"

<<(xfinb+L*sin(theta))/10.0<<","<<(yfint+L*cos(theta))/10.0<<","<<zf
int/10.0<<">,<"

<<(xfinb+L*sin(theta))/10.0<<","<<(yfint+L*cos(theta))/10.0<<","<<zfinb
/10.0<<">pigment {color rgb<1,1,0>}}";
for(int i=0;i<=n;i++)
{
myfile<<"sphere { <"<<endl;
myfile<<x[i]/10.0<<","<<y[i]/10.0<<","<<z[i]/10.0<<">,"<<endl;
myfile<<sphere_size/15.<<endl;
blue=1.0;
green=1.0;
red=1.0;
myfile<<"texture {pigment"<<endl;
myfile<<"{color rgb<"<<red<<","<<green<<","<<blue<<">}"<<endl;
myfile<<"finish{specular 1}} }"<<endl;
}
myfile.close();
}
//###############################################################
}


Calculate and write out the number of snowflakes in the simulation

ofstream myfile2;
myfile2.open("final_snowfall.res");
for(int i=0;i<=99;i++)
{
for(int j=0;j<=99;j++)
{
xleft=float(i)*acell/100.0;
xright=float(i+1)*acell/100.0;
ybot=float(j)*bcell/100.0;
ytop=float(j+1)*bcell/100.0;
count=0.0;
for(int k=1;k<=npart;k++)
{
if(x[k]>=xleft&&x[k]<xright&&y[k]>=ybot&&y[k]<ytop)
{
count=count+1.0;
}
}
myfile2<<(xleft+xright)/2.0<<" "<<(ytop+ybot)/2.0<<" "<<count<<endl;
}
}
```

```
return (0);
}
```

As a note: the computer simulation discussed here represents a very simple physical phenomenon. I am under a winter storm warning as I am writing this chapter of the book and, looking out my window, I realize how simple and direct nature is, in contrast to the effort we need to go through for our simulation, and an imperfect one at that. It's food for thought. As for the July bit, I had you wait until the end of the story to explain: the original work done on this project was done over a summertime, and began, as requested through consultation in the month following June.

6.6 How Fluids Move through Porous Media

6.6.1 General Thoughts

So far, we have dealt with fluid flow in the presence of boundaries. What happens when the fluid is actually flowing through something? In the interest of not only addressing the fundamental physical problem but also applying our model to a practical system, in this section we seek to describe contaminant flow in a confined aquifer with an impermeable rectangular boundary, in collaboration with a group overseeing a system of wells at the University of Northern Iowa. Moreover, there are contaminants that are actually flowing into a nearby river! With the option of inserting a constant-head river within the aquifer, the model at hand can be thought of as that of a semiconfined aquifer. The approach described in this section consists of two steps. First, the behavior of the physical flow of water is modeled analytically with the porous medium flow equation. Second, the water velocity field obtained from the first step is used in a finite-difference approach to solving the advection-dispersion equation [2–6]. The *potentiometric head* can be thought of as the water level with respect to some reference point for an unconfined aquifer but is generally the pressure height (height an equivalent column of water would have for the given pressure) at a certain point in the water table even if it is confined.

6.6.2 Groundwater Flow and Contamination in a Rectangle

The example C++ program Hyd_Contam.cpp included in this chapter is a basic model of transient behavior of a groundwater system including contaminant flow. Important parameters in the simulation are shown in Table 6.4. Let's go through it, see what I/O it entails, and see how it works. I have also provided an Excel spreadsheet with which you can make other input files for this program. Although this book is printed in grayscale I commonly refer to the use of color in visualization, which is nearly essential when dealing with water table and contaminant data.

First all the niceties...

Table 6.4. Parameters Important in Modeling the Groundwater Flow

Parameter	Symbol
Length of aquifer	A
Width of aquifer	B
Hydraulic conductivity	K
Specific storage	S_s
Number of terms in width (horizontal) sum	n_x
Number of terms in depth (vertical) sum	n_y
Hydrodynamic dispersivity of Cl^-	
Initial concentration of Cl^-	C_{10}

Source: E. Maldonado, M.W. Roth, *Journal of Applied Fluid Mechanics*, 5(3), 71–78, 2012 [7].

```
//A C++ Program that calculates water level and contaminant flow in a
rectangular aquifer

#include<cstdlib>
#include<iostream>
#include<fstream>
#include<math.h>
#include<iomanip>
#include<stdio.h>

using namespace std;
```

Now we begin the program and declare our variables.

```
//------Declare Variables------

int main()

{
const int nsteps=1000000,nevery=50000
const int nx_max=100,ny_max=100;
const float dt=1.0;
const float hint=0.0;
const float Stor=1.0;
float hmax,hmin;
float cmax,cmin;
float red,green,blue;
float h[nx_max][ny_max],hp1[nx_max][ny_max];
float hxder,hyder,hdummy;
float dx,dy;
float x1,x2,y1,y2;
float cond[nx_max][ny_max],cp1[nx_max][ny_max];
float dummy,cdummy;
float a,b;
float vx,vy,disp;
float kxavg,kyavg;
int idummy,nx,ny;
int activity[nx_max][ny_max];
int ifile;
float cont[nx_max][ny_max];
char filename [7];
```

Deal with I/O.

```
ifstream myfile2;
myfile2.open("input_contam.txt");
```

Read in the grid and aquifer dimensions.

```
myfile2>>idummy;
nx=idummy;
myfile2>>idummy;
ny=idummy;
myfile2>>dummy;
a=dummy;
myfile2>>dummy;
b=dummy;
myfile2>>disp;
```

Here is some screen I/O if you want to use it.

```
//   cout<<"Horizontal Dimension Pleeeease\n>";
//   cin>>a;
```

```
//   cout<<"Vertical Dimension or I'll tell a Roth joke\n>";
//   cin>>b;
```

Initialize the file counter.

```
ifile=0;
```

Initialize the arrays to hold them in memory.

```
for(int i=0;i<nx_max;i++)
{
for(int j=0;j<ny_max;j++)
{
h[i][j]=0.0;
hp1[i][j]=0.0;
activity[i][j]=1;
cond[i][j]=0.0;
cont[i][j]=0.0;
}
}
```

Initialize maxima, minima, and grid spacings.

```
cmin=0.0;
cmax=0.0;
hmin=0.0;
hmax=0.0;

dx=a/float(nx-1);
dy=b/float(ny-1);
```

Read in the activity, initial head, conductivity, and initial contaminant concentration arrays.

```
for(int k=0;k<ny;k++)
{
for(int j=0;j<nx;j++)
{
myfile2>>activity[j][k];
}
}

for(int k=0;k<ny;k++)
{
for(int j=0;j<nx;j++)
{
myfile2>>h[j][k];
}
}

for(int k=0;k<ny;k++)
{
for(int j=0;j<nx;j++)
{
myfile2>>cond[j][k];
}
}
```

```
for(int k=0;k<ny;k++)
{
for(int j=0;j<nx;j++)
{
myfile2>>cont[j][k];
}
}
```

Now start the "time" loop and keep track of each step.

```
for(int k=0; k<nsteps; k++)
{
cout<<k<<endl;
```

Loop around the aquifer and begin the head calculations. The head can be calculated from Equation 6.17 written in update form as used in the code, and it's a homework problem for you to write it that way.

$$\left[\frac{\partial^2[Kh(x,y,t)]}{\partial x^2}+\frac{\partial^2[Kh(x,y,t)]}{\partial y^2}\right]=S_s\frac{\partial h(x,y,t)}{\partial t}, \tag{6.26}$$

$$K\left\{\begin{matrix}\frac{h_{i+1,j}^n-2h_{i,j}^n+h_{i-1,j}^n}{(\Delta x)^2}\\+\frac{h_{i,j+1}^n-2h_{i,j}^n+h_{i,j-1}^n}{(\Delta y)^2}\end{matrix}\right\}+\left\{\begin{matrix}\frac{h_{i+1,j}^n-h_{i-1,j}^n}{2\Delta x}\frac{K_{i+1,j}^n-K_{i-1,j}^n}{2\Delta x}\\+\frac{h_{i,j+1}^n-h_{i,j-1}^n}{2\Delta y}\frac{K_{i,j+1}^n-K_{i,j-1}^n}{2\Delta y}\end{matrix}\right\}=S\left\{\frac{h_{i,j}^{n+1}-h_{i,j}^n}{\Delta t}\right\} \tag{6.27}$$

```
for(int i=0;i<nx;i++)
{
for(int j=0;j<ny;j++)
{
```

Consider only active cells.

```
if(activity[i][j]==1)
{
```

Start with the interior.

```
if(i>0&&i<(nx-1)&&j>0&&j<(ny-1))
{
kxavg=(cond[i+1][j]+cond[i][j]+cond[i-1][j])/3.0;
kyavg=(cond[i][j+1]+cond[i][j]+cond[i][j-1])/3.0;
hdummy=kxavg*(h[i+1][j]-2.0*h[i][j]+h[i-1][j])/(dx*dx)+
kyavg*(h[i][j+1]-2.0*h[i][j]+h[i][j-1])/(dy*dy)+
(cond[i+1][j]-cond[i-1][j])*(h[i+1][j]-h[i-1][j])/(4.0*dx*dx)+
(cond[i][j+1]-cond[i][j-1])*(h[i][j+1]-h[i][j-1])/(4.0*dy*dy);
dummy=h[i][j]+hdummy*dt/Stor;
hp1[i][j]=dummy;
}
```

And now deal with the boundary.

```
if(i==0&&j>0&&j<(ny-1))
{
```

```
hdummy=(h[i][j+1]+h[i][j-1]+h[i+1][j])/3.0;
hp1[i][j]=hdummy;
}
if(i==(nx-1)&&j>0&&j<(ny-1))
{
hdummy=(h[i][j+1]+h[i][j-1]+h[i-1][j])/3.0;
hp1[i][j]=hdummy;
}
if(i>0&&i<(nx-1)&&j==0)
{
hdummy=(h[i+1][j]+h[i-1][j]+h[i][j+1])/3.0;
hp1[i][j]=hdummy;
}
if(i>0&&i<(nx-1)&&j==(ny-1))
{
hdummy=(h[i][j-1]+h[i-1][j]+h[i+1][j])/3.0;
hp1[i][j]=hdummy;
}
if(i==0&&j==0)
{
hdummy=(h[i+1][j]+h[i][j+1])/2.0;
hp1[i][j]=hdummy;
}
if(i==(nx-1)&&j==0)
{
hdummy=(h[i-1][j]+h[i][j+1])/2.0;
hp1[i][j]=hdummy;
}
if(i==0&&j==(ny-1))
{
hdummy=(h[i+1][j]+h[i][j-1])/2.0;
hp1[i][j]=hdummy;
}
if(i==(nx-1)&&j==(ny-1))
{
hdummy=(h[i-1][j]+h[i][j-1])/2.0;
hp1[i][j]=hdummy;
}
}
```

Transfer data for inactive cells.

```
if(activity[i][j]==0)
{
hdummy=h[i][j];
hp1[i][j]=hdummy;
}
```

Close the aquifer loop.

```
}
}
```

Transfer data from the temporary head array into the main one.

```
for(int i=0;i<nx;i++)
{
for(int j=0;j<ny;j++)
```

```
{
hdummy=hp1[i][j];
h[i][j]=hdummy;
}
}
```

Now find the maximum and minimum head and concentrations for coloring purposes.

```
cmax=0.0;
hmax=0.0;
hmin=1.0e+06;
cmin=1.0e+06;
for(int i=0;i<nx;i++)
{
for(int j=0;j<ny;j++)
{
if(h[i][j]<=hmin)
{
hmin=h[i][j];
}
if(h[i][j]>=hmax)
{
hmax=h[i][j];
}
if(cont[i][j]<=cmin)
{
cmin=cont[i][j];
}
if(cont[i][j]>=cmax)
{
cmax=cont[i][j];
}
}
}
```

Now loop around the aquifer and start the contaminant concentration calculations.

```
//////////////////////////////////////////////////
for(int i=0;i<nx;i++)
{
for(int j=0;j<ny;j++)
{
```

Here is the finite difference advection-dispersion equation in 3D, which we adapt for 2D below (see Figure 6.5).

$$D\nabla^2 C(x,y,t) + \vec{v}\cdot\vec{\nabla}C(x,y,t) = \frac{\partial C(x,y,t)}{\partial t}. \tag{6.28}$$

$$D\left(\frac{C_{i+1,j,k}-2C_{i,j,k}+C_{i-1,j,k}}{(\Delta x)^2} + \frac{C_{i,j+1,k}-2C_{i,j,k}+C_{i,j-1,k}}{(\Delta y)^2}\right)$$
$$+ v_x\left(\frac{C_{i+1,j,k}-C_{i-1,j,k}}{2\Delta x}\right) + v_y\left(\frac{C_{i,j+1,k}-C_{i,j-1,k}}{2\Delta y}\right) = \frac{C_{i,j,k+1}-C_{i,j,k}}{\Delta t}. \tag{6.29}$$

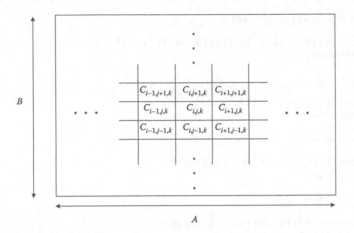

Figure 6.5. Example of finite difference layout for contaminant concentration (E. Maldonado, M.W. Roth, *Journal of Applied Fluid Mechanics*, 5(3), 71–78, 2012.) [7].

First the interior:

```
if(i>0&&i<(nx-1)&&j>0&&j<(ny-1))
{
kxavg=(cond[i+1][j]+cond[i][j]+cond[i-1][j])/3.0;
kyavg=(cond[i][j+1]+cond[i][j]+cond[i][j-1])/3.0;
vx=-kxavg*(h[i+1][j]-h[i-1][j])/(2.0*dx);
vy=-kyavg*(h[i][j+1]-h[i][j-1])/(2.0*dy);
cdummy=cont[i][j]+dt*
(disp*(cont[i+1][j]-2.0*cont[i][j]+cont[i-1][j])/(dx*dx)+
disp*(cont[i][j+1]-2.0*cont[i][j]+cont[i][j-1])/(dy*dy)-
vx*(cont[i+1][j]-cont[i-1][j])/(2.0*dx)-
vy*(cont[i][j+1]-cont[i][j-1])/(2.0*dy));
if(cdummy<0.0)
{
cdummy=0.0;
}
cp1[i][j]=cdummy;
}
```

Deal with updating the boundary.

```
if(i==0&&j>0&&j<(ny-1))
{
cdummy=(cont[i][j+1]+cont[i][j-1]+cont[i+1][j])/3.0;
cp1[i][j]=cdummy;
}
if(i==(nx-1)&&j>0&&j<(ny-1))
{
cdummy=(cont[i][j+1]+cont[i][j-1]+cont[i-1][j])/3.0;
cp1[i][j]=cdummy;
}
if(i>0&&i<(nx-1)&&j==0)
{
cdummy=(cont[i+1][j]+cont[i-1][j]+cont[i][j+1])/3.0;
cp1[i][j]=cdummy;
}
```

```
if(i>0&&i<(nx-1)&&j==(ny-1))
{
cdummy=(cont[i][j-1]+cont[i-1][j]+cont[i+1][j])/3.0;
cp1[i][j]=cdummy;
}
if(i==0&&j==0)
{
cdummy=(cont[i+1][j]+cont[i][j+1])/2.0;
cp1[i][j]=cdummy;
}
if(i==(nx-1)&&j==0)
{
cdummy=(cont[i-1][j]+cont[i][j+1])/2.0;
cp1[i][j]=cdummy;
}
if(i==0&&j==(ny-1))
{
cdummy=(cont[i+1][j]+cont[i][j-1])/2.0;
cp1[i][j]=cdummy;
}
if(i==(nx-1)&&j==(ny-1))
{
cdummy=(cont[i-1][j]+cont[i][j-1])/2.0;
cp1[i][j]=cdummy;
}
```

Close the loop over the aquifer.

```
}
}
```

Transfer data from the temporary concentration array into the main one.

```
for(int i=0;i<nx;i++)
{
for(int j=0;j<ny;j++)
{
cdummy=cp1[i][j];
cont[i][j]=cdummy;
}
}
```

Now write out the pov files.

```
/////////////////////////////////////////////////////
if(float(k/nevery)==float(k)/float(nevery))
{
ifile=ifile+1;
sprintf(filename,"head%d.pov",ifile);
ofstream myfile;
myfile.open(filename);

myfile<<"#version 3.6"<<endl;
myfile<<"#include "<<"\""<<"colors.inc"<<"\""<<endl;
myfile<<"global_settings"<<endl;
myfile<<"{"<<endl;
myfile<<"assumed_gamma 1.0"<<endl;
myfile<<"}"<<endl;
```

```
myfile<<"camera"<<endl;
myfile<<"{"<<endl;
myfile<<"location"<<"<"<<a/2.0<<","<<b/2.0<<","<<-
3.0*max(a,b)<<">"<<endl;
myfile<<"direction 1.5*z"<<endl;
myfile<<"right -x"<<endl;
myfile<<"sky -y"<<endl;
myfile<<"look_at  <"<<a/2.0<<","<<b/2.0<<","<<0.0<<">"<<endl;
myfile<<"}"<<endl;
myfile<<"background { color red 0.5 green 0.5 blue 0.5 }"<<endl;
myfile<<"light_source"<<endl;
myfile<<"{"<<endl;
myfile<<"0*x"<<endl;
myfile<< "color red"<<endl;
myfile<<"1.0  green 1.0  blue 1.0"<<endl;
myfile<<"translate <-30, 30, -30>"<<endl;
myfile<<"}"<<endl;
myfile<<"light_source"<<endl;
myfile<<"{"<<endl;
myfile<<"0*x"<<endl;
myfile<<"color red"<<endl;
myfile<< "0.5  green 0.5  blue 0.5"<<endl;
myfile<<"translate <30, 30, -30>"<<endl;
myfile<<"}"<<endl;

for(int i=0;i<nx;i++)
{
x1=float(i)*dx;
x2=float(i+1)*dx;
for(int j=0;j<ny;j++)
{
y1=float(j)*dy;
y2=float(j+1)*dy;
```

Color from blue to red.

```
if(h[i][j]>=hmin&&h[i][j]<=(hmin+(hmax-hmin/3.0)))
{
red=0.0;
green=(h[i][j]-hmin)/((hmax-hmin)/3.0);
blue=((hmin+(hmax-hmin)/3.0)-h[i][j])/((hmax-hmin)/3.0);
}


if(h[i][j]>=(hmin+(hmax-hmin)/3.0)&&h[i][j]<=(hmin+2.0*(hmax-hmin/3.0)))
{
red=(h[i][j]-(hmin+(hmax-hmin)/3.0))/(hmin+(hmax-hmin)/3.0);
green=(h[i][j]-(hmin+(hmax-hmin)/3.0))/(hmin+(hmax-hmin)/3.0);
green=green+((hmin+2.0*(hmax-hmin)/3.0)-h[i][j])/(hmin+(hmax-hmin)/3.0);
blue=0.0;
}


if(h[i][j]>=(hmin+2.0*(hmax-hmin)/3.0)&&h[i][j]<=hmax)
{
red=((hmax-h[i][j])/(hmax-(2.0*(hmax-hmin))/3.0));
green=((hmax-h[i][j])/(hmax-(2.0*(hmax-hmin))/3.0));
red=(h[i][j]-(hmax-(2.0*(hmax-hmin))/3.0))/(hmax-((hmax-(2.0*(hmax-
hmin))/3.0)));
blue=0.0;
}
```

Color lenses purple.

```
if (cond[i][j]==0.0)
{
red=1.0;
blue=1.0;
green=0.0;
}

myfile<<"plane{z,"<<0.0<<" pigment{color
rgb<"<<red<<","<<green<<","<<blue<<">}"<<endl;
myfile<<"clipped_by {box {<"<<x1<<","<<y1<<","<<-0.001<<">,<"<<x2<<",
"<<y2<<","<<0.001<<">}}}"<<endl;

}
}

  sprintf(filename,"cont%d.pov",ifile);

  ofstream myfile5;
  myfile5.open(filename);

hmin=cmin;
hmax=cmax;

myfile5<<"#version 3.6"<<endl;
myfile5<<"#include "<<"\""<<"colors.inc"<<"\""<<endl;
myfile5<<"global_settings"<<endl;
myfile5<<"{"<<endl;
myfile5<<"assumed_gamma 1.0"<<endl;
myfile5<<"}"<<endl;
myfile5<<"camera"<<endl;
myfile5<<"{"<<endl;
myfile5<<"location"<<"<"<<a/2.0<<","<<b/2.0<<","<<-
3.0*max(a,b)<<">"<<endl;
myfile5<<"direction 1.5*z"<<endl;
myfile5<<"right -x"<<endl;
myfile5<<"sky -y"<<endl;
myfile5<<"look_at  <"<<a/2.0<<","<<b/2.0<<","<<0.0<<">"<<endl;
myfile5<<"}"<<endl;
myfile5<<"background { color red 0.5 green 0.5 blue 0.5 }"<<endl;
myfile5<<"light_source"<<endl;
myfile5<<"{"<<endl;
myfile5<<"0*x"<<endl;
myfile5<< "color red"<<endl;
myfile5<<"1.0  green 1.0  blue 1.0"<<endl;
myfile5<<"translate <-30, 30, -30>"<<endl;
myfile5<<"}"<<endl;
myfile5<<"light_source"<<endl;
myfile5<<"{"<<endl;
myfile5<<"0*x"<<endl;
myfile5<<"color red"<<endl;
myfile5<< "0.5  green 0.5  blue 0.5"<<endl;
myfile5<<"translate <30, 30, -30>"<<endl;
myfile5<<"}"<<endl;

for(int i=0;i<nx;i++)
{
x1=float(i)*dx;
x2=float(i+1)*dx;
for(int j=0;j<ny;j++)
{
```

```
y1=float(j)*dy;
y2=float(j+1)*dy;
```

Figure out blue to red colors again.

```
if(cont[i][j]>=hmin&&cont[i][j]<=(hmin+(hmax-hmin/3.0)))
{
red=0.0;
green=(cont[i][j]-hmin)/((hmax-hmin)/3.0);
blue=((hmin+(hmax-hmin)/3.0)-cont[i][j])/((hmax-hmin)/3.0);
}

if(cont[i][j]>=(hmin+(hmax-hmin)/3.0)&&cont[i][j]<=(hmin+2.0*(hmax-
hmin/3.0)))
{
red=(cont[i][j]-(hmin+(hmax-hmin)/3.0))/(hmin+(hmax-hmin)/3.0);
green=(cont[i][j]-(hmin+(hmax-hmin)/3.0))/(hmin+(hmax-hmin)/3.0);
green=green+((hmin+2.0*(hmax-hmin)/3.0)-cont[i][j])/(hmin+(hmax-
hmin)/3.0);
blue=0.0;
}

if(cont[i][j]>=(hmin+2.0*(hmax-hmin)/3.0)&&cont[i][j]<=hmax)
{
red=((hmax-cont[i][j])/(hmax-(2.0*(hmax-hmin))/3.0));
green=((hmax-cont[i][j])/(hmax-(2.0*(hmax-hmin))/3.0));
red=(cont[i][j]-(hmax-(2.0*(hmax-hmin))/3.0))/(hmax-((hmax-(2.0*(hmax-
hmin))/3.0)));
blue=0.0;
}
```

Lenses are purple here too.

```
if (cond[i][j]==0.0)
{
red=1.0;
blue=1.0;
green=0.0;
}

myfile5<<"plane{z,"<<0.0<<" pigment{color rgb<"<<red<<","<<green<<","<
<blue<<">}"<<endl;
myfile5<<"clipped_by {box {<"<<x1<<","<<y1<<","<<-0.001<<">,<"<<x2<<
","<<y2<<","<<0.001<<">}}"<<endl;


}
}
}
if(k==nsteps-1)
{
```

Write out the final arrays in case you want to use a different graphics program.

```
ofstream myfile3;
myfile3.open("final_arrays.res");
for(int i=0;i<ny;i++)
{
y1=float(i)*dy;
y2=float(i+1)*dy;
for(int j=0;j<nx;j++)
```

```
{
x1=float(j)*dx;
x2=float(j+1)*dx;

myfile3<<(x1+x2)/2.0<<" "<<(y1+y2)/2.0<<" "<<h[j][i]<<" "<<cont[j]
[i]<<endl;
}
}
}
```

Close the time loop.

```
}
```

And close the program.

```
return(0);
}
```

Results from this program for two different scenarios are shown in Figure 6.6a–f using input_contam_trial.txt and Figure 6.7a–f using input_contam_wells.txt as input_con-tam.txt.

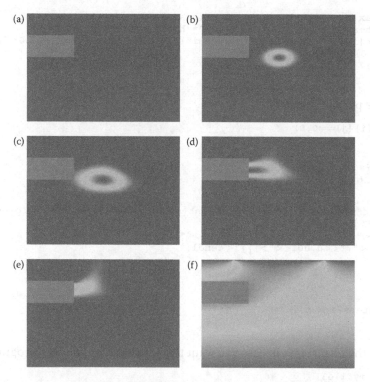

Figure 6.6. Sample transient-state contamination plume evolution (a–e) and potentiometric head (f) contour maps from blue (low) to red (high). Shading helps distinguish the building and fins in the grayscale printing and the electronic version is in color, where purple regions are inactive. Note that you can use smoothing programs for less choppy visuals or of course use a finer grid in the simulation.

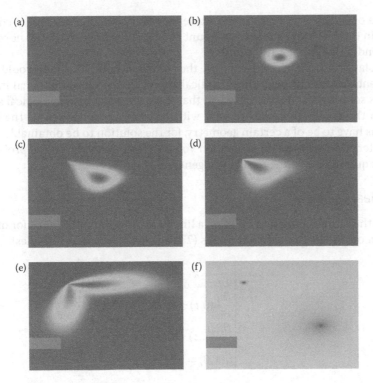

Figure 6.7. Transient-state contamination plume evolution (a–e) and potentiometric head (f) contour maps for a dipole well system (one pumping and one injection). Shading helps distinguish the contamination concentration in the grayscale printing and the electronic version is in color, where purple regions are inactive.

I strongly suggest looking at Freeze and Cherry [10] for ideas of various groundwater systems you might consider simulating to either validate or extend to a project.

6.6.3 More Math, More Insight, Less Versatility

Just as is true with the case for studying the mathematical structure of sound, there is an analytical approach for solving groundwater flow that works in certain cases and, when it does there can be a great wealth of mathematical insight afforded. Appendix B online shows solutions of analytical problems for rectangular and circular aquifers. The rectangular ones with impermeable boundaries have solutions of the form

$$h(x,y,t) = \sum_{j=0}^{\infty}\sum_{k=0}^{\infty} a_{lk}(t)\cos\left(\frac{j\pi x}{A}\right)\cos\left(\frac{k\pi y}{B}\right) \tag{6.30}$$

and analogous circular ones look like

$$h(r,\phi,t) = \sum_{m=0}^{\infty}\sum_{n=1}^{\infty} J_m\left(k_{mn}r\right)\{a_{mn}(t)\cos(m\phi) + b_{mn}(t)\sin(m\phi)\}. \tag{6.31}$$

The name of the game is then to find expressions for the time-dependent coefficients, and that in turn tells you about the amount and behavior of each spatial normal mode. See Appendix B online for full details.

Now, far be it from me to discourage the use of mathematics, but I would be remiss not to mention that although the analytical approach gives more physical insight, the method is somewhat more restrictive in that there are a group of geological structures for which the conductivity has to vary with space in certain ways, or the boundary conditions have to be of a certain geometry, for the solution to be obtainable. It is up to you to determine the method of solution you want to pursue, given the problem setup and the requirements of the results you generate.

6.7 The Heat Equation

Consider the 1D heat equation IVPB on a line $[0,L]$ describing the behavior of the temperature u on wire insulated on the ends (T is also used in other examples):

$$\begin{cases} \dfrac{\partial u}{\partial t} = k \dfrac{\partial^2 u}{\partial x^2} \\[2mm] \dfrac{\partial u}{\partial x}(0,t) = 0 \\[2mm] \dfrac{\partial u}{\partial x}(L,t) = 0 \\[2mm] u(x,0) = f(x) \end{cases} \tag{6.32}$$

$$k\left\{ \frac{u_{i+1}^n - 2u_i^n + u_{i-1}^n}{(\Delta x)^2} \right\} = S\left\{ \frac{u_i^{n+1} - u_i^n}{\Delta t} \right\} \tag{6.33}$$

Extending to higher dimensions, $k\nabla^2 u = \partial u / \partial t$ + boundary conditions + initial conditions and expressions for the Laplacian in common 3D coordinate systems is shown in Table 6.5.

An example of a 2D rectangular system n $[0,L]$ X $[0,H]$ is

$$\begin{cases} \dfrac{\partial u}{\partial t} = k \dfrac{\partial^2 u}{\partial x^2} \\[2mm] \dfrac{\partial u}{\partial y}(x,0,t) = 0 \\[2mm] \dfrac{\partial u}{\partial y}(x,H,t) = 0 \\[2mm] \dfrac{\partial u}{\partial x}(0,y,t) = 0 \\[2mm] \dfrac{\partial u}{\partial x}(L,y,t) = 0 \\[2mm] u(x,y,0) = f(x,y) \end{cases} \tag{6.34}$$

Which, when discretized becomes

$$k\left\{ \frac{u_{i+1,j}^n - 2u_{i,j}^n + u_{i-1,j}^n}{(\Delta x)^2} + \frac{u_{i,j+1}^n - 2u_{i,j}^n + u_{i,j-1}^n}{(\Delta y)^2} \right\} = S\left\{ \frac{u_{i,j}^{n+1} - u_{i-1,j}^n}{\Delta t} \right\}, \tag{6.35}$$

Table 6.5. Expressions for the Laplacian in Commonly Used Coordinate Systems

Coordinate System	Laplacian
Cartesian (x,y,z)	$\dfrac{\partial^2 u}{\partial x^2}+\dfrac{\partial^2 u}{\partial y^2}+\dfrac{\partial^2 u}{\partial z^2}$
Plane Polar (r,θ,z)	$\dfrac{1}{r}\dfrac{\partial}{\partial r}\left(r\dfrac{\partial u}{dr}\right)+\dfrac{1}{r^2}\dfrac{\partial^2 u}{\partial\theta^2}$
Spherical Polar (r,θ,ϕ)	$\dfrac{1}{r^2}\dfrac{\partial}{\partial r}\left(r^2\dfrac{\partial u}{dr}\right)+\dfrac{1}{r^2\sin\theta}\dfrac{\partial}{\partial\theta}\left(\sin\theta\dfrac{\partial u}{d\theta}\right)+\dfrac{1}{r^2}\dfrac{\partial^2 u}{\partial\phi^2}$

and here is what the Laplacian looks like elsewhere.

In Cartesian coordinates the problem is easily extended to 3D. Let's see what a basic program simulating transient-state 3D temperature fields looks like. Please see 3DHeat.cpp and a simpler 2D version RectTemp.cpp for relevant code.

First, call libraries.

```
#include<cstdlib>
#include<iostream>
#include<fstream>
#include<math.h>
```

Start coding for the program.

```
int main()
{

using namespace std;
```

Declare and in some cases assign values to variables.

```
const int nx=20,ny=20,nz=20,nstep=10000,nevery=100;
const float R=1.5,r=1.0,h=2.0,pi=3.141592;
const float a=20.0,b=20.0,c=20.0;
const float dt=0.001;

float x,y,z,T[nx][ny][nz],Tnew[nx][ny][nz];
float dx,dy,dz,updatex,updatey,updatez;
float Vod,Vid,xcenter,ycenter,zcenter;

dx=a/float(nx-1);
dy=b/float(ny-1);
dz=c/float(nz-1);
Vod=pi*(R*R)*h;
Vid=pi*(r*r)*h;
xcenter=nx/2;
ycenter=ny/2;
zcenter=nz/2;
```

Assign values to the initial temperature field.

```
for(int i=0;i<=nx-2;i++)
{
for(int j=0;j<=ny-2;j++)
{
for(int k=0;k<=nz-2;k++)
{
if ((x-xcenter)+(z-zcenter)<=Vid && (x-xcenter)+(z-zcenter)>=Vod);
{
if (y-ycenter<=h/2);
{
T[i][j][k]=100.0;
Tnew[i][j][k]=0.0;
}
}
}
}
}
```

I/O preliminaries:

```
ofstream myfile;
myfile.open("Heat_Transfer_Example.res");

for(int itime=1;itime<=nstep;itime++)
{
```

Boundary Conditions:

```
for(int i=0;i<=nx-1;i++)
{
for(int j=0;j<=ny-1;j++)
{
T[i][j][0]=0.0;
}
}

for(int i=0;i<=nx-1;i++)
{
for(int j=0;j<=ny-1;j++)
{
T[i][j][nz-1]=0.0;
}
}

for(int j=0;j<=ny-1;j++)
{
for(int k=0;k<=nz-1;k++)
{
T[0][j][k]=0.0;
}
}

for(int j=0;j<=ny-1;j++)
{
for(int k=0;k<=nz-1;k++)
{
```

```
T[nx-1][j][k]=0.0;
}
}

for(int i=0;i<=nx-1;i++)
{
for(int k=0;k<=nz-1;k++)
{
T[i][0][k]=0.0;
}
}

for(int i=0;i<=nx-1;i++)
{
for(int k=0;k<=nz-1;k++)
{
T[i][ny-1][k]=0.0;
}
}
```

Temperature Update:

```
for (int i=1;i<=nx-2;i++)
{
for (int j=1;j<=ny-2;j++)
{
for (int k=1;k<=nz-2;k++)
{
if ((x-xcenter)+(z-zcenter)<=Vid && (x-xcenter)+(z-zcenter)>=Vod);
{
if (y-ycenter<=h/2);
{
updatex=(T[i+1][j][k]-2.0*T[i][j][k]+T[i-1][j][k])/pow(dx,2);
updatey=(T[i][j+1][k]-2.0*T[i][j][k]+T[i][j-1][k])/pow(dy,2);
updatez=(T[i][j][k+1]-2.0*T[i][j][k]+T[i][j][k-1])/pow(dz,2);
Tnew[i][j][k]=T[i][j][k]+(updatex+updatey+updatez)*dt;
}
}
}
}
}
```

Transfer temporary array into permanent array.

```
for(int i=1;i<=nx-1;i++)
{
for(int j=1;j<=ny-1;j++)
{
for(int k=1;k<=nz-1;k++)
{
if ((x-xcenter)+(z-zcenter)<=Vid && (x-xcenter)+(z-zcenter)>=Vod);
{
if (y-ycenter<=h/2);
{
T[i][j][k]=Tnew[i][j][k];
}
}
}
}
}
```

Write out the temperature array as a surface every nevery steps.

```
if(float(itime/nevery)==float(itime)/float(nevery))
{
myfile<<float(itime)*dt<<endl;
for(int i=1;i<=nx-2;i++)
{
for(int j=1;j<=ny-2;j++)
{
for(int k=1;k<=nz-2;k++)
{
myfile<<float(i-1)*dx<<"        "<<float(j-1)*dy<<"        "<<float(k-1)*dz<<"
"<<T[i][j][k]<<endl;
}
}
}
}
}
```

Wrap it all up.

```
}
```

I suggest your paging through Haberman [11] who gives a wide range of example heat equation problems that you could use for either validation or extension to a project.

PROBLEMS

NOTE: There are several programs provided for you which you can use as starting points for homework. Please feel free to take them for a spin!

6.1. Identify something that is classified in a way (solid, fluid, liquid, gas) that doesn't fit with your intuition and discuss.

6.2. There is a certain type of faulting in geology that can be demonstrated with the card deck example discussed earlier. Research the type of faulting and discuss.

6.3. What changes would need to be made to Equation 6.1 if we were studying a compressible fluid? Is any fluid truly incompressible? Discuss, including your thoughts on when it is OK to call something "incompressible".

6.4. Show that the change in momentum density is equal to force density for the term in the left hand side of Equation 6.1.

6.5. Verify that the units match for the middle and last terms on the right hand side of Equation 6.1.

6.6. Show that Equations 6.1 and 6.2 become Equations 6.3 through 6.5 and 6.6.

6.7. Explain why the zero velocity boundary conditions match the pressure boundary conditions in Table 6.2.

6.8. Add comments to the variable declaration part of the program in Section 6.3.

6.9. Why does the no slip boundary condition for the ceiling of the computational cell make life easier?

6.10. Discretize Equations 6.3 through 6.5 to obtain 6.7 through 6.9.

6.11. Why do we need Equation 6.6? That is, why must the wind velocity field lack divergence? What would it mean if the velocity field did diverge?

6.12. Explain and justify the initial conditions mentioned in Section 6.2.2.

6.13. Describe the boundary conditions in Table 6.1 with words.

6.14. Show how to get Equations 6.17–6.19 from Equation 6.16.

6.15. Discretize Equations 6.10 to get Equation 6.11.

6.16. In all the simulations we did, we did not build in vortices. Considering the snowfall simulation, build in some vortices with either boundary conditions or sources and run the simulation.

6.17. In reality the effect of the snowflakes on the air velocity can be important. Is it acceptable to neglect it here? Are there any pats of the simulation for which that interaction would matter most?

6.18. As an advanced topic, incorporate interactions with snowflakes on the velocity field and run the simulation. Discuss any difference in results.

6.19. Modify the snowstorm code to include an intrusive shape on the ground that the snow has to accumulate on. Run the program and compare to results without the shape there.

6.20. Take a look at Freeze and Cherry [10] and get ideas on various groundwater systems you can model. Select one and run the contamination transport simulations for various initial and boundary conditions.

6.21. The groundwater system modeled in the contaminant transport program is really only an abstraction. Select one of the two scenarios provided (where figure sequences are shown) and draw the corresponding 3D structure being simulated.

6.22. All the programs here give different results for different grid resolutions. Modify and run the code for either of the two figure sequences shown for 5×5, 20×20 (I did this one), 50×50, 100×100 and as large as you can practically go. Discuss the differences, and what resolution you feel the grid has to have in order for your results to be "good."

6.23. Expanding on Table 6.4, there are geometries other than rectangular and circular in which it is easy ("easy") to write down, discretize, and solve the flow through porous medium equation. Do some research and briefly list and discuss those geometries, along with types of aquifers in reality that might apply.

6.24. Expand the derivatives in 6.26 using the product rule and write the result. Discretize and compare to the code.

6.25. For the groundwater contamination simulation, identify and discuss a few simplifying assumptions we make.

6.26. Write Equation 6.27 as an update expression.

6.27. After Equation 6.33 we also had to calculate the velocity of the system from the head values. Write the analytical expression and show how we got to the expression used in the program.

6.28. Extend the 2D contaminant transport model to either 3D or another coordinate system of your choice.

6.29. Write Equation 6.35 as an update expression as similar in form to Equations 6.7–6.9. Outline a possible algorithm for an update loop.

6.30. I strongly suggest looking at Haberman, Reference 11, for ideas of heat equation problems you might consider simulating to either validate or extend to a project. Look into what you have identified and discuss as appropriate with a class instructor or colleagues.

References

1. H.M. Schey, *Div Grad Curl and all that: An Informal Text on Vector Calculus*. 4th Edition, W. W. Norton & Company, New York, NY, 2005.

2. R.B. Bird, W.E. Stewart, E.N. Lightfoot, *Transport Phenomena*, 2nd Edition, Wiley and Sons, New York, NY, 2002.

3. F.M. White, *Viscous Fluid Flow*, 3rd Edition, McGraw-Hill, Singapore, 1974.

4. G.K. Batchelor, *An Introduction to Fluid Dynamics*, Cambridge University Press, Cambridge, United Kingdom, 1967.

5. L.D. Landau, E.M. Lifshitz, *Fluid Mechanics. Course of Theoretical Physics*, Second edition. Pergamon Press, Ontario, Canada, 1987.

6. S. Nazarenko, *Fluid Dynamics Via Examples and Solutions*, CRC Press (Taylor & Francis group), Boca Raton, FL, 2014.

7. E. Maldonado, M.W. Roth, Direct two-phase numerical simulation of snowdrift remediation using three-dimensional deflection fins, *Journal of Applied Fluid Mechanics*, 5(3), 71–78, 2012.

8. M. Matyka, Solution to two-dimensional incompressible Navier-Stokes Equations with SIMPLE, SIMPLER, and Vorticity-Stream, Function Approaches Driven-Lid Cavity Problem: Solution and Visualization, Preprint.

9. J.R. Claeyssen, R.B. Platte, E. Bravo, Simulation in primitive variables of incompressible flow with pressure neumann condition, *International Journal for Numercial Methods in Fluids*, 30, 1009–1026, 1999.

10. G.A. Freeze, J.A. Cherry, *Groundwater*, Prentice Hall, Upper Saddle River, NJ, 1979.

11. R. Haberman. *Applied Partial Differential Equations with Fourier Series and Boundary Value Problems*, Fifth edition, Pearson, Upper Saddle River, NJ, 2013.

SECTION III

Beyond Everyday Phenomena

CHAPTER 7

///

One of the Most Versatile Simulation Tools Around

7.1 Introduction

One of the most versatile simulations around is the Material Point Method (MPM) [1] with an algorithm developed, evaluated, and utilized by Z. Chen and others [2]. In MPM we will see that we model system components as being composed of groups of particles (material points) and we then advance it in time. MPM is advantageous because it doesn't require a complex algorithm to simulate collisions or material failure. Instead, MPM maps the mass and momentum of groups of particles in the simulation to a background grid and utilizes the conservation equation of mass as well as conservation of momentum to ultimately advance the system through time. The use of the background grid also avoids entanglement associated with other contact-based algorithms. Because MPM can accommodate a large number of particles, it can also provide realistic visuals and was utilized by Disney, for example in simulating snow in the feature cartoon *Frozen*.

7.2 Theory Behind the Material Point Method

7.2.1 The MPM Algorithm

We have to start somewhere, and so in any dynamical simulation the initial position and velocities of the particles must be defined. So, we start with a collection of particles having initial positions $\{\vec{x}_p^0\}$ and velocities $\{\vec{v}_p^0\}$. Moreover, we need to specify any edge, surface, or interior boundary conditions in effect. The initial conditions for the simulation should be chosen so as to reflect the important physics of the actual system as much as possible. Next, the background grid is defined, where computational space is separated into a defined number of grid cells in which nodes are placed in each corner of each cell. Because there is important physics (and calculations) going on at the nodes, we need to distribute the particle's mass and momentum to the nodes in a weighted fashion so that nodes closer to the particle get a greater fraction. At each time t then, the mass M_p of each particle p is mapped to each node i of the grid cells based on the shape function $N_i(\vec{\mathbf{x}}_p^t)$, resulting in the mass m_i^t of node i at time t:

$$m_i^t = \sum_{p=1}^{Np} M_p N_i(\vec{\mathbf{x}}_p^t). \tag{7.1}$$

The same is done for the momentum $(Mv)_p^t$ of the particles:

$$(m\mathbf{v})_i^t = \sum_{p=1}^{Np} (M\mathbf{v})_p^t N_i(\mathbf{x}_p^t) \tag{7.2}$$

The shape functions $N_i(\vec{\mathbf{x}}_p^t)$ are responsible for weighting the particle's mass and momentum so that nodes closer to the particle get the lion's share of its physical properties. To calculate shape functions, the coordinates of the cell containing a particle p are normalized so that they take on extreme values of 0 or 1 at vertices of the box. So, if a particular box has a vertex at (x_j, y_j, z_j) and another vertex at $(x_j + \Delta x, y_j + \Delta y, z_j + \Delta z)$ across its diagonal, the normalized coordinates are

Table 7.1. Expressions for the Eight Shape Functions Used to Map Particle Properties to Each of the Eight Vertices of the Grid Cube It Resides In

$$\zeta = \frac{(x_p^t - x_j)}{\Delta x}$$

$$\eta = \frac{(y_p^t - y_j)}{\Delta y} \tag{7.3}$$

$$\chi = \frac{(z_p^t - z_j)}{\Delta z}$$

Node Coordinates	Shape Function
(0,0,0)	$(1-\xi)(1-\eta)(1-\chi)$
(1,0,0)	$\xi(1-\eta)(1-\chi)$
(1,1,0)	$\xi\eta(1-\chi)$
(0,1,0)	$(1-\xi)\eta(1-\chi)$
(0,0,1)	$(1-\xi)(1-\eta)\chi$
(1,0,1)	$\xi(1-\eta)\chi$
(1,1,1)	$\xi\eta\chi$
(0,1,1)	$(1-\xi)\eta\chi$

Source: C. Massina, M.W. Roth, P.A. Gray, Material Point Method Computer Simulations of Grazing Impacts on the Martian Surface, *14th Annual Sigma Xi Student Research Conference*, April 18, 2007 [26].

and the eight shape functions used in the particle-to-node mappings are calculated from the normalized coordinates, as shown in Table 7.1, so that they are equal to unity for a vertex (node) where the particle is at and is zero for a given node if the particle is at any other node. Once the mappings in Equations 7.1 and 7.2 are completed then the forces at each node must be calculated so we will know how the node momentum is going to behave. There are many ways to categorize forces and in this case we can speak of internal and external force. The external force (if implemented) is then calculated at each node as

$$\vec{f}_i^{ext} = \vec{c}_i^t + M_p \vec{g} m_i^t. \tag{7.4}$$

Here, the first term incorporates any applied force at a particular time you may choose to apply and the second term is for force due to gravity, which is just the weight of the node. But wait, it gets better.

The internal force (f_i^t) arises from elastic deformation of the object and is calculated as

$$(\vec{f}_i^t)^{\text{int}} = -\sum_{p=1}^{Np} \mathbf{G}_i(\mathbf{x}_p^t) \cdot \mathbf{s}_p^t \frac{M_p}{\rho_p^t}. \tag{7.5}$$

Here $\mathbf{G}_i(\mathbf{x}_p^t)$ is the gradient of the shape functions, $\mathbf{s}_p^t$ is the stress tensor and ρ_p^t is density, all of which are evaluated for particle p at a given time t in the simulation. In the next section I will go through elasticity theory and stress–strain tensor relationships but since we are covering MPM theory, let's finish that and then dive into the stress–strain stuff.

After the forces are calculated the nodal accelerations are obtained using $a_i^t = f_i^t / m_i^t$ where the total force is given by $f_i^t = (f_i^t)^{\text{int}} + (f_i^t)^{\text{ext}}$. A force is the same as a time rate of change of momentum and so the node momenta are then updated in time

$$(mv)_i^{t+\Delta t} = (mv)_i^t + f_i^t \Delta t \tag{7.6}$$

Now, we know how the nodes are behaving but we have to map the kinematic variables of the nodes back to the particles in the system, so we can update them and track their behavior. After all, it's the collection of particles that is reflecting the actual system symmetry and boundary conditions—not the nodes per se.

$$\vec{a}_p^t = \sum_{i=1}^{Nn} \frac{\vec{f}_i^t}{m_i^t} N_i(\vec{x}_p^t) \tag{7.7}$$

$$\vec{v}_p^t = \sum_{i=1}^{Nn} \frac{(m\vec{v})_i^{t+\Delta t}}{m_i^t} N_i(\vec{x}_p^t) \tag{7.8}$$

$$\vec{x}_p^{t+\Delta t} = \vec{x}_p^t + \vec{v}_p^{t+\Delta t} \Delta t \tag{7.9}$$

The new particle kinematic variables in Equations 7.7 through 7.9 are mapped back onto the nodes using Equations 7.1 and 7.2 to obtain updated nodal velocities.

7.2.2 Elasticity Theory and Stress–Strain Relationships

Because the material being modeled may deform throughout the course of the simulation, there are internal forces generated that have to be calculated and incorporated in the algorithm discussed in the previous section. If we are working in 1D we don't have to go fishing very long for what we need to describe elasticity with: Hooke's Law. In higher dimensions (2D, 3D) the story becomes more complicated. Imagine a cube of gelatin

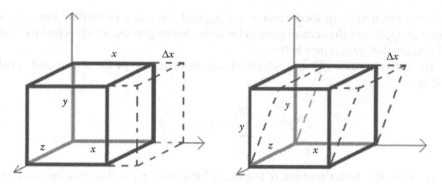

Figure 7.1. Different types of material deformation. Normal compression or extension (left) results in a volume change for the material and is related to diagonal elements of the strain tensor; shear strain (right) conserves volume and relates to off – diagonal elements. (Adapted from http://folk.ntnu.no/stoylen/strainrate/Basic_concepts.html.) [3].

dessert. If I hold it and squeeze it from left to right it bulges from top to bottom as well as front to back. This means we have responses (strain) that are *not in the same direction as the driving force* (stress) and, whenever we have such phenomena in physics we need tensor algebra calculus to describe the mechanics going on. Strain is generally broken down into two types as illustrated in Figure 7.1: [3] compressive (non-volume-conserving) and shear (volume-conserving). So here comes some higher math, but it's great preparation if you are thinking about going into theoretical physics or an engineering related field.

In calculating internal forces due to material deformation in Equation 7.5, the *rate of strain tensor* $\ddot{e}$ is calculated from the velocity gradients $e_{ij} = \frac{1}{2}(\partial_j x_i' + \partial_i x_j')$. Here the primed variables represent velocity components and, explicitly in 3D we have

$$\ddot{e} = \begin{bmatrix} \dfrac{\partial u}{\partial x} & \dfrac{1}{2}\left(\dfrac{\partial u}{\partial y}+\dfrac{\partial v}{\partial x}\right) & \dfrac{1}{2}\left(\dfrac{\partial u}{\partial z}+\dfrac{\partial w}{\partial x}\right) \\[2ex] \dfrac{1}{2}\left(\dfrac{\partial v}{\partial y}+\dfrac{\partial v}{\partial x}\right) & \dfrac{\partial v}{\partial y} & \dfrac{1}{2}\left(\dfrac{\partial v}{\partial z}+\dfrac{\partial w}{\partial y}\right) \\[2ex] \dfrac{1}{2}\left(\dfrac{\partial u}{\partial z}+\dfrac{\partial w}{\partial x}\right) & \dfrac{1}{2}\left(\dfrac{\partial u}{\partial y}+\dfrac{\partial u}{\partial y}\right) & \dfrac{\partial w}{\partial z} \end{bmatrix}. \tag{7.10}$$

Subsequently, the 3D elastic stress–strain relationship is utilized to calculate the updated stress tensor $\ddot{s}^{t+\Delta t} = \ddot{s}^t + \Delta\ddot{s}$, where $\Delta s_{ij} = \sigma_{ij} = \sum_{ijkl} C_{ijkl}\varepsilon_{kl}$ and $\ddot{C}$ is the *elastic tensor* discussed in the following section. The stress tensor is visualized in Figure 7.2 [4] and I strongly encourage you to work through the mathematical relationships introduced in this section. There are several special cases of the elastic tensor as well as strain tensor which will be discussed. Compressive stress creates pressure differences, and pressure may be calculated as the sum of the diagonal elements (the trace) of the stress tensor: $P = -\frac{1}{3}Tr\,\ddot{s} = -\frac{1}{3}(s_{11} + s_{22} + s_{33})$.

There is a lot of beauty in the symmetry of tensor mathematics, and when studying material properties the form of the elastic tensor is related to the symmetry properties

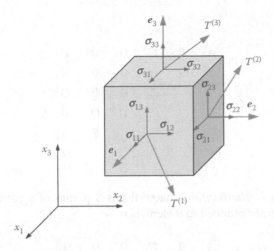

Figure 7.2. Visualization of material stress. (Adapted from https://en.wikipedia.org/wiki/Cauchy_stress_tensor). [4] The e_i are unit outward normals to the cube of material and the elements of the stress tensor σ_{ij} are the stresses in direction j acting on the face of the cube spanned by the tangent vectors $T^{(i)}$ normal to the unit vectors e_i with coordinate $x_i =$ const.

of the material being modeled. If you are not familiar with tensors or matrix mathematics, I suggest that you grab a good tensor calculus or tensor algebra and linear algebra reference.

In the most general case, the material we are dealing with does not exhibit any symmetry at all. In that case, the elastic tensor requires 21 unique elements to describe the elastic response of the material to the six unique stress elements. The tensor is necessarily symmetric and so the lower diagonal block is omitted but implied:

$$\overset{\leftrightarrow}{C} = \begin{vmatrix} A & B & C & D & E & F \\ & G & H & I & J & K \\ & & L & M & N & O \\ & & & P & Q & R \\ & & & & S & T \\ & & & & & U \end{vmatrix}. \tag{7.11}$$

Now suppose the material has one plane of symmetry. It is termed *monoclinic* and the number of elements of the elastic tensor needed are now only 13. This arises from the fact that the material properties do not change upon a mirror reflection about, say the (x,y) plane so the stresses are identical for the strains $\overset{\leftrightarrow}{\varepsilon}$ and $\overset{\leftrightarrow}{\varepsilon}' = \overset{\leftrightarrow}{L}^T \overset{\leftrightarrow}{\varepsilon} L$, where

$$\overset{\leftrightarrow}{L} = \begin{vmatrix} 1 & 0 & 0 \\ 0 & 1 & 0 \\ 0 & 0 & -1 \end{vmatrix}.$$

The result is:

$$\ddot{C} = \begin{bmatrix} A & B & C & 0 & 0 & F \\ & G & H & 0 & 0 & K \\ & & L & 0 & 0 & O \\ & & & P & Q & 0 \\ & & & & S & 0 \\ & & & & & U \end{bmatrix}. \tag{7.12}$$

An *orthotropic (orthorhombic)* material has 3 planes of symmetry and then the elastic tensor is further reduced to 9 elements:

$$\ddot{C} = \begin{bmatrix} A & B & C & 0 & 0 & 0 \\ & G & H & 0 & 0 & 0 \\ & & L & 0 & 0 & 0 \\ & & & P & 0 & 0 \\ & & & & S & 0 \\ & & & & & U \end{bmatrix}. \tag{7.13}$$

Now, further assume that the material tetragonal, having 3 planes of symmetry and one is isotropic. Then only 5 independent constants are needed

$$\ddot{C} = \begin{bmatrix} A & B & C & 0 & 0 & 0 \\ & A & C & 0 & 0 & 0 \\ & & L & 0 & 0 & 0 \\ & & & P & 0 & 0 \\ & & & & P & 0 \\ & & & & & (A-B) \end{bmatrix}. \tag{7.14}$$

Now suppose that the material is entirely *isotropic*. Then the number of constants reduces to two:

$$\ddot{C} = \begin{bmatrix} A & B & B & 0 & 0 & 0 \\ & A & B & 0 & 0 & 0 \\ & & A & 0 & 0 & 0 \\ & & & (A-B)/2 & 0 & 0 \\ & & & & (A-B)/2 & 0 \\ & & & & & (A-B)/2 \end{bmatrix}. \tag{7.15}$$

In our simulations we assume isotropic media and so the elastic tensor takes for form equation 7.15 and the stress–strain relationship can be expressed explicitly in terms of relevant material properties:

$$
\begin{bmatrix} \sigma_x \\ \sigma_y \\ \sigma_z \\ \sigma_{xy} \\ \sigma_{yz} \\ \sigma_{zx} \end{bmatrix} = \frac{E}{(1+v)(1-2\nu)} \begin{bmatrix} 1-v & v & v & 0 & 0 & 0 \\ v & 1-v & v & 0 & 0 & 0 \\ v & v & 1-v & 0 & 0 & 0 \\ 0 & 0 & 0 & \frac{1-2v}{2} & 0 & 0 \\ 0 & 0 & 0 & 0 & \frac{1-2v}{2} & 0 \\ 0 & 0 & 0 & 0 & 0 & \frac{1-2v}{2} \end{bmatrix} \begin{bmatrix} \varepsilon_x \\ \varepsilon_y \\ \varepsilon_z \\ \varepsilon_{xy} \\ \varepsilon_{yz} \\ \varepsilon_{xz} \end{bmatrix}.
\tag{7.16}
$$

Here E is the elastic modulus and v is Poisson's ratio. In the case where a fluid is being modeled, sheer forces are completely absent and the elastic tensor is written as

$$
\overset{\leftrightarrow}{C} = \begin{bmatrix} A & 0 & 0 & 0 & 0 & 0 \\ & A & 0 & 0 & 0 & 0 \\ & & A & 0 & 0 & 0 \\ & & & 0 & 0 & 0 \\ & & & & 0 & 0 \\ & & & & & 0 \end{bmatrix}.
\tag{7.17}
$$

There are also times when the geometry of structures being considered has a coordinate whose behavior can be ignored. This is usually the case, for example, when the dimensions of an object on one dimension are much larger than in two others. The results is plane strain, where the strain tensor takes the form

$$
\overset{\leftrightarrow}{\varepsilon} = \begin{bmatrix} A & B & 0 \\ C & D & 0 \\ 0 & 0 & 0 \end{bmatrix},
\tag{7.18}
$$

And for certain symmetries there may be *antiplane strain* where

$$
\overset{\leftrightarrow}{\varepsilon} = \begin{bmatrix} 0 & 0 & c \\ 0 & 0 & f \\ g & h & 0 \end{bmatrix}.
\tag{7.19}
$$

For antiplane strain the response of the material is purely shear.

7.3 A Material Point Method Program

7.3.1 The Code

Make sure the needed libraries are called.

```
#include<cstdlib>
#include<iostream>
#include<fstream>
#include<math.h>

using namespace std;
```

Shape Function Calculations.

```
float chooseshape(int a,float b,float c,float d)

{

float shape;

if (a==1)
{
shape=(1.0-b)*(1.0-c)*(1.0-d);
}
if (a==2)
{
shape=(b)*(1.0-c)*(1.0-d);
}
if (a==3)
{
shape=(c)*(b)*(1.0-d);
}
if (a==4)
{
shape=(1.0-b)*(c)*(1.0-d);
}
if (a==5)
{
shape=(1.0-b)*(1.0-c)*(d);
}
if (a==6)
{
shape=(b)*(1.0-c)*(d);
}
if (a==7)
{
shape=(c)*(b)*(d);
}
if (a==8)
{
shape=(1.0-b)*(c)*(d);
}
return (shape);
}
```

Shape function x gradient.

```
float choosegradx(int a,float b,float c,float d,float e,float f,float g)
{
float gradx;
if (a==1){
gradx=-(1.0-c)*(1.0-d)/e;
}
if (a==2){
gradx=(1.0-c)*(1.0-d)/e;
}
if (a==3){
gradx=(c)*(1.0-d)/e;
}
if (a==4){
gradx=-(c)*(1.0-d)/e;
}
if (a==5){
gradx=-(1.0-c)*(d)/e;
}
if (a==6){
  gradx=(1.0-c)*(d)/e;
}
if (a==7){
gradx=(c)*(d)/e;
}
if (a==8){
gradx=-(c)*(d)/e;
}
return (gradx);
}
```

Shape function y gradient.

```
float choosegrady(int a,float b,float c,float d,float e,float f,float g)
{
float grady;
if (a==1){
grady=-(1.0-b)*(1.0-d)/f;
}
if (a==2){
grady=-(b)*(1.0-d)/f;
}
if (a==3){
grady=(b)*(1.0-d)/f;
}
if (a==4) {
grady=(1.0-b)*(1.0-d)/f;
}
if (a==5){
grady=-(1.0-b)*(d)/f;
}
if (a==6){
grady=-(b)*(d)/f;
}
if (a==7){
grady=(b)*(d)/f;
}
if (a==8){
grady=(1.0-b)*(d)/f;
}
return (grady);
```

```
}
```

Shape function z gradient.

```
float choosegradz(int a,float b,float c,float d,float e,float f,float g)
{
float gradz;
if (a==1) {
gradz=-(1.0-b)*(1.0-c)/g;
}
if (a==2) {
gradz=-(b)*(1.0-c)/g;
}
if (a==3) {
gradz=-(b)*(c)/g;
}
if (a==4) {
gradz=-(1.0-b)*(c)/g;
}
if (a==5) {
gradz=(1.0-b)*(1.0-c)/g;
}
if (a==6) {
gradz=(b)*(1.0-c)/g;
}
if (a==7) {
gradz=(b)*(c)/g;
}
if (a==8) {
gradz=(1.0-b)*(c)/g;
}
return (gradz);
}
```

Now the main program starts.

```
int main()
{

using namespace std;
```

Declare variables. This particular code was translated from FORTRAN to C++. Can you explain my choice of array dimensioning based on that? Because the program is larger, there is a variable list in Section 7.3.2.

```
const int npmax=5000,nstep=40000,nevery=2000;
const int nodesx=50,nodesy=50,nodesz=50;
const float part_mass=1.0,part_dens=1.0;
const float agrid=10.0,bgrid=10.0,cgrid=10.0;
const float delta_t=0.00005;
const float sphere_size=0.15,cyl_rad=0.1;
const int nvel=1;
const float nu=0.3;
const float centx1=4.25,centy1=4.25,centz1=4.25;
const float centx2=5.75,centy2=5.75,centz2=5.75;
const float rsph1=1.0,rsph2=1.0;
char fileanim[10];
int ii,ifr,np;
int inode,jnode,knode;
```

```
int elementnum[npmax+1];
int surfacetag[3][npmax+1];
int nsph1,nsph2;
float mi[nodesx+1][nodesy+1][nodesz+1];
float vxi[nodesx+1][nodesy+1][nodesz+1],vyi[nodesx+1][nodesy+1]
[nodesz+1];
float vzi[nodesx+1][nodesy+1][nodesz+1],pxi[nodesx+1][nodesy+1]
[nodesz+1];
float pyi[nodesx+1][nodesy+1][nodesz+1],pzi[nodesx+1][nodesy+1]
[nodesz+1];
float vxp[npmax+1],vyp[npmax+1],vzp[npmax+1];
float xp[npmax+1],yp[npmax+1],zp[npmax+1];
float lp[4][4][npmax+1];
float
fxi[nodesx+1][nodesy+1][nodesz+1],fyi[nodesx+1][nodesy+1]
[nodesz+1],fzi[nodesx+1][nodesy+1][nodesz+1];
float axp[npmax+1],ayp[npmax+1],azp[npmax+1];
float vbxp[npmax+1],vbyp[npmax+1],vbzp[npmax+1];
float evec[7],ecum[7][npmax+1];
float shape_funct,left,right,back,front,bottom,top;
float xi,eta,chi;
float s[4][4][npmax+1];
float e[4][4][npmax+1],cstiff[7][7][npmax+1],cfactor;
float dmax;
float grad_x,grad_y,grad_z;
float xtry,ytry,ztry,rtest;
float deltx,delty,deltz;
float press[npmax+1];
float red,green,blue,pmax,pmin;
float xc1,xc2,yc1,yc2,zc1,zc2;
float rmin1,rmax1,rmin2,rmax2;
float radius;
float vxc1,vxc2,vyc1,vyc2,vzc1,vzc2,vdotr;
float vmax,vel;
float ekl,ekr,ul,ur,duml[7][npmax+1],dumr[7][npmax+1];
float centx,centy,centz;
float dotx,doty,dotz,econst;
float dummy;

//Manage I/O...energy files
ofstream energies;
energies.open("energy.res");
```

Initialize constants, variables, and all arrays used in the simulation.

```
deltx=agrid/float(nodesx-1);
delty=bgrid/float(nodesy-1);
deltz=cgrid/float(nodesz-1);
red=0.0;
green=0.0;
blue=0.0;
pmax=0.0;
pmin=0.0;
rmin1=0.0;
rmax1=0.0;
rmin2=0.0;
rmax2=0.0;
nsph1=0;
nsph2=0;
xc1=0.0;
```

```
xc2=0.0;
yc1=0.0;
yc2=0.0;
zc1=0.0;
zc2=0.0;
radius=0.0;
vxc1=0.0;
vxc2=0.0;
vyc1=0.0;
vyc2=0.0;
vzc1=0.0;
vzc2=0.0;
vdotr=0.0;
vel=0.0;
ekl=0.0;
ekr=0.0;
ul=0.0;
ur=0.0;

for(int i=0;i<=6;i++)
{
evec[i]=0.0;
for(int j=0;j<=npmax;j++)
{
ecum[i][j]=0.0;
duml[i][j]=0.0;
dumr[i][j]=0.0;
}
}

for(int i=0;i<=3;i++)
{
for(int j=0;j<=3;j++)
{
for(int k=0;k<=npmax;k++)
{
lp[i][j][k]=0.0;
s[i][j][k]=0.0;
e[i][j][k]=0.0;
}
}
}

for(int i=0;i<=nodesx;i++)
{
for(int j=0;j<=nodesy;j++)
{
for(int k=0;k<=nodesz;k++)
{
vxi[i][j][k]=0.0;
vyi[i][j][k]=0.0;
vzi[i][j][k]=0.0;
pxi[i][j][k]=0.0;
pyi[i][j][k]=0.0;
pzi[i][j][k]=0.0;
fxi[i][j][k]=0.0;
fyi[i][j][k]=0.0;
fzi[i][j][k]=0.0;
mi[i][j][k]=0.0;
}
```

```
}
}

for(int j=0;j<=npmax;j++)
{
vxp[j]=0.0;
vyp[j]=0.0;
vzp[j]=0.0;
xp[j]=0.0;
yp[j]=0.0;
zp[j]=0.0;
axp[j]=0.0;
ayp[j]=0.0;
azp[j]=0.0;
vbxp[j]=0.0;
vbyp[j]=0.0;
vbzp[j]=0.0;
press[j]=0.0;
elementnum[j]=0;
for(int k=0;k<=2;k++)
{
surfacetag[k][j]=0;
}
}
ifr=0;
```

Create initial conditions—specific to each type of simulation.

Create sphere initial configurations.

```
np=0;
for(int i=1;i<=npmax;i++)
{
elementnum[i]=0;
}
```

Set up one sphere. (Note: you can use random numbers and have things look smoother.)

```
for(int i=0;i<=nodesx-1;i++)
{
for(int j=0;j<=nodesy-1;j++)
{
for(int k=0;k<=nodesz-1;k++)
{
centx=float(i)*deltx;
centy=float(j)*delty;
centz=float(k)*deltz;
for(int ix=1;ix<=1;ix++)
{
for(int iy=1;iy<=1;iy++)
{
for(int iz=1;iz<=1;iz++)
{
xtry=centx+0.5*deltx*float(ix);
ytry=centy+0.5*delty*float(iy);
ztry=centz+0.5*deltz*float(iz);
rtest=sqrt(pow(xtry-centx1,2)+pow(ytry-centy1,2)+pow(ztry-centz1,2));
if(rtest<rsph1)
{
np=np+1;
```

```
elementnum[np]=1;
if(np>npmax)
{
cout<<"Warning: maximum nunber of particles exceeded"<<endl;
break; goto line9975;
}
xp[np]=xtry;
yp[np]=ytry;
zp[np]=ztry;
vxp[np]=0.5;
vyp[np]=0.5;
vzp[np]=0.5;
```

Tag the particles near the sphere's surface for visualization purposes. This can also apply to a slice through the center of the sphere(s) or any other subset of particles you want to visualize. We will find extrema and scale colors for visualization later.

```
if(abs(rtest-rsph1)<=0.2)
{
surfacetag[1][np]=1;
}
}
}
}
}
}
}
}
```

Now set up another sphere:

```
for(int i=0;i<=nodesx-1;i++)
{
for(int j=0;j<=nodesy-1;j++)
{
for(int k=0;k<=nodesz-1;k++)
{
centx=float(i)*deltx;
centy=float(j)*delty;
centz=float(k)*deltz;
for(int ix=1;ix<=1;ix++)
{
for(int iy=1;iy<=1;iy++)
{
for(int iz=1;iz<=1;iz++)
{
xtry=centx+0.5*deltx*float(ix);
ytry=centy+0.5*delty*float(iy);
ztry=centz+0.5*deltz*float(iz);
rtest=sqrt(pow(xtry-centx2,2)+pow(ytry-centy2,2)+pow(ztry-centz2,2));
if(rtest<rsph2) {
np=np+1;
elementnum[np]=2;
if(np>npmax)
{
cout<<"Warning: maximum nunber of particles exceeded"<<endl;
```

```
break; goto line9975;
}
xp[np]=xtry;
yp[np]=ytry;
zp[np]=ztry;
vxp[np]=-0.5;
vyp[np]=-0.5;
vzp[np]=-0.5;
```

Tag the particles near the sphere's surface for visualization purposes.

```
if(abs(rtest-rsph2)<=0.2)
{
surfacetag[2][np]=1;
}
}
}
}
}
}
}
}
```

Create elasticity tensor. We may define different stiffnesses for different materials. Set up stiffness matrix C (3D spring constant).

```
for(int i=1;i<=np;i++)
{
if(elementnum[i]==1)
{
econst=100.0;
}
if(elementnum[i]==2)
{
econst=100.0;
}

cfactor=econst/((1.0+nu)*(1.0-2.*nu));
cstiff[1][1][i]=cfactor*(1.0-nu);
cstiff[1][2][i]=cfactor*nu;
cstiff[1][3][i]=cfactor*nu;
cstiff[1][4][i]=0.0;
cstiff[1][5][i]=0.0;
cstiff[1][6][i]=0.0;
cstiff[2][1][i]=cfactor*nu;
cstiff[2][2][i]=cfactor*(1.0-nu);
cstiff[2][3][i]=cfactor*nu;
cstiff[2][4][i]=0.0;
cstiff[2][5][i]=0.0;
cstiff[2][6][i]=0.0;
cstiff[3][1][i]=cfactor*nu;
cstiff[3][2][i]=cfactor*nu;
cstiff[3][3][i]=cfactor*(1.0-nu);
cstiff[3][4][i]=0.0;
cstiff[3][5][i]=0.0;
cstiff[3][6][i]=0.0;
cstiff[4][1][i]=0.0;
cstiff[4][2][i]=0.0;
cstiff[4][3][i]=0.0;
```

```
cstiff[4][4][i]=cfactor*((1.0-2.0*nu)/2.0);
cstiff[4][5][i]=0.0;
cstiff[4][6][i]=0.0;
cstiff[5][1][i]=0.0;
cstiff[5][2][i]=0.0;
cstiff[5][3][i]=0.0;
cstiff[5][4][i]=0.0;
cstiff[5][5][i]=cfactor*((1.0-2.0*nu)/2.0);
cstiff[5][6][i]=0.0;
cstiff[6][1][i]=0.0;
cstiff[6][2][i]=0.0;
cstiff[6][3][i]=0.0;
cstiff[6][4][i]=0.0;
cstiff[6][5][i]=0.0;
cstiff[6][6][i]=cfactor*((1.0-2.0*nu)/2.0);
}
```

Start time loop.

```
for(int it=1;it<=nstep;it++)
{
cout<<it<<endl;
```

Zero out the nodes.

```
for(int i=1;i<=nodesx;i++)
{
for(int j=1;j<=nodesy;j++)
{
for(int k=1;k<=nodesz;k++)
{
mi[i][j][k]=0.0;
pxi[i][j][k]=0.0;
pyi[i][j][k]=0.0;
pzi[i][j][k]=0.0;
fxi[i][j][k]=0.0;
fyi[i][j][k]=0.0;
fzi[i][j][k]=0.0;
}
}
}
```

Map the particle mass and momenta to the nodes.

```
for(int k=1;k<=np;k++)
{
```

Determine the node closest to particle k.

```
inode=int(xp[k]/deltx)+1;
if(inode<0)
{
inode=0;
}
if (inode==nodesx)
{
inode=nodesx-1;
}
jnode=int(yp[k]/delty)+1;
```

```
if(jnode<0)
{
jnode=0;
}
if (jnode==nodesy)
{
jnode=nodesy-1;
}
knode=int(zp[k]/deltz)+1;
if(knode<0)
{
knode=0;
}
if (knode==nodesz)
{
knode=nodesz-1;
}
```

Calculate the positions of the corners of the box the particle resides in.

```
left=float(inode-1)*deltx;
right=float(inode)*deltx;
front=float(jnode)*delty;
back=float(jnode-1)*delty;
top=float(knode)*deltz;
bottom=float(knode-1)*deltz;
```

Calculate internal (normalized) coordinates.

```
xi=(xp[k]-left)/deltx;
eta=(yp[k]-back)/delty;
chi=(zp[k]-bottom)/deltz;
```

Loop around the eight corners of the box (cube), calculate the portion of the particle's momentum that belongs to that corner, and put it there.

```
for(int iiz=knode;iiz<=knode+1;iiz++)
{
for(int iiy=jnode;iiy<=jnode+1;iiy++)
{
for(int iix=inode;iix<=inode+1;iix++)
{
if(iix==inode&&iiy==jnode&&iiz==knode)
{
ii=1;
}
if(iix==inode+1&&iiy==jnode&&iiz==knode)
{
ii=2;
}
if(iix==inode+1&&iiy==jnode+1&&iiz==knode)
{
ii=3;
}
if(iix==inode&&iiy==jnode+1&&iiz==knode)
{
ii=4;
}
```

```
if(iix==inode&&iiy==jnode&&iiz==knode+1)
{
ii=5;
}
if(iix==inode+1&&iiy==jnode&&iiz==knode+1)
{
ii=6;
}
if(iix==inode+1&&iiy==jnode+1&&iiz==knode+1)
{
ii=7;
}
if(iix==inode&&iiy==jnode+1&&iiz==knode+1)
{
ii=8;
}
shape_funct = chooseshape(ii,xi,eta,chi);
dummy=mi[iix][iiy][iiz]+part_mass*shape_funct;
mi[iix][iiy][iiz]=dummy;
dummy=pxi[iix][iiy][iiz]+vxp[k]*part_mass*shape_funct;
pxi[iix][iiy][iiz]=dummy;
dummy=pyi[iix][iiy][iiz]+vyp[k]*part_mass*shape_funct;
pyi[iix][iiy][iiz]=dummy;
dummy=pzi[iix][iiy][iiz]+vzp[k]*part_mass*shape_funct;
pzi[iix][iiy][iiz]=dummy;
}
}
}
}
```

Calculate internal forces due to stress v/strain increments.

```
for(int k=1;k<=np;k++)
{
inode=int(xp[k]/deltx)+1;
if(inode<0)
{
inode=0;
}
if (inode==nodesx)
{
inode=nodesx-1;
}
jnode=int(yp[k]/delty)+1;
if(jnode<0)
{
jnode=0;
}
if (jnode==nodesy)
{
jnode=nodesy-1;
}
knode=int(zp[k]/deltz)+1;
if(knode<0)
{
knode=0;
}
if (knode==nodesz)
{
knode=nodesz-1;
```

```
}
left=float(inode-1)*deltx;
right=float(inode)*deltx;
front=float(jnode)*delty;
back=float(jnode-1)*delty;
top=float(knode)*deltz;
bottom=float(knode-1)*deltz;
xi=(xp[k]-left)/deltx;
eta=(yp[k]-back)/delty;
chi=(zp[k]-bottom)/deltz;
```

Calculate the shape function gradients and use them to get the force on each node due to object deformation.

```
for(int iiz=knode;iiz<=knode+1;iiz++)
{
for(int iiy=jnode;iiy<=jnode+1;iiy++)
{
for(int iix=inode;iix<=inode+1;iix++)
{
if(iix==inode&&iiy==jnode&&iiz==knode)
{
ii=1;
}
if(iix==inode+1&&iiy==jnode&&iiz==knode)
{
ii=2;
}
if(iix==inode+1&&iiy==jnode+1&&iiz==knode)
{
ii=3;
}
if(iix==inode&&iiy==jnode+1&&iiz==knode)
{
ii=4;
}
if(iix==inode&&iiy==jnode&&iiz==knode+1)
{
ii=5;
}
if(iix==inode+1&&iiy==jnode&&iiz==knode+1)
{
ii=6;
}
if(iix==inode+1&&iiy==jnode+1&&iiz==knode+1)
{
ii=7;
}
if(iix==inode&&iiy==jnode+1&&iiz==knode+1)
{
ii=8;
}
grad_x = choosegradx(ii,xi,eta,chi,deltx,delty,deltz);
grad_y = choosegrady(ii,xi,eta,chi,deltx,delty,deltz);
grad_z = choosegradz(ii,xi,eta,chi,deltx,delty,deltz);
dotx=grad_x*s[1][1][k]+grad_y*s[2][1][k]+grad_z*s[3][1][k];
dummy=fxi[iix][iiy][iiz]-dotx*part_mass/part_dens;
fxi[iix][iiy][iiz]=dummy;
doty=grad_x*s[1][2][k]+grad_y*s[2][2][k]+grad_z*s[3][2][k];
dummy=fyi[iix][iiy][iiz]-doty*part_mass/part_dens;
```

```
fyi[iix][iiy][iiz]=dummy;
dotz=grad_x*s[1][3][k]+grad_y*s[2][3][k]+grad_z*s[3][3][k];
dummy=fzi[iix][iiy][iiz]-dotz*part_mass/part_dens;
fzi[iix][iiy][iiz]=dummy;
}
}
}
}
```

Use $\vec{F} = \Delta\vec{p}/\Delta t$ to update the node momenta due to the internal forces.

```
for(int i=1;i<=nodesx;i++)
{
for(int j=1;j<=nodesy;j++)
{
for(int k=1;k<=nodesz;k++)
{
dummy=pxi[i][j][k]+fxi[i][j][k]*delta_t;
pxi[i][j][k]=dummy;
dummy=pyi[i][j][k]+fyi[i][j][k]*delta_t;
pyi[i][j][k]=dummy;
dummy=pzi[i][j][k]+fzi[i][j][k]*delta_t;
pzi[i][j][k]=dummy;
}
}
}
```

Calculate shape functions and map kinematic variables from the nodes back onto the particles to calculate their acceleration.

```
for(int k=1;k<=np;k++)
{
axp[k]=0.0;
ayp[k]=0.0;
azp[k]=0.0;
vbxp[k]=0.0;
vbyp[k]=0.0;
vbzp[k]=0.0;
inode=int(xp[k]/deltx)+1;
if(inode<0)
{
inode=0;
}
if (inode==nodesx)
{
inode=nodesx-1;
}
jnode=int(yp[k]/delty)+1;
if(jnode<0)
{
jnode=0;
}
if (jnode==nodesy)
{
jnode=nodesy-1;
}
knode=int(zp[k]/deltz)+1;
if(knode<0)
{
```

```
knode=0;
}
if (knode==nodesz)
{
knode=nodesz-1;
}
left=float(inode-1)*deltx;
right=float(inode)*deltx;
front=float(jnode)*delty;
back=float(jnode-1)*delty;
top=float(knode)*deltz;
bottom=float(knode-1)*deltz;
xi=(xp[k]-left)/deltx;
eta=(yp[k]-back)/delty;
chi=(zp[k]-bottom)/deltz;
for(int iiz=knode;iiz<=knode+1;iiz++)
{
for(int iiy=jnode;iiy<=jnode+1;iiy++)
{
for(int iix=inode;iix<=inode+1;iix++)
{
if(iix==inode&&iiy==jnode&&iiz==knode)
{
ii=1;
}
if(iix==inode+1&&iiy==jnode&&iiz==knode)
{
ii=2;
}
if(iix==inode+1&&iiy==jnode+1&&iiz==knode)
{
ii=3;
}
if(iix==inode&&iiy==jnode+1&&iiz==knode)
{
ii=4;
}
if(iix==inode&&iiy==jnode&&iiz==knode+1)
{
ii=5;
}
if(iix==inode+1&&iiy==jnode&&iiz==knode+1)
{
ii=6;
}
if(iix==inode+1&&iiy==jnode+1&&iiz==knode+1)
{
ii=7;
}
if(iix==inode&&iiy==jnode+1&&iiz==knode+1)
{
ii=8;
}
if(mi[iix][iiy][iiz]!=0.0)
{
shape_funct = chooseshape(ii,xi,eta,chi);
dummy=axp[k]+fxi[iix][iiy][iiz]*shape_funct/mi[iix][iiy][iiz];
axp[k]=dummy;
dummy=ayp[k]+fyi[iix][iiy][iiz]*shape_funct/mi[iix][iiy][iiz];
ayp[k]=dummy;
dummy=azp[k]+fzi[iix][iiy][iiz]*shape_funct/mi[iix][iiy][iiz];
```

```
azp[k]=dummy;
dummy=vbxp[k]+pxi[iix][iiy][iiz]*shape_funct/mi[iix][iiy][iiz];
vbxp[k]=dummy;
dummy=vbyp[k]+pyi[iix][iiy][iiz]*shape_funct/mi[iix][iiy][iiz];
vbyp[k]=dummy;
dummy=vbzp[k]+pzi[iix][iiy][iiz]*shape_funct/mi[iix][iiy][iiz];
vbzp[k]=dummy;
}
}
}
}
```

Update position and velocity in time for particle k.

```
dummy=vxp[k]+axp[k]*delta_t;
vxp[k]=dummy;
dummy=vyp[k]+ayp[k]*delta_t;
vyp[k]=dummy;
dummy=vzp[k]+azp[k]*delta_t;
vzp[k]=dummy;
dummy=xp[k]+vbxp[k]*delta_t;
xp[k]=dummy;
dummy=yp[k]+vbyp[k]*delta_t;
yp[k]=dummy;
dummy=zp[k]+vbzp[k]*delta_t;
zp[k]=dummy;
```

Now, enforce particle boundary conditions. Keep all particles inside the computational box. Implement reflection off any boundaries by reversing the component of the particle's velocity normal to the boundary's surface at the point of contact.

```
if(xp[k]<0.0)
{
xp[k]=0.0;
if(vxp[k]<0.0)
{
vxp[k]=-vxp[k];
}
}
if(xp[k]>agrid)
{
xp[k]=agrid;
if(vxp[k]>0.0)
{
vxp[k]=-vxp[k];
}
}
if(yp[k]<0.0)
{
yp[k]=0.0;
if(vyp[k]<0.0)
{
vyp[k]=-vyp[k];
}
}
if(yp[k]>bgrid)
{
yp[k]=bgrid;
if(vyp[k]>00.)
{
```

```
vyp[k]=-vyp[k];
}
}
if(zp[k]<0.0)
{
zp[k]=0.0;
{
if(vzp[k]<0.0)
vzp[k]=-vzp[k];
}
}
if(zp[k]>cgrid)
{
zp[k]=cgrid;
if(vzp[k]>0.0)
{
vzp[k]=-vzp[k];
}
}
}
```

Update node masses and momenta again.

```
for(int i=1;i<=nodesx;i++)
{
for(int j=1;j<=nodesy;j++)
{
for(int k=1;k<=nodesz;k++)
{
pxi[i][j][k]=0.0;
pyi[i][j][k]=0.0;
pzi[i][j][k]=0.0;
mi[i][j][k]=0.0;
}
}
}

for(int k=1;k<=np;k++)
{
inode=int(xp[k]/deltx)+1;
if(inode<0)
{
inode=0;
}
if (inode==nodesx)
{
inode=nodesx-1;
}
jnode=int(yp[k]/delty)+1;
if(jnode<0)
{
jnode=0;
}
if (jnode==nodesy)
{
jnode=nodesy-1;
}
knode=int(zp[k]/deltz)+1;
if(knode<0)
{
knode=0;
```

```
}
if (knode==nodesz)
{
knode=nodesz-1;
}
left=float(inode-1)*deltx;
right=float(inode)*deltx;
front=float(jnode)*delty;
back=float(jnode-1)*delty;
top=float(knode)*deltz;
bottom=float(knode-1)*deltz;
xi=(xp[k]-left)/deltx;
eta=(yp[k]-back)/delty;
chi=(zp[k]-bottom)/deltz;

for(int iiz=knode;iiz<=knode+1;iiz++)
{
for(int iiy=jnode;iiy<=jnode+1;iiy++)
{
for(int iix=inode;iix<=inode+1;iix++)
{
if(iix==inode&&iiy==jnode&&iiz==knode)
{
ii=1;
}
if(iix==inode+1&&iiy==jnode&&iiz==knode)
{
ii=2;
}
if(iix==inode+1&&iiy==jnode+1&&iiz==knode)
{
ii=3;
}
if(iix==inode&&iiy==jnode+1&&iiz==knode)
{
ii=4;
}
if(iix==inode&&iiy==jnode&&iiz==knode+1)
{
ii=5;
}
if(iix==inode+1&&iiy==jnode&&iiz==knode+1)
{
ii=6;
}
if(iix==inode+1&&iiy==jnode+1&&iiz==knode+1)
{
ii=7;
}
if(iix==inode&&iiy==jnode+1&&iiz==knode+1)
{
ii=8;
}
shape_funct = chooseshape(ii,xi,eta,chi);
dummy=pxi[iix][iiy][iiz]+part_mass*vxp[k]*shape_funct;
pxi[iix][iiy][iiz]=dummy;
dummy=pyi[iix][iiy][iiz]+part_mass*vyp[k]*shape_funct;
pyi[iix][iiy][iiz]=dummy;
dummy=pzi[iix][iiy][iiz]+part_mass*vzp[k]*shape_funct;
pzi[iix][iiy][iiz]=dummy;
```

```
dummy=mi[iix][iiy][iiz]+part_mass*shape_funct;
mi[iix][iiy][iiz]=dummy;
}
}
}
}
```

Use $\vec{p} = m\vec{v}$ to update nodal velocities.

```
for(int i=1;i<=nodesx;i++)
{
for(int j=1;j<=nodesy;j++)
{
for(int k=1;k<=nodesz;k++)
{
if(mi[i][j][k]!=0.0)
{
vxi[i][j][k]=pxi[i][j][k]/mi[i][j][k];
vyi[i][j][k]=pyi[i][j][k]/mi[i][j][k];
vzi[i][j][k]=pzi[i][j][k]/mi[i][j][k];
}
}
}
}
```

Zero rate of strain tensor.

```
for(int i=1;i<=3;i++)
{
for(int j=1;j<=3;j++)
{
for(int k=1;k<=np;k++)
{
lp[i][j][k]=0.0;
}
}
}
```

Update the rate of strain tensor.

```
for(int k=1;k<=np;k++)
{
inode=int(xp[k]/deltx)+1;
if(inode<0)
{
inode=0;
}
if (inode==nodesx)
{
inode=nodesx-1;
}
jnode=int(yp[k]/delty)+1;
if(jnode<0)
{
jnode=0;
}
```

```
if (jnode==nodesy)
{
jnode=nodesy-1;
}
knode=int(zp[k]/deltz)+1;
if(knode<0)
{
knode=0;
}
if (knode==nodesz)
{
knode=nodesz-1;
}
left=float(inode-1)*deltx;
right=float(inode)*deltx;
front=float(jnode)*delty;
back=float(jnode-1)*delty;
top=float(knode)*deltz;
bottom=float(knode-1)*deltz;
xi=(xp[k]-left)/deltx;
eta=(yp[k]-back)/delty;
chi=(zp[k]-bottom)/deltz;
for(int iiz=knode;iiz<=knode+1;iiz++)
{
for(int iiy=jnode;iiy<=jnode+1;iiy++)
{
for(int iix=inode;iix<=inode+1;iix++)
{
if(iix==inode&&iiy==jnode&&iiz==knode)
{
ii=1;
}
if(iix==inode+1&&iiy==jnode&&iiz==knode)
{
ii=2;
}
if(iix==inode+1&&iiy==jnode+1&&iiz==knode)
{
ii=3;
}
if(iix==inode&&iiy==jnode+1&&iiz==knode)
{
ii=4;
}
if(iix==inode&&iiy==jnode&&iiz==knode+1)
{
ii=5;
}
if(iix==inode+1&&iiy==jnode&&iiz==knode+1)
{
ii=6;
}
if(iix==inode+1&&iiy==jnode+1&&iiz==knode+1)
{
ii=7;
}
if(iix==inode&&iiy==jnode+1&&iiz==knode+1)
{
ii=8;
}
```

Calculate shape function gradients.

```
grad_x = choosegradx(ii,xi,eta,chi,deltx,delty,deltz);
grad_y = choosegrady(ii,xi,eta,chi,deltx,delty,deltz);
grad_z = choosegradz(ii,xi,eta,chi,deltx,delty,deltz);
```

Use them to update the rate of strain tensor.

```
dummy=lp[1][1][k]+grad_x*vxi[iix][iiy][iiz];
lp[1][1][k]=dummy;
dummy=lp[1][2][k]+grad_x*vyi[iix][iiy][iiz];
lp[1][2][k]=dummy;
dummy=lp[1][3][k]+grad_x*vzi[iix][iiy][iiz];
lp[1][3][k]=dummy;
dummy=lp[2][1][k]+grad_y*vxi[iix][iiy][iiz];
lp[2][1][k]=dummy;
dummy=lp[2][2][k]+grad_y*vyi[iix][iiy][iiz];
lp[2][2][k]=dummy;
dummy=lp[2][3][k]+grad_y*vzi[iix][iiy][iiz];
lp[2][3][k]=dummy;
dummy=lp[3][1][k]+grad_z*vxi[iix][iiy][iiz];
lp[3][1][k]=dummy;
dummy=lp[3][2][k]+grad_z*vyi[iix][iiy][iiz];
lp[3][2][k]=dummy;
dummy=lp[3][3][k]+grad_z*vzi[iix][iiy][iiz];
lp[3][3][k]=dummy;
}
}
}
}
```

Use rate of strain tensor and its symmetry properties to update the strain tensor.

```
for(int i=1;i<=3;i++)
{
for(int j=1;j<=3;j++)
{
for(int k=1;k<=np;k++)
{
e[i][j][k]=0.5*(lp[i][j][k]+lp[j][i][k])*delta_t;
}
}
}
```

Use 3D linear elastic relationships (3D Hooke's law) to update stress tensor.

```
for(int k=1;k<=np;k++)
{
evec[1]=e[1][1][k];
evec[2]=e[2][2][k];
evec[3]=e[3][3][k];
evec[4]=e[2][3][k];
evec[5]=e[1][3][k];
evec[6]=e[1][2][k];
for(int i=1;i<=6;i++)
```

Now update the stress tensor using $\Delta s_{ij} = \sigma_{ij} = \sum_{ijkl} C_{ijkl} \varepsilon_{kl}$.

```
{
dummy=s[1][1][k]+cstiff[1][i][k]*evec[i];
s[1][1][k]=dummy;
dummy=s[2][2][k]+cstiff[2][i][k]*evec[i];
s[2][2][k]=dummy;
dummy=s[3][3][k]+cstiff[3][i][k]*evec[i];
s[3][3][k]=dummy;
dummy=s[2][3][k]+cstiff[4][i][k]*evec[i];
s[2][3][k]=dummy;
dummy=s[1][3][k]+cstiff[5][i][k]*evec[i];
s[1][3][k]=dummy;
dummy=s[1][2][k]+cstiff[6][i][k]*evec[i];
s[1][2][k]=dummy;
}
dummy=s[1][2][k];
s[2][1][k]=dummy;
dummy=s[2][3][k];
s[3][2][k]=dummy;
dummy=s[1][3][k];
s[3][1][k]=dummy;
}
```

Calculate pressure using $P = -\frac{1}{3}Tr\vec{\vec{s}} = -\frac{1}{3}(s_{11} + s_{22} + s_{33})$. Find pressure extrema for visualization purposes.

```
for(int k=1;k<=np;k++)
{
press[k]=-(s[1][1][k]+s[2][2][k]+s[3][3][k])/3.0;
}
for(int k=1;k<=np;k++)
{
if(k==1)
{
pmin=press[k];
pmax=press[k];
}
if(abs(xp[k]-yp[k])<=0.2)
{
if(press[k]<pmin)
{
pmin=press[k];
}
if(press[k]>pmax)
{
pmax=press[k];
}
}
}
```

Now calculate maximum and minimum sphere radii and maximum speed which can be used later for scaling colors in visualization.

```
rmin1=0.0;
rmax1=0.0;
rmin2=0.0;
rmax2=0.0;
nsph1=0;
nsph2=0;
```

```
xc1=0.0;
yc1=0.0;
zc1=0.0;
xc2=0.0;
yc2=0.0;
zc2=0.0;
vxc1=0.0;
vyc1=0.0;
vzc1=0.0;
vxc2=0.0;
vyc2=0.0;
vzc2=0.0;
for(int k=1;k<=np;k++)
{
if(elementnum[k]==1)
{
xc1=xc1+xp[k];
yc1=yc1+yp[k];
zc1=zc1+zp[k];
vxc1=vxc1+vxp[k];
vyc1=vyc1+vyp[k];
vzc1=vzc1+vzp[k];
nsph1=nsph1+1;
}
if(elementnum[k]==2)
{
xc2=xc2+xp[k];
yc2=yc2+yp[k];
zc2=zc2+zp[k];
vxc2=vxc2+vxp[k];
vyc2=vyc2+vyp[k];
vzc2=vzc2+vzp[k];
nsph2=nsph2+1;
}
xc1=xc1/float(nsph1);
yc1=yc1/float(nsph1);
zc1=zc1/float(nsph1);
xc2=xc2/float(nsph2);
yc2=yc2/float(nsph2);
zc2=zc2/float(nsph2);
vxc1=vxc1/float(nsph1);
vyc1=vyc1/float(nsph1);
vzc1=vzc1/float(nsph1);
vxc2=vxc2/float(nsph2);
vyc2=vyc2/float(nsph2);
vzc2=vzc2/float(nsph2);
for(int k=1;k<=np;k++)
{
vel=sqrt(pow(vxp[k]-vxc1,2)+pow(vyp[k]-vyc1,2)+pow(vzp[k]-vzc1,2));
if(k==1)
{
rmin1=rsph1;
rmax1=rsph1;
rmin2=rsph2;
rmax2=rsph2;
vmax=vel;
}
if(surfacetag[1][k]==1)
{
radius=sqrt(pow(xp[k]-xc1,2)+pow(yp[k]-yc1,2)+pow(zp[k]-zc1,2));
if(radius<rmin1)
```

```
{
rmin1=radius;
}
if(radius>rmax1)
{
rmax1=radius;
}
}
if(surfacetag[2][k]==1)
{
radius=sqrt(pow(xp[k]-xc2,2)+pow(yp[k]-yc2,2)+pow(zp[k]-zc2,2));
if(radius<rmin2)
{
rmin2=radius;
}
if(radius>rmax2)
{
rmax2=radius;
}
}
if(vel>vmax)
{
vmax=vel;
}
}
```

Here, we calculate energies of the two spheres in our simulation. This section can also clearly be adapted to components comprising the particular system you choose to work on.

Zero everything out.

```
ekl=0.0;
ekr=0.0;
ul=0.0;
ur=0.0;
for(int i=1;i<=6;i++)
{
for(int j=1;j<=npmax;j++)
{
duml[i][j]=0.0;
dumr[i][j]=0.0;
}
}
```

First calculate the kinetic energy.

```
for(int k=1;k<=np;k++)
{
if(elementnum[k]==1)
{
ekl=ekl+0.5*part_mass*(pow(vxp[k],2)+pow(vyp[k],2)+pow(vzp[k],2));
}
if(elementnum[k]==2)
{
ekr=ekr+0.5*part_mass*(pow(vxp[k],2)+pow(vyp[k],2)+pow(vzp[k],2));
}
}
```

And now, the elastic potential energy.

```
for(int k=1;k<=np;k++)
{
ecum[1][k]=ecum[1][k]+e[1][1][k];
ecum[2][k]=ecum[2][k]+e[2][2][k];
ecum[3][k]=ecum[3][k]+e[3][3][k];
ecum[4][k]=ecum[4][k]+e[2][3][k];
ecum[5][k]=ecum[5][k]+e[1][3][k];
ecum[6][k]=ecum[6][k]+e[1][2][k];

for(int i=1;i<=6;i++)
{
for(int j=1;j<=6;j++)
{
if(elementnum[k]==1)
{
duml[i][k]=duml[i][k]+cstiff[i][j][k]*ecum[j][k];
}
if(elementnum[k]==2)
{
dumr[i][k]=dumr[i][k]+cstiff[i][j][k]*ecum[j][k];
}
}
}
for(int i=1;i<=6;i++)
{
if(elementnum[k]==1)
{
ul=ul+duml[i][k]*ecum[i][k];
}
if(elementnum[k]==2)
{
ur=ur+dumr[i][k]*ecum[i][k];
}
}
}
if (floor(float(it)/float(nvel))==float(it)/float(nvel)){
energies<<it<<" "<<ekl<<" "<<ekr<<" "<<ul<<" "<<ur<<endl;
}
```

Now, write out to a POV file for visualization.

```
for(int i=1;i<=np;i++)
{
if(it==1)
if(i==1)
{
dmax=abs(xp[i]);
}
if(abs((xp[i]))>dmax)
{
dmax=abs(xp[i]);
}
if(abs(yp[i])>dmax)
{
dmax=abs(yp[i]);
}
if(abs(zp[i])>dmax)
{
dmax=abs(zp[i]);
```

```
}
}
```

Write things out every so often (technically speaking).

```
if (float(it/nevery)==float(it)/float(nevery))
{
ifr=ifr+1;
sprintf(fileanim,"%d.pov",ifr);
ofstream myfile;
myfile.open(fileanim);
myfile<<"#version 3.6"<<endl;
myfile<<"#include "<<"\""<<"colors.inc"<<"\""<<endl;
myfile<<"global_settings"<<endl;
myfile<<"{"<<endl;
```

Here are some specifics such as camera position, viewing angle, lighting, etc. I encourage you to look into these parameters and see what effect changing them produces.

```
myfile<<"assumed_gamma 1.0"<<endl;
myfile<<"}"<<endl;
myfile<<"camera"<<endl;
myfile<<"{"<<endl;
myfile<<"location  <"<<0.04*agrid<<","<<0.6*agrid<<","<<0.08*cgrid<<">
"<<endl;
myfile<<"direction 1.5*z"<<endl;
myfile<<"right     4/3*x"<<endl;
myfile<<"look_at   <"<<0.04*agrid<<","<<0.05*bgrid<<","<<0.08*cgrid<<"
>"<<endl;
myfile<<"}"<<endl;
myfile<<"background { color red 0 green 0 blue 0 }"<<endl;
myfile<<"light_source"<<endl;
myfile<<"{"<<endl;
myfile<<"0*x"<<endl;
myfile<<"color red"<<endl;
myfile<<"1.0  green 1.0  blue 1.0"<<endl;
myfile<<"translate <-30, 30, 30>"<<endl;
myfile<<"}"<<endl;
myfile<<"light_source"<<endl;
myfile<<"{"<<endl;
myfile<<"0*x"<<endl;
myfile<<"color red"<<endl;
myfile<<"0.5  green 0.5  blue 0.5"<<endl;
myfile<<"translate <30, 30, 30>"<<endl;
myfile<<"}"<<endl;
myfile<<"light_source"<<endl;
myfile<<"{"<<endl;
myfile<<"0*x"<<endl;
myfile<<"color red"<<endl;
myfile<<"0.1  green 0.1  blue 0.1"<<endl;
myfile<<"translate <30, -30, 30>"<<endl;
myfile<<"}"<<endl;
```

Now, actually place the spheres that comprise the object.

```
for(int i=1;i<=np;i++)
```

```
{
myfile<<"sphere { <"<<xp[i]/10.0<<","<<yp[i]/10.0<<","<<zp[i]/10.0<<">
,"<<endl;
myfile<<sphere_size/10.0;
if(elementnum[i]==1)
{
blue=1.0;
green=0.0;
red=0.0;
}
if(elementnum[i]==2)
{
blue=0.0;
green=0.0;
red=1.0;
}
//      if(surfacetag[1][i]==1)
//      {
//      radius=sqrt(pow(xp(i)-xc1,2)+pow(yp(i)-yc1,2)+pow(zp(i)-zc1,2));
//      red=(rmax1-radius)/(rmax1-rmin1)+(radius-rmin1)/(rmax1-rmin1);
//      green=(radius-rmin1)/(rmax1-rmin1);
//      blue=0.0;
//      }
//      if(surfacetag[2][i]==1)
//      {
//      radius=sqrt(pow(xp(i)-xc2,2)+pow(yp(i)-yc2,2)+pow(zp(i)-zc2,2));
//      red=0.0;
//      green=(radius-rmin2)/(rmax2-rmin2);
//      blue=(rmax2-radius)/(rmax2-rmin2);
//      }
//      if(surfacetag[1][i]==0&&surfacetag[2][i]==0)
//      {
//      red=1.0;
//      blue=1.0;
//      green=1.0;
//      }
myfile<<"texture {pigment{color
rgb<"<<red<<","<<green<<","<<blue<<">}"<<endl;
myfile<<"finish{specular 1}} }"<<endl;
}
```

Draw the computational cell if you want.

```
right=float(nodesx)*deltx;
left=float(0)*deltx;
front=float(nodesy)*delty;
back=float(0)*delty;
top=float(nodesz)*deltz;
bottom=float(0)*deltz;
myfile<<"cylinder { <"<<left/10.0<<","<<back/10.0<<","<<0.00/10.0<<">,
<"<<left/10.0<<","<<front/10.0<<","<<0.00/10.0<<">,"<<endl;
myfile<<cyl_rad/10.0<<endl;
myfile<<"texture {pigment {color rgb<0,0,1>}"<<endl;
myfile<<"finish{specular 1}} }"<<endl;
myfile<<"cylinder { <"<<right/10.0<<","<<back/10.0<<","<<0.00/10.0<<">,
<"<<right/10.0<<","<<front/10.0<<","<<0.00/10.0<<">,"<<endl;
myfile<<cyl_rad/10.0<<endl;
myfile<<"texture {pigment {color rgb<0,0,1>}"<<endl;
```

```
myfile<<"finish{specular 1}} }"<<endl;
myfile<<"cylinder { <"<<left/10.0<<","<<back/10.0<<","<<0.00/10.0<<">,
<"<<right/10.0<<","<<back/10.0<<","<<0.00/10.0<<">,"<<endl;
myfile<<cyl_rad/10.0<<endl;
myfile<<"texture {pigment {color rgb<0,0,1>}"<<endl;
myfile<<"finish{specular 1}} }"<<endl;
myfile<<"cylinder { <"<<left/10.0<<","<<front/10.0<<","<<0.00/10.0<<">,
<"<<right/10.0<<","<<front/10.0<<","<<0.00/10.0<<">,"<<endl;
myfile<<cyl_rad/10.0<<endl;
myfile<<"texture {pigment {color rgb<0,0,1>}"<<endl;
myfile<<"finish{specular 1}} }"<<endl;
myfile<<"cylinder { <"<<left/10.0<<","<<back/10.0<<","<<top/10.0<<">,
<"<<left/10.0<<","<<front/10.0<<","<<top/10.0<<">,"<<endl;
myfile<<cyl_rad/10.0<<endl;
myfile<<"texture {pigment {color rgb<0,0,1>}"<<endl;
myfile<<"finish{specular 1}} }"<<endl;
myfile<<"cylinder { <"<<right/10.0<<","<<back/10.0<<","<<top/10.0<<">,
<"<<right/10.0<<","<<front/10.0<<","<<top/10.0<<">,"<<endl;
myfile<<cyl_rad/10.0<<endl;
myfile<<"texture {pigment {color rgb<0,0,1>}"<<endl;
myfile<<"finish{specular 1}} }"<<endl;
myfile<<"cylinder { <"<<left/10.0<<","<<back/10.0<<","<<top/10.0<<">,
<"<<right/10.0<<","<<back/10.0<<","<<top/10.0<<">,"<<endl;
myfile<<cyl_rad/10.0<<endl;
myfile<<"texture {pigment {color rgb<0,0,1>}"<<endl;
myfile<<"finish{specular 1}} }"<<endl;
myfile<<"cylinder { <"<<left/10.0<<","<<front/10.0<<","<<top/10.0<<">,
<"<<right/10.0<<","<<front/10.0<<","<<top/10.0<<">,"<<endl;
myfile<<cyl_rad/10.0<<endl;
myfile<<"texture {pigment {color rgb<0,0,1>}"<<endl;
myfile<<"finish{specular 1}} }"<<endl;
myfile<<"cylinder { <"<<left/10.0<<","<<back/10.0<<","<<bottom/10.0<<"
>,<"<<left/10.0<<","<<back/10.0<<","<<top/10.0<<">,"<<endl;
myfile<<cyl_rad/10.0<<endl;
myfile<<"texture {pigment {color rgb<0,0,1>}"<<endl;
myfile<<"finish{specular 1}} }"<<endl;
myfile<<"cylinder { <"<<right/10.0<<","<<back/10.0<<","<<bottom/10.0<<
">,<"<<right/10.0<<","<<back/10.0<<","<<top/10.0<<">,"<<endl;
myfile<<cyl_rad/10.0<<endl;
myfile<<"texture {pigment {color rgb<0,0,1>}"<<endl;
myfile<<"finish{specular 1}} }"<<endl;
myfile<<"cylinder { <"<<left/10.0<<","<<front/10.0<<","<<bottom/10.0<<
">,<"<<left/10.0<<","<<front/10.0<<","<<top/10.0<<">,"<<endl;
myfile<<cyl_rad/10.0<<endl;
myfile<<"texture {pigment {color rgb<0,0,1>}"<<endl;
myfile<<"finish{specular 1}} }"<<endl;
myfile<<"cylinder { <"<<right/10.0<<","<<front/10.0<<","<<bottom/10.0<
<">,<"<<right/10.0<<","<<front/10.0<<","<<top/10.0<<">,"<<endl;
myfile<<cyl_rad/10.0<<endl;
myfile<<"texture {pigment {color rgb<0,0,1>}"<<endl;
myfile<<"finish{specular 1}} }"<<endl;
myfile.close();
}
```

The visualization section is all over; now wrap things up!

```
} //End time loop

    line9975:;
```

```
     energies.close();

     return (0);
     }
```

7.3.2 Variable List and Descriptions for the MPM Code

The following is a variable list for the program in the previous section. I present it this way because, the way the program is written and variables are grouped, a separate list provides much more organization.

```
npmax: Maximum number of particles
nstep: Number of Steps
nevery: Writes out pov files every this many steps
nodesx: Number of nodes (grid boxes) in the x direction
nodesy: Number of nodes (grid boxes) in the y direction
nodesz: Number of nodes (grid boxes) in the z direction
part_mass: Mass of MPM particles
part_dens: Density of MPM particles
agrid: Total box size in the x direction
bgrid: Total box size in the y direction
cgrid: Total box size in the z direction
delta_t: Time step
sphere_size: Size of MPM spheres for animation purposes only
cyl_rad: Radius of cylinders making box for animation purposes only
nvel: Writes out velocity info every this many steps
nu: Poisson ratio
centx1,centy1,centz1: Initial coordinates of large sphere 1
centx2,centy2,centz2: Initial coordinates of large sphere 2
rsph1,rsph2: Radii of two large colliding spheres
fileanim: Character string for file names
ii,ifr,np: Dummy index for node sums, dummy file index, number of particles
inode,jnode,knode. Dummy indices for the node closet to a particle
elementnum: Element number to distinguish material types
surfacetag: Used for animation to tell if something is on a surface
nsph1,nsph2: Number of particles in each sphere
mi Mass of nodes:
vxi,vyi: x and y node velocities
vzi,pxi z :node velocity and x node momentum
pxi,pyi,pzi : x, y and z node momenta
vxp,vyp,vzp: x, y and z particle velocities
xp,yp,zp: x, y and z particle positions
lp: Rate of strain tensor
fxi,fyi,fzi: x, y and z components of force on nodes
axp,ayp,azp: x, y and z components of particle accelerations
vbxp,vbyp,vbzp: x, y and z average particle velocities
evec,ecum: instantaneous and cumulative strain vector tensors
shape_funct: Shape functions for property mapping
left,right,back,front,bottom,top: Limits of small grid box that a particle
is in
xi,eta,chi: Reduced coordinates of particle within box
s: Stress tensor
e,cstiff,cfactor: Strain tensor, stiffness tensor and stiffness constant
dmax: Maximum distance dimension for visualization purposes
grad_x,grad_y,grad_z: x, y and z shape function gradients.
xtry,ytry,ztry,rtest: Trial coordinates and trial radius
deltx,delty,deltz: Small grid box dimensions
press: Pressure
red,green,blue,pmax,pmin: Colors and max/min pressures
xc1,xc2,yc1,yc2,zc1,zc2: Sphere centers for energy calculations
```

```
rmin1,rmax1,rmin2,rmax2: Max/min dummy variables
radius: Dummy radius
vxc1,vxc2,vyc1,vyc2,vzc1,vzc2,vdotr: Dummy x, y, z velocities and dot
product for forces
vmax,vel: Dummy velocities
ekl,ekr,ul,ur,duml,dumr: Obsolete; dummy and final energies
centx,centy,centz: x, y and z sphere dummy centers
float dotx,doty,dotz,econst: dot products for velocities and elastic constant
dummy: REALLY a dummy variable!!
```

7.4 Applications of the Material Point Method Simulation

7.4.1 Blood Flow

You can't get blood out of a turnip but apparently it's no problem for computers. Understanding the behavior of human blood is of fundamental importance for the advancement of health care, as well as scientific and medical education at all levels. Blood presents an interesting confluence of fluid flow in irregularly-shaped networks and dynamics of particles having varied shapes and sizes. Even though there has been a wealth of experimental exploration of human blood, on both gross and microscopic levels, computer simulations can afford perspectives on blood behavior not offered in the laboratory. Moreover, as science grows more interdisciplinary in the twenty-first century, computer explorations of biological systems become more meaningful. There exist a wealth of computer simulations regarding blood flow, including gross flow calculations [5], flow inside realistic arterial and capillary networks [6–13], flow related to aneurisms [14], as well as detailed simulation of the behavior of rigid and elastic blood corpuscles [15,16] and their interactions [16–21]. All this to say that there is a huge amount of very interesting work done on blood flow simulations, but there is always room for something new.

One significant benefit of computer simulations is that parameters provided by nature and technology can be systematically altered and results can be analyzed. On the other hand, any computer model necessarily cannot perfectly reproduce natural reality and so the results of computer simulations must be understood within their limitations. Even so, more reasonable and reliable models have increased relevance and can therefore be very useful in understanding the behavior of physical systems. The three major constituents of human blood are platelets (approximated as small spheres), leukocytes (protean-shaped cells) and erythrocytes (elastic disks), which flow through the body with plasma as their vehicle. In healthy humans there are about 5,000,000 erythrocytes per mm^3 of blood, as well as about 150,000 to 400,000 platelets and about 4000 to 11,000 leukocytes; clearly platelets and erythrocytes are the major constituents of blood and their dynamics on a microscopic level are of interest.

The work on blood and blood flow presented so far focuses on biologically realistic conditions in vivo, but in the near future it is conceivable that human blood will flow through artificial conduits for treatment, storage, microtechnological, or educational purposes. So, it is potentially useful to study how blood corpuscles behave as they flow through such channels. Moreover, when artificial barriers are introduced in the stream, the coupling of the particles to the plasma velocity field as well as to each other could result in interesting, counterintuitive, and potentially useful effects such as selective

Table 7.2. Details of the Runs Done in this Study

	Particle Number	Cell Number	Cell Density
Low Density Erythrocytes	3000	60	8571
High Density Erythrocytes	15,000	300	42,857

Source: C. Massina, M.W. Roth, P.A. Gray, Material Point Method Computer Simulations of Grazing Impacts on the Martian Surface, *14th Annual Sigma Xi Student Research Conference*, April 18, 2007 [26].

sieving, blocking, and redirection. So even though there is a wealth of work done on blood flow in realistic settings it can be useful to take the path less traveled.

My student and I decided to develop and use a two-phase MPM computer simulation involving elastic discs (modeling erythrocytes) flowing through plasma. Table 7.2 gives some details of the simulations.

Figure 7.3 illustrates the behavior of the elastic erythrocytes encountering vertical (wall) barriers for both densities studied. The high-density case clearly illustrates that the cells aggregate on the high-pressure side of the barrier and then pass over it, almost completely avoiding the wake created by the wall. The vertical distribution on the downstream side of the barrier is essentially bimodal, with a small number of cells moving along the floor boundary layer and the preponderance above the geometrical center of the box. The low-density case shows the same general dynamics, but the aggregation on the

Low density equilibrium cell flow:
Vertical obstructions

High density equilibrium cell flow:
Vertical obstructions

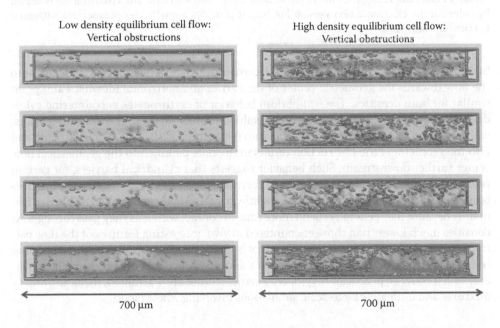

←——————————→
700 μm

←——————————→
700 μm

Figure 7.3. Snapshots of equilibrium erythrocyte flow for low density (left panel) and high-density (right panel) simulations with vertical barriers. The background shows the plasma velocity field, illustrating the effects of the boundaries and barriers. The visualization is scaled by color for the electronic text version but here it provides a qualitative perspective without a particular shade scale.

Low density equilibrium cell flow: High density equilibrium cell flow:
Cylindrical obstructions   Cylindrical obstructions

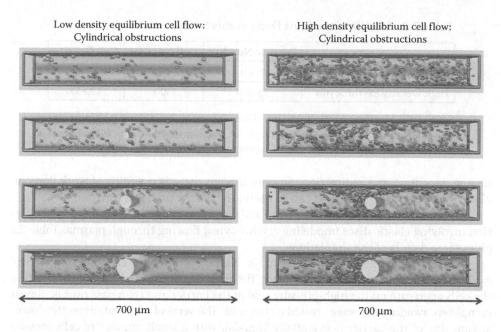

700 μm   700 μm

Figure 7.4. Snapshots of equilibrium erythrocyte flow for low density (left panel) and high-density (right panel) simulations with cylindrical barriers. The background shows the plasma velocity field, illustrating the effects of the boundaries and barriers. The visualization is scaled by color for the electronic text version but here it provides a qualitative perspective without a particular shade scale.

high-pressure (upstream) side is absent. Downstream the particles are drawn back into the flow towards the geometric center of the barrier at a horizontal (downstream) point similar for both densities. The equilibrium behavior of erythrocytes encountering cylindrical barriers is shown in Figure 7.4. The results are similar to those seen for wall barriers, with the exception that the cell flow is cut into two symmetric branches downstream, and also the geometry of the obstruction results in the cells pulling into the geometrical flow center further downstream. Such behavior suggests that cylindrical barriers, for certain ranges of plasma velocities could be very effective in separating erythrocyte flow into two branches. In addition, the cells show a slightly greater propensity to exit the barrier flat, mainly because they generally spend more time in contact with it as they pass through. At densities much lower than those encountered in vivo, interesting features of the flow patterns emerge and the simulations suggest that both barriers but especially cylindrical ones result in different behavior for erythrocytes and platelets downstream from the blockage, which could be practically exploited. Further studies are warranted, including binary mixtures and ultimately large-scale simulations involving whole blood.

7.4.2 Planetary Impact

Valles Marineris is the deepest trench known to exist on a terrestrial body in the solar system. If you carefully compare Valles to a typical water-carved structure on earth (like the Grand Canyon), you will notice the lack of the tortuous network typical for rivers and tributaries in Valles. Moreover, Martian surface gravitational acceleration is about 1/3 that of the Earth and comparing the two structures, it just doesn't make sense that water could have formed the structure on Mars. It seems as though water or dry sand [22] may have run down the sides of Valles in the past, but didn't have a role in the channel's formation [23]. Even if Valles is related to a fault in the Martian crust [24] the channel's formation could still be of different origin. A student and I had the tantalizing thought: Could it be that a grazing impact created the trench? Thinking even further... could the irregularly shaped Martian satellites Phobos and Deimos be evidence of such a grazing impact?

After some discussion, we decided to simulate grazing impacts on a terrestrial planetary surface [25] and determine if we could reproduce the general morphology of Valles and end up with bodies orbiting Mars whose orbital elements are close to those of Phobos and Deimos. If you are not familiar with the term "orbital elements" and want to follow up, I suggest you take a look at a celestial mechanics reference. The MPM method is perfectly suited to study this problem because of the presence of collisions. There are times, however, at the beginning and end of the simulation where gravitationally-governed dynamics is the main physics going on. To maximize efficiency, we married MPM with classical Newtonian planetary dynamics. Before the impact, a two-particle center of mass attraction algorithm is just what the doctor ordered! This gravitational method is employed until the particles enter a zone around the planet where collisions can happen and MPM takes center stage. We spent a good deal of time constructing our program and validating the code against energy and momentum conservation for various two-dimensional collisions between balls and energy conservation in the collision of three dimensional balls. We had a ball doing this work, I must add.

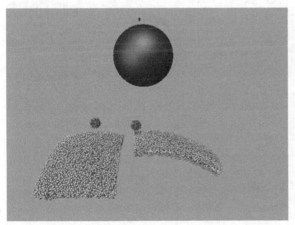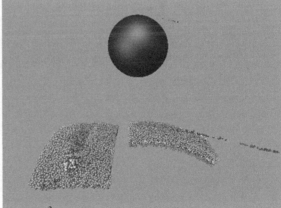

Figure 7.5. Initial configuration for a grazing impact simulation (left) and a typical configuration shortly after impact (right). The view of the whole planet (left) shows particles only governed my classical Newtonian planetary dynamics, and so the impacting body disappears after impact and the surface patch is never visible. The impacting body as well as Martian topsoil, ice layer and subsoil (progressively deeper) are shaded slightly differently and are colored in the electronic version of the text (C. Massina, M.W. Roth, P.A. Gray, Material Point Method Computer Simulations of Grazing Impacts on the Martian Surface, *14th Annual Sigma Xi Student Research Conference*, April 18, 2007) [26].

Table 7.3. Useful Simulation Parameters

Parameter	Value
Planet Mass	1 Martian Mass
Planet Radius	1 Martian Radius
Impacting Body Mass	0.011 Martian Masses
Impacting Body Speed	10 km/sec
Impacting Body Radius	4.7 Phoebian Radii
Number of Particles in Impacting Body	111
Number of Particles on Simulated Surface	9889
Longest Simulated Time	1 Phoebian Mo. = 7.6 h
Time Step	5×10^{-5} Phoebian Mo. = 1.37 sec.

Source: C. Massina, M.W. Roth, P.A. Gray, Material Point Method Computer Simulations of Grazing Impacts on the Martian Surface, *14th Annual Sigma Xi Student Research Conference*, April 18, 2007 [26].

The impact zone was constructed using anywhere from about 10,000 to 120,000 small spheres (material points) placed on the surface of a larger reflecting sphere. We consider a topsoil layer, subsurface ice layer, and soil layer beneath the ice. The initial condition for the simulation can be seen in the upper panel of Figure 7.5 and relevant physical data is given in Table 7.3.

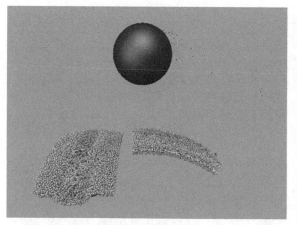 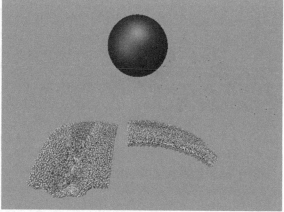

Figure 7.6. Typical configurations from the intermediate stage of impact (left panel), after the rapidly moving horizontal ejecta has left, and late phase of impact where the trench has been carved out and forward momentum of objects inside the trench continue to carve it out to some extent (right panel). The impacting body as well as Martian topsoil, ice layer and subsoil (progressively deeper) are shaded slightly differently and are colored in the electronic version of the text (C. Massina, M.W. Roth, P.A. Gray, Material Point Method Computer Simulations of Grazing Impacts on the Martian Surface, *14th Annual Sigma Xi Student Research Conference*, April 18, 2007.) [26].

Just like with the making of a good movie, we needed many runs with the initial conditions being adjusted so that a reasonable grazing collision was obtained. Figure 7.6 shows the initial progress of the simulation during the first phase of impact, where the colliding body elongates and breaks apart, and high-speed ejecta from the planet leaves the surface. Figure 7.7 shows the next distinct dynamical phase where slower moving ejecta from the planet as well as parts of the impacting body begin to noticeably curve in towards the surface. In addition, there is a large number of fragments orbiting the planet. On the planet's surface the trench continues to be cut out, exposing the sub surface ice due to forward horizontal momentum delivered by the impacting body. In some simulations, we even saw a temporary partial ring around the planet, which extinguished with time. In the last phase of the simulation (Figure 7.8), there were at most two or three bodies in stable orbits about the planet, with orbital properties somewhat similar to Mars' natural satellites. We also examined collisions where the impacting body had a core (Figure 7.8) as well as large-scale simulations with 120,000

Figure 7.7. Some simulations yielded a partial ring of debris during the intermediate stage of impact (C. Massina, M.W. Roth, P.A. Gray, Material Point Method Computer Simulations of Grazing Impacts on the Martian Surface, *14th Annual Sigma Xi Student Research Conference*, April 18, 2007.) [26].

particles. In all our simulations the general morphology of the trench is similar to that of Valles and the ice layer in the trench is exposed. The larger simulations will be required if we desire to model the canal with more structure and fine details.

So what do you think?? Our simulations presented here seem to show that it is possible to carve trenches on planetary surfaces with morphologies similar to that of

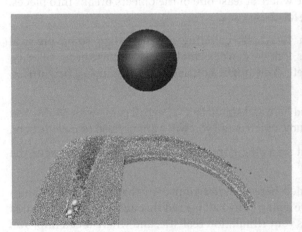 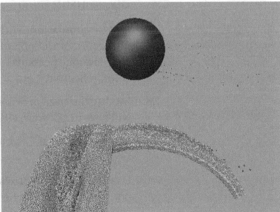

Figure 7.8. Configurations from simulations with 120,000 particles on the planet's impact site and a double spherical core for the impacting body. The core of the impacting body, the material surrounding it, the Martian topsoil, ice layer and subsoil (progressively deeper) are all shaded slightly differently and are colored in the electronic version of the text (C. Massina, M.W. Roth, P.A. Gray, Material Point Method Computer Simulations of Grazing Impacts on the Martian Surface, *14th Annual Sigma Xi Student Research Conference*, April 18, 2007.) [26].

Valles: narrow on the ends and wider in the middle. It also looks like subsurface ice could have been exposed. It's even possible to wind up with orbiting fragments that could explain the irregular satellites that Mars has! There is something very important missing from our simulations here: an equation of state (EOS) that would allow vaporization and melting to take place and allowing necessary modeling of the changes in rock and ice that took place during the impact.

PROBLEMS

7.1.   Translate the MPM code given here into FORTRAN. Run the new program and validate.

7.2.   Run the MPM code provided here and make sure that it is validated in terms of energy conservation.

7.3.   Run the code provided for this chapter and try some new visualization techniques, one being two deformable balls colliding. You will have to carefully think about how you will determine the shape of the two objects you are using.

7.4.   Change the initial conditions of the simulation so that the objects colliding are ellipsoids. Run the simulation and discuss the differences between these new simulations and the one with spheres.

7.5.   Change the simulation so that the spheres are different masses. Run a variety of simulations including one where at least one of the objects breaks into pieces. I suggest UNIX scripting for these runs.

7.6.   Change the simulation so that the spheres are different sizes. Run a variety of simulations including one where at least one of the objects breaks into pieces. I suggest UNIX scripting for these runs.

7.7.   Run a simulation with two balls colliding and visualize some physical properties of the objects that we can't normally see, such as speed, velocity, acceleration, pressure, etc. You might consider methods using brightness, color, reflectivity, etc...

7.8.   Modify the code provided here and simulate a volleyball game with MPM. You will have to ask yourself how much detail you will want or need to include here.

7.9.   Same question as 7.8, but for a ping-pong game. What's different between the two?

7.10. Use the program provided here or write your own that simulates a "ping-pong ball cannon," where a ping pong ball is fired through an aluminum soda can. Discuss what makes your simulation realistic, and where it can improve.

7.11. Run a simulation that generates a series of pov files and write a UNIX script to render them.

7.12. Think of your own project—some system exhibiting material failure, collisions, or bending and modify the program provided here to construct an appropriate simulation. Run it and discuss the results.

7.13. Find the form of the elastic tensor for three other types of symmetry including hexagonal, cubic, two types of trigonal and monoclinic with respect to the x and y planes.

7.14. Plan to fail! Take some time to make the MPM code either crash or give nonphysical (aka useless) results by changing only numbers. Discuss what you see and especially if any patterns emerged.

References

1. Z. Chen, R. Brannon, An evaluation of the material point method, SAND Report SAND2002-0482 1–42, 2002.

2. Z. Chen, W. Hu, L. Shen, X. Xin, R. Brannon, An evaluation of the MPM for simulating dynamic failure with damage diffusion, *Engineering Fracture Mechanics*, 69(17), 1873–1890, 2002.

3. http://folk.ntnu.no/stoylen/strainrate/Basic_concepts.html

4. https://en.wikipedia.org/wiki/Cauchy_stress_tensor

5. N. Jamshidi, J.S. Edwards, T. Fahland, G.M. Church, B.O. Palsson, Dynamic simulation of the human red blood cell metabolic network, *Bioinformatics Applications Note*, 17(3), 286–287, 2001.

6. A.R. Pries, T.W. Secomb, P. Gaehtgens, J.F. Gross, Blood flow in microvascular networks, *Experiments and Simulation in Circulation Research*, 67, 826–834, 1990.

7. R. Krams, J.J. Wentzel, J.A.F. Oomen, R. Vinke, J.C.H. Schuurbiers, P.J. de Feyter, P.W. Serruys, C.J. Slager, Evaluation of endothelial shear stress and 3D geometry as factors determining the development of atherosclerosis and remodeling in human coronary arteries in vivo, *Arteriosclerosis, Thrombosis, and Vascular Biology*, 17, 2061–2065, 1997.

8. J.S. Milner, J.A. Moore, B.K. Rutt, D.A. Steinman, Hemodynamics of human carotid artery bifurcations: computational studies with models reconstructed from magnetic resonance imaging of normal subjects, *Journal of Vascular Surgery*, 28(1), 143–156, 1998.

9. S.Z. Zhao, X.Y. Xu, A.D. Hughes, S.A. Thom, Blood flow and vessel mechanics in a physiologically realistic model of a human carotid arterial bifurcation, *Journal of Biomechanics*, 33(8), 975–984, 2000.

10. D.A. Steinman, J.B. Thomas, H.M. Ladak, J.S. Milner, B.K. Rutt, J.D. Spence, Reconstruction of carotid bifurcation hemodynamics and wall thickness using computational fluid dynamics and MRI, *Magnetic Resonance in Medicine*, 47(1), 149–159, 2001.

11. J.B. Grotberg, O.E. Jensen, Biofluid mechanics in flexible tubes, *Annual Review of Fluid Mechanics*, 36, 121–147, 2004.

12. C.A. Figueroa, I.E. Vignon-Clementel, K.E. Jansen, T.J.R. Hughes, C.A. Taylor, A coupled momentum method for modeling blood flow in three-dimensional deformable arteries, *Computer Methods in Applied Mechanics and Engineering*, 195, 5685–5706, 2006.

13. I.E. Vignon-Clementel, C.A. Figueroa, K.E. Jansen, C.A. Taylor, Outflow boundary conditions for three-dimensional finite element modeling of blood flow and pressure in arteries, *Computer Methods in Applied Mechanics and Engineering*, 195(29–32), 3776–3796, 2006.

14. K. Perktold, R. Peter, M. Resch, Pulsatile non-Newtonian blood flow simulation through a bifurcation with an aneurysm, *Biorheology*, 26(6), 1011–1030, 1989.

15. R.M. Macmeccan, J.R. Clausen, G.P. Neitzel, C.K. Aidun, Simulating deformable particle suspensions using a coupled lattice-Boltzmann and finite-element method, *Journal of Fluid Mechanics*, 618, 13–39, 2009.

16. M.M. Dupin, I. Halliday, C.M. Care, L.L. Munn, Lattice Boltzmann modelling of blood cell dynamics, *International Journal of Computational Fluid Dynamics*, 22(7), 481–492, 2008.

17. C. Bui, V. Lleras, O. Pantz, *Dynamics of red blood cells in 2D*, ESAIM Proceedings, 28, 182–194, 2009.

18. J.K.W. Chesnutt, J.S. Marshall, Blood cell transport and aggregation using discrete ellipsoidal particles, *Computers and Fluids*, 38(9), 1782–1794, 2009.

19. P. Peyla, Rheology and dynamics of a deformable object in a microfluidic configuration: A numerical study, *Europhysical Letters*, 80(3), 2007.

20. H. Fang, Z. Wang, Z. Lin, M. Liu, Lattice Boltzmann method for simulating the viscous flow in large distensible blood vessels, *Physical Revue E*, 65, 051925–051935, 2002.

21. Y. Liu, W.K. Liu, Rheology of red blood cell aggregation by computer simulation, *Journal of Computational Physics*, 220(1), 139–154, 2006.

22. C. Quantin, P. Allemand, C. Delacourt, Morphology and geometry of Valles Marineris landslides, *Planetary and Space Science*, 52(11), 1011–1022, 2004.

23. N. Mangold, C. Quantin, V. Ansan, C. Delacourt, P. Allemand, *Science*, 305(5680), 78–81, 2004.

24. http://www.nasa.gov/multimedia/imagegallery/image_feature_83.html

25. C.J. Massina, M.W. Roth, P.A. Gray, Deterministic computer simulations of grazing impacts on planetary surfaces, *American Journal of Undergraduate Research.*, 8(1), 15–22, 2009.

26. C. Massina, M.W. Roth, P.A. Gray, Material Point Method Computer Simulations of Grazing Impacts on the Martian Surface, *14th Annual Sigma Xi Student Research Conference*, April 18, 2007.

CHAPTER 8

//

Simulations that Explore Atoms and Planets

8.1 Introduction to Molecular Dynamics Computer Simulations

If you recall Newton's second law of motion, $F = ma$, you know that we can calculate acceleration if we know the (net external) force on an object and its mass. If you don't recall Newton's second law we can still do the calculation, but it would be a good thing to review Newton's work. Newton's second law is just what its name suggests, a law, so it applies in a global sense—to everything, or at least everything classical. The "classical" caveat has to be added there because quantum mechanics involves the world of the very small, special relativity involves the world of the very fast, and general relativity deals with the very massive. Classical physics breaks down in such places right along with our intuition, but it is such thoughts that captivate the imagination, and I strongly recommend external reading [1–6]. Molecular Dynamics (MD) is a wonderful starting point for students to learn atomic and molecular simulation because it is just a direct application of Newton's second law. Put very simply, we calculate forces on atoms, divide by the mass to get acceleration, and then step the system through time. As the simulation progresses, it has a *thermalization* phase for equilibration from its initial condition, and subsequently a *production* phase where observables (important averages) are calculated as time averages. There are issues like temperature control, periodic boundary conditions, and other details which we're going to talk about, but in essence the heart and soul of MD is simple and straightforward. In the next section, we are going to go through the development of an MD program that simulates particles (atoms), and then we'll up the ante and talk about molecules and much more complicated systems.

8.2 Molecular Dynamics Simulation of a System of Particles

8.2.1 Initial Thoughts

I think that the best way to learn MD is to *do* MD. In this book, I often present theory and then sample programs but sometimes, when the program tends to be modular enough and also contains data analysis, I will intermingle theory and the sample program. I am choosing the latter approach in this section regarding a basic MD program that simulates a system of particles. I tend to think that such conversations are best had in coffee shops and other relaxed venues, so I suggest that you draw your drink of preference as you begin learning about MD.

The next section steps through the construction of an MD simulation of a box of particles. I know that doesn't sound like the epitomy of excitement, but when constructing computer simulations it is crucial to start from a known place and move into the unknown from there. The particles we're talking about here are called "Lennard-Jonesium." They interact with each other through the *Lennard-Jones* potential [7] (Equation 8.5), which is a basic standard in atomic and molecular simulation. Moreover, the behavior of such systems has been thoroughly studied and so there is a wealth of data available when you go to validate your computer simulation—that is, to make sure it works correctly.

As you go through the program below, I want you to keep in mind that there is a certain transition students make in going from coding to actually living in the simulation. To do the latter, it is important to get a feeling for natural quantities and units in the simulation, which is to say that you can take the potential parameters ε (well depth) and σ (collision diameter) in Equation 8.5 and construct times, lengths, masses, and such whose values aren't unduly large or small numerically.

Table 8.1 shows expressions for some of the *reduced units* for a Lennard-Jones system and it's a homework problem for you to derive them!

Although "Lennard-Jonesium" is an abstraction, experimental measurements can be made to determine Lennard-Jones potential parameters for various systems. The noble gases He, Ne, Ar, Kr, and Xe form a nice comparative group and, although the parameter values depend a bit on the experiment they are being derived from, representative ones we have used in previous simulations are shown in Table 8.2, along with those for carbon [8].

Table 8.1. Expression for Reduced Units in a Lennard-Jones System

Quantity	Expression
Length	σ
Mass	m
Time	$\sigma\sqrt{\dfrac{m}{\epsilon}}$
Velocity	$\sqrt{\dfrac{\varepsilon}{m}}$
Force	ε/σ
Temperature	ε/k_B
Pressure	ε/σ^3
Energy	ε

Note: Here, $k_B = 1.3806488 \times 10^{-23}$ m^2 kg/Ks^2 Boltzmann's constant.

8.2.2 Constructing the Program

As with any program, we first need to get through the niceties of including necessary C++ libraries. Please see the program MD.cpp provided for relevant code.

```
#include<cstdlib>
#include<iostream>
#include<fstream>
#include<math.h>
using namespace std;
```

Table 8.2. Values for Lennard-Jones Potential Parameters for Various Noble Gases, as well as for Carbon

Species	ε(K)	σ (Å)
He–He	10.80	2.57
Ne–Ne	36.68	2.79
Ar–Ar	120.0	3.38
Kr–Kr	171.0	3.60
Xe–Xe	221.0	4.10
C–C	34.839	3.805

Source: From M.K. Balasubramanya et al. *J. Comput. Theor. Nanosci.* 5, 627–634, 2008 [8].

The next step is to declare the variables in the program.

```
const int Nmax=1000; //Number of particles
const int Ndist=1000; //Number of points used in calculating
distributions.
const int Nstep=10; //Number of steps in the simulation
const float dt=0.004; // Time step
const float Mass=1.00 //Mass of the particles
const float eps=1.0,sig=1.0; //Lennard-Jones potential parameters
const float rho=0.5; //Density of the system
const float rclose=0.98; //The closest the particles can be initially
const float Cutoff=1.5; //Cutoff for calculating interactions
const float rad=0.3; //Radius of particles for visualization in pov file
char filename[7]; //Character string that holds the file name for pov files
int ifile,irij,iUt;//file counter, distribution index and internal energy
distribution index
int seed = 3; //Random number seed
float abox,bbox,cbox; //Computational cell (box) parameters
float x[Nmax],y[Nmax],z[Nmax]; //Position vector
float vx[Nmax],vy[Nmax],vz[Nmax]; //Velocity vector
float vxold[Nmax],vyold[Nmax],vzold[Nmax]; //Previous velocity vector
float Fx[Nmax],Fy[Nmax],Fz[Nmax]; //Force vector
float Fxold[Nmax],Fyold[Nmax],Fzold[Nmax]; //Previous force vector
float Prij[Ndist],PUt[Ndist]; //Particle separation and total energy dists.
float red,green,blue; //Color components for visualization
float pi,rij,theta,phi; //pi, separation and angle variables
float EKavg,EKpre,EKnew,Tpre,Tpreavg; //Energy and temperature variables
float Vmag,vxdum,vydum,vzdum,vel; //Velocity and speed placeholder variables
float dx,dy,dz; //Coordinate separations
float Uc,Ucavg,Ut,Utavg,Utavg2,Utavg4; //Energy variables and averages
float VL; //Binder's Fourth Cumulant
float Press,Pavg; //Pressures
float Utmax,Utmin,dUt; Internal energies

float drij; //Step in separation used for creating Prij distribution
```

Now start the program.

```
int main()
{
```

Open the file that will have the simulation results.

```
ofstream myfile;
myfile.open("MD_Cpp.res");
```

Initialize arrays and any other variables as needed.

```
for(int i=1;i<=Nmax;i++)
{
x[i]=0.0;
y[i]=0.0;
z[i]=0.0;
vx[i]=0.0;
vy[i]=0.0;
vz[i]=0.0;
Fxold[i]=0.0;
Fyold[i]=0.0;
Fzold[i]=0.0;
Fx[i]=0.0;
Fy[i]=0.0;
Fz[i]=0.0;
vxold[i]=0.0;
vyold[i]=0.0;
vzold[i]=0.0;
}
for(int i=1;i<=Ndist;i++)
{
Prij[i]=0.0;
PUt[i]=0.0;
}
```

We now will calculate constants in the simulation here. In this program, the number density ρ and number of particles N_{max} are taken to be fixed values and so we have the relationship $N_{max} = abc\rho$, where a, b, anc c are rectangular box dimensions. For a cubic box, however, $a = b = c = L$ and we then have $N_{max} = \rho L^3$, and solving for L we get

$$a = b = c = L = \left(\frac{N_{max}}{\rho}\right)^{1/3}. \tag{8.1}$$

We could use the "pow" statement to code in the box parameters but we chose to use exp statements and logs instead (it's a homework problem to study the difference). So then Equation 8.1 becomes

$$a = b = c = L = \exp\left\{\log\left[\left(\frac{N_{max}}{\rho}\right)^{1/3}\right]\right\}, \tag{8.2}$$

which reduces to

$$a = b = c = L = \exp\left\{\frac{1}{3}\log\left[\frac{N_{max}}{\rho}\right]\right\}. \tag{8.3}$$

```
pi=4.0*atan(1.0);
abox=exp(0.3333*log(Nmax/rho));
bbox=exp(0.3333*log(Nmax/rho));
cbox=exp(0.3333*log(Nmax/rho));
```

```
drij=Cutoff/float(Ndist);
Utmin=-10000.0;
Utmax=-10.0;
dUt=(Utmax-Utmin)/float(Ndist);
```

Now that we have built the computational box, we need to place the particles at initial random positions within it. We are starting the loop here over all the particles in the simulation.

```
for(int i=1;i<=Nmax;i++)
{
```

First, we try an initial position. Thinking ahead, we know that we may be using periodic boundary conditions (I will discuss it later), which means we will copy the box on all sides of itself to mimic the behavior of a larger system. If we were to have particles on the boundaries of the box, then they could overlap when the boundary conditions are implemented. There are many ways to take care of such a problem, but here we simply make sure that we don't place the particles any closer than "rclose" to the faces or edges. So, to determine a trial initial position, pick a random spot that is no closer than rclose to any of the faces or edges:

```
x[i]=rclose+0.001*(rand()%1000)*(abox-2.0*rclose);
y[i]=rclose+0.001*(rand()%1000)*(bbox-2.0*rclose);
z[i]=rclose+0.001*(rand()%1000)*(cbox-2.0*rclose);
```

Note that the precision of the rand function can be adjusted, and in this case 1000 numbers in the interval between 0 and 1 is very reasonable. There can be another source of overlap when more than one particle is in the box, and so if we just tried to place the second (or later) articles in the box, we have to make sure that the new trial position is not too close to any of the other existing ones:

```
if(i>1)
{
line250:
for(int j=1;j<=i-1;j++)
{
```

We check the distance here.

```
rij=sqrt(pow(x[i]-x[j],2)+pow(y[i]-y[j],2)+
pow(z[i]-z[j],2));
```

And if the new particle is too close to any of the others, try again and go back to do the same check.

```
if(rij<rclose)
{
x[i]=rclose+0.001*(rand()%1000)*(abox-2.*rclose);
y[i]=rclose+0.001*(rand()%1000)*(bbox-2.*rclose);
z[i]=rclose+0.001*(rand()%1000)*(cbox-2.*rclose);
goto line250;
}
}
}
```

Things can slow down here when the box gets crowded and so we ask the computer to give us a shout out as to how things are going.

```
    cout<<"Atom "<<i<<" out of "<<Nmax<<" rare gas atoms just placed
in the box"<<endl;
    }
```

Now that we have the particles placed inside the computational cell (box) we need to assign initial velocities. This can be done in many ways, including incorporating a Gaussian or Boltzmann velocity distribution, but in this case we start out with *speeds* appropriate for a given temperature

$$v = \sqrt{\frac{3k_B T}{m}}, \tag{8.4}$$

and the *velocity* directed randomly over the unit sphere. In Equation 8.4 k_B is Boltzmann's constant, T is temperature and m is the particle's mass. The velocities of physical systems exhibit the Maxwell–Boltzmann distribution and I refer you to a good physical thermodynamics text for more background [9].

```
    for(int i=1;i<=Nmax;i++)
    {
    Vmag=sqrt(3.0*Temp/(Mass));
    theta=0.001*(rand()%1000)*pi;
    phi=0.001*(rand()%1000)*2.0*pi;
    vx[i]=Vmag*sin(theta)*cos(phi);
    vy[i]=Vmag*sin(theta)*sin(phi);
    vz[i]=Vmag*cos(theta);
    }
```

Now write out the initial configuration to the results file.

```
    myfile<<"    "<<endl;
    myfile<<"Initial Configuration (x,y,z):"<<endl;
    for(int i=1;i<=Nmax;i++)
    {
      myfile<<x[i]<<" "<<y[i]<<" "<<z[i]<<endl;
    }
```

as well as the initial velocities.

```
    myfile<<"    "<<endl;
    myfile<<"Initial Velocities (x,y,z):"<<endl;
    for(int i=1;i<=Nmax;i++)
    {
    myfile<<vx[i]<<" "<<vy[i]<<" "<<vz[i]<<endl;
    }
```

Together, the positions and velocities of the particles in the system are referred to as the system's *classical state* and I would point you towards theoretical mechanics texts if you want to pursue the theoretical background here further [10–12].

Now, we are going to write out instantaneous and average temperatures, energies, and pressures, so let's write a header in the results file so we can find the data there easily:

```
myfile<<" "<<endl;
myfile<<"Insantaneous and average prescaled T,Uc,U,P:"<<endl;
```

We are also going to write out multiple pov files for visualization and animation, and each file will have a number in its name so let's initialize an integer we will use later that ratchets up by one every so often (which we also get to determine—and there's your technical speak for the day!). Moreover, let's also initialize thermal averages (observables) before the time loop starts.

```
ifile=0;
Ucavg=0.0;
Tpreavg=0.0;
Utavg=0.0;
Utavg2=0.0;
Utavg4=0.0;
Pavg=0.0;
```

Now, we start the time loop and write out the time step to the screen, so we can monitor the progress of the simulation.

```
for(int it=1;it<=Nstep;it++)
{
cout<<it<<endl;
```

Initialize the force vectors and assign the correct values to the previous forces used in the integration algorithm.

```
for(int i=1;i<=Nmax;i++)
{
if(it==1)
{
Fxold[i]=0.0;
Fyold[i]=0.0;
Fzold[i]=0.0;
}
if (it>1)
{
Fxold[i]=Fx[i];
Fyold[i]=Fy[i];
Fzold[i]=Fz[i];
}
Fx[i]=0.0;
Fy[i]=0.0;
Fz[i]=0.0;
}
```

Now, we proceed with calculation of particle-particle quantities including pressure and configurational energies. We will begin by initializing the variables before the loop starts summing over the particles.

```
Press=0.0;
Uc=0.0;
```

The energy present when two particles interact can take on many forms but a very long-standing and useful one is the Lennard-Jones potential. There is an attractive term that arises from oscillating charge distributions in the two particles interacting and there is a repulsive term present due to the Pauli exclusion principle—that two electrons can't occupy the same space at the same time. For further history on the Lennard-Jones potential I refer you out [7].

$$ u_{LJ}(r_{ij}) = \varepsilon_{ij} \left[\left(\frac{\sigma_{ij}}{r_{ij}} \right)^{12} - \left(\frac{\sigma_{ij}}{r_{ij}} \right)^{6} \right]. \tag{8.5} $$

When we calculate the total interaction energy, we have to sum over all particle pairs in the system. We must be careful not to overcount so we can take an identical double sum and divide by two, or make the restriction that one particle index always be greater than the other, as in Equation 8.6.

$$ U = \left\langle \sum_{i=1}^{N} \sum_{j>i} 4\varepsilon_{ij} \left[\left(\frac{\sigma_{ij}}{r_{ij}} \right)^{12} - \left(\frac{\sigma_{ij}}{r_{ij}} \right)^{6} \right] \right\rangle \tag{8.6} $$

Now, the force on particle (i) is equal to the negative of the spatial gradient of the potential energy it experiences

$$ \vec{F}^i = -\vec{\nabla} U^i \tag{8.7} $$

and Equation 8.7 takes the following form for the three individual components of the force on particle (i):

$$ F_x^i = -\frac{\partial U^i}{\partial x_i} = 48 \sum_{\substack{j=1 \\ j \neq i}}^{N} \epsilon_{ij} \left[\left(\frac{\sigma_{ij}}{r_{ij}} \right)^{14} - \frac{1}{2} \left(\frac{\sigma_{ij}}{r_{ij}} \right)^{8} \right] (x_i - x_j) \tag{8.8} $$

$$ F_y^i = -\frac{\partial U^i}{\partial y_i} = 48 \sum_{\substack{j=1 \\ j \neq i}}^{N} \epsilon_{ij} \left[\left(\frac{\sigma_{ij}}{r_{ij}} \right)^{14} - \frac{1}{2} \left(\frac{\sigma_{ij}}{r_{ij}} \right)^{8} \right] (y_i - y_j) \tag{8.9} $$

$$ F_z^i = -\frac{\partial U^i}{\partial z_i} = 48 \sum_{\substack{j=1 \\ j \neq i}}^{N} \epsilon_{ij} \left[\left(\frac{\sigma_{ij}}{r_{ij}} \right)^{14} - \frac{1}{2} \left(\frac{\sigma_{ij}}{r_{ij}} \right)^{8} \right] (z - z_j) \tag{8.10} $$

The pressure also involves particle-particle interactions and can be calculated from the *Virial*, which is expressed as

$$P = \frac{48\rho}{6N}\left\langle \sum_{i=1}^{N}\sum_{j>i} \varepsilon_{ij} r_{ij}\left[\left(\frac{\sigma_{ij}}{r_{ij}}\right)^{14} - \frac{1}{2}\left(\frac{\sigma_{ij}}{r_{ij}}\right)^{8}\right]\right\rangle. \tag{8.11}$$

The next part ensures that periodic boundary conditions are enforced in our calculations. They help mimic a larger system but can also introduce periodicities that are artificial and unduly stabilize the system. For further discussion on periodic boundary conditions, I refer you to external reading [13].

```
for(int i=1;i<=Nmax;i++)
{
for(int j=1;j<=Nmax;j++)
{
for(int kx=1;kx<=3;kx++)
{
dx=x[i]-x[j]-abox*float(kx-2);
for(int ky=1;ky<=3;ky++)
{
dy=y[i]-y[j]-bbox*float(ky-2);
for(int kz=1;kz<=3;kz++)
{
dz=z[i]-z[j]-cbox*float(kz-2);
rij=sqrt(pow(dx,2)+pow(dy,2)+
pow(dz,2));
if(rij!=0.0)
{

if(rij<=Cutoff)
{
Fx[i]=Fx[i]+48.0*(eps/pow(sig,2))*dx*(pow(sig/rij,14)
-0.5*pow(sig/rij,8));
Fy[i]=Fy[i]+48.0*(eps/pow(sig,2))*dy*(pow(sig/rij,14)
-0.5*pow(sig/rij,8));
Fz[i]=Fz[i]+48.0*(eps/pow(sig,2))*dz*(pow(sig/rij,14)
-0.5*pow(sig/rij,8));
Press=Press-(rho/(6.0*float(Nmax)))*rij*
48.0*(eps/pow(sig,2))*(pow(sig/rij,14)
-0.5*pow(sig/rij,8));
irij=int(rij/drij);
Prij[irij]=Prij[irij]+0.5;
Uc=Uc+4.*eps*(pow(sig/rij,12)-pow(sig/rij,6));
}
}
}
}
}
}
}
Press=Press/2.0;
Press=rho*Temp-Press;
```

Calculate running averages.

```
Pavg=(float(it-1)*Pavg+Press)/float(it);
Ucavg=0.5*(float(it-1)*Ucavg+Uc)/float(it);
```

Now that we have the current force, as well as the force from the previous time step, we can get the corresponding accelerations and take the first step in the Velocity Verlet integration algorithm

$$v^n = v^{n-1} + \frac{1}{2}(a^n + a^{n-1})\Delta t \qquad (8.12)$$

which looks like the following to the computer:

```
for(int i=1;i<=Nmax;i++)
{
vxold[i]=vx[i];
vyold[i]=vy[i];
vzold[i]=vz[i];
if (it==1)
{
vxdum=vx[i]+Fx[i]*dt/Mass;
vydum=vy[i]+Fy[i]*dt/Mass;
vzdum=vz[i]+Fz[i]*dt/Mass;
}
if(it!=1)
{
vxdum=vx[i]+0.5*(Fx[i]+Fxold[i])*dt/Mass;
vydum=vy[i]+0.5*(Fy[i]+Fyold[i])*dt/Mass;
vzdum=vz[i]+0.5*(Fz[i]+Fzold[i])*dt/Mass;
}
vx[i]=vxdum;
vy[i]=vydum;
vz[i]=vzdum;

}
```

But now that we have updated the velocities it could be that the temperature changed, because the temperature is related to the system's average kinetic energy:

$$T = \frac{1}{3N\,k_B} \sum_{i=1}^{N_m} m_i v_i^2. \qquad (8.13)$$

There is no adequate way to correctly keep temperature constant in an MD simulation and there are many different types of thermostats I would encourage you to read up on [14,15], but here we employ good old homespun velocity rescaling. That is, we will multiply all the atomic/molecular speeds by a factor so that the average kinetic energy is appropriate for the temperature we are trying to target. We can rescale every step or choose to allow some drift and rescale every so often (there's that technical speak again) with a frequency we specify. In code speak we have

```
EKavg=0.0;
for(int i=1;i<=Nmax;i++)
{
vel=sqrt(pow(vx[i],2)+pow(vy[i],2)+pow(vz[i],2));
EKavg=EKavg+
0.5*Mass*pow(vel,2);
}
EKpre=EKavg;
Tpre=2.0*EKpre/3.0;
```

```
Tpreavg=(float(it-1)*Tpreavg+Tpre)/float(it);
EKnew=0.0;
for(int i=1;i<=Nmax;i++)
{
vxdum=vx[i]*
sqrt((1.5*Temp)/EKavg);
vydum=vy[i]*
sqrt((1.5*Temp)/EKavg);
vzdum=vz[i]*
sqrt((1.5*Temp)/EKavg);
vx[i]=vxdum;
vy[i]=vydum;
vz[i]=vzdum;
vel=sqrt(pow(vx[i],2)+pow(vy[i],2)+pow(vz[i],2));
EKnew=EKnew+
0.5*Mass*pow(vel,2);
}
```

After the velocities are rescaled we then can take the second step in the Velocity Verlet integration

$$x^n = x^{n-1} + \frac{1}{2}(v^{n-1} + v^n)\Delta t. \tag{8.14}$$

which when coded looks like

```
for(int i=1;i<=Nmax;i++)
{
x[i]=x[i]+0.5*(vxold[i]+vx[i])*dt;
y[i]=y[i]+0.5*(vyold[i]+vy[i])*dt;
z[i]=z[i]+0.5*(vzold[i]+vz[i])*dt;
}
```

We have to be careful that we didn't send a particle outside the box in the previous step, and so the following code makes sure that all particles remain within the box:

```
for(int i=1;i<=Nmax;i++)
{
if(x[i]<0.0)
{
x[i]=x[i]+abox;
}
if(x[i]>abox)
{
x[i]=x[i]-abox;
}
if(y[i]<0.0)
{
y[i]=y[i]+bbox;
}
if(y[i]>bbox)
{
y[i]=y[i]-bbox;
}
if(z[i]<0.0)
{
z[i]=z[i]+cbox;
```

```
}
if(z[i]>cbox)
{
z[i]=z[i]-cbox;
}
}
```

Now, calculate the total internal energy, which is a sum of kinetic and potential energy terms, and also begin to build up the internal energy distribution.

```
Ut=Uc+1.5*float(Nmax)*Temp;
iUt=int((Ut-Utmin)/dUt);
if(iUt<=Ndist)
{
PUt[iUt]=PUt[iUt]+1.0;
}
```

Calculate averages of important quantities and write them out to the results file:

```
Utavg=(float(it-1)*Utavg+Ut)/float(it);
Utavg2=(float(it-1)*Utavg2+pow(Ut,2))/float(it);
Utavg4=(float(it-1)*Utavg4+pow(Ut,4))/float(it);
myfile<< float(it)*dt<<" "<<Tpre<<" "<<Tpreavg<<" "<<Uc<<"
"<<Ucavg<<" "<<Ut<<" "<<Utavg<<" "<<Press<<
" "<<Pavg<<endl;
```

Now, ratchet up the pov file counter by one, write the new file name and open it for writing:

```
ifile=ifile+1;
sprintf(filename,"%d.pov",ifile);
ofstream myfile2;
myfile2.open(filename);
```

Start writing a new .pov file for visualization.

```
myfile2<<"#version 3.6"<<endl;
myfile2<<"#include "<<"\""<<"colors.inc"<<"\""<<endl;
myfile2<<"global_settings"<<endl;
myfile2<<"{"<<endl;
myfile2<<"assumed_gamma 1.0"<<endl;
myfile2<<"}"<<endl;
myfile2<<"camera"<<endl;
myfile2<<"{"<<endl;
myfile2<<"location"<<"<"<<2.5*abox<<","<<2.5*bbox<<","<<2.5*cbox<<">"
<<endl;
myfile2<<"direction 1.5*z"<<endl;
myfile2<<"right 4/3*z"<<endl;
myfile2<<"look_at  <"<<abox/2.0<<","<<bbox/2.0<<","<<cbox/2.0<<">"
<<endl;
myfile2<<"}"<<endl;
myfile2<<"background { color red 0.5 green 0.5 blue 0.5 }"<<endl;
myfile2<<"light_source"<<endl;
myfile2<<"{"<<endl;
myfile2<<"0*x"<<endl;
myfile2<< "color red"<<endl;
myfile2<<"1.0  green 1.0  blue 1.0"<<endl;
myfile2<<"translate <-30, 30, -30>"<<endl;
```

```
myfile2<<"}"<<endl;
myfile2<<"light_source"<<endl;
myfile2<<"{"<<endl;
myfile2<<"0*x"<<endl;
myfile2<<"color red"<<endl;
myfile2<< "0.5  green 0.5  blue 0.5"<<endl;
myfile2<<"translate <30, 30, -30>"<<endl;
myfile2<<"}"<<endl;
for(int k=0;k<=Nmax;k++)
{
red=1.0;
green=0.0;
blue=0.0;
myfile2<<"sphere { <"<<x[k]<<","<<y[k]<<","<<z[k]<<">,"<<rad<<endl;
myfile2<<"texture {pigment {color
rgb<"<<red<<","<<green<<","<<blue<<">}"<<endl;
myfile2<<"finish{specular 1}} }"<<endl;
}
```

Close the pov file,

```
myfile2.close();
}
```

and end the MD time loop.

```
        }
```

In studying the thermodynamics of various systems it is useful to calculate Binder's fourth cumulant

$$V_L = 1 - \frac{<U^4>}{3<U^2>^2},$$ (8.15)

which is useful in determining the modality of the system's energy distribution. Why on earth is that useful, you ask? It's useful because phase transitions in systems can be sharp (first-order, like ice to water and vice versa) or more continuous. When the phase transitions are sharp, the system jumps between two distinct phases bracketing the transition and so its energy distribution is bimodal, but in more continuous transitions (second order and higher) there is no hopping and no bimodality either. It is left as a homework problem for you to look deeper into Binder's fourth cumulant and how it indicates bimodality.

```
VL=1.0-Utavg4/(3.0*pow(Utavg,2));

    myfile<<" "<<endl;
    myfile<<"Binder's 4th Cumulant:"<<VL<<endl;
    myfile<<"" ""<<endl;
    myfile<<"Pair Distribution Function"<<endl;
    for(int i=1;i<=Ndist;i++)
    {
    myfile<<Cutoff*0.5*(float(i-1)+float(i))/float(Ndist)<<
    " "<<Prij[i]<<endl;
    }
    myfile<<" "<<endl;
    myfile<<"Internal Energy Distribution Function"<<endl;
```

```
                    for(int  i=1;i<=Nmax;i++)
                    {
                    myfile<<Utmin+dUt*0.5*(float(i-1)+float(i))<<
            " "<<PUt[i]<<endl;
                    }
myfile.close();
return(0);
}
```

8.2.3 Thermostats: Nobody Does Temperature Control Like Nature Does

Fact of the matter is that nobody knows how to perfectly control temperature in an MD simulation. This is partly because temperature is a quantity that is defined formally in the thermodynamic limit (for an infinite system). There are many ways to make sure that the system's average kinetic energy has the correct value based on the targeted temperature but it's not always certain that, for example, velocities will exhibit the Maxwell–Boltzmann distribution which is required and understood through statistical physics. Each thermostat has benefits and drawbacks and a few of the more widely used thermostats are presented in Table 8.3.

Table 8.3. Various Thermostats Used in Molecular Dynamics Simulations along with Pro's and Con's

	Action	Pro's	Con's
Thermostat			
Velocity Rescaling	Rescale kinetic energies so the temperature has the target value	Straightforward, good for initial investigations	Doesn't target any ensemble, not time-reversible, system not necessarily ergodic
Nose Hoover	Extra term to equation of motion related to heat bath coupiling	Easy to implement and use, deterministic and time reversible. Consistent with canonical	Doesn't handle instantaneous temperature jumps well
Lengevin	Constant friction and random force	Correct canonical ensemble, ergodic, can use larger time step	Difficult to implement drag for nonspherical particles: related toparticle radius; momentum transfer lost: cannot compute diffusion coefficient
Andersen	Assign velocity of particle to Maxweell distribution	Allows sampling from canonical ensemble	Algorithm randomly decorrelates velocity: dynamics are not physical
Berendsen	Velocity rescaling Weak coupling to heat bath	Very efficient	Not consistent with canonical ensemble
Gaussian	Strong coupling to heat bath	Canonical	Non-hamiltonian perturbations

(Continued)

Table 8.3. (*Continued*) Various Thermostats Used in Molecular Dynamics Simulations along with Pro's and Con's

	Action	Pro's	Con's
Integration Algorithm			
Verlet	Position is calculated from previous two time steps	Straightforward easy on memory, easy implementation, stable	Not the most precise, not self starting
Verlet Leapfrog	Position and velocities calculated at alternating half integer time steps	Velocities are explicitly calculated, easy implimentation, stable	Vel and pos not simultaneously calculated, not self-started
Velocity Verlet	Velocity and position calculated at the same time with a force update in between.	Improved accuracy compared to standard, Verlet, easy implementation, stable	Not the most precise
Beeman's	Predictor-corrector variation of Verlet. Identical positions as Verlet.	More accurate expression for the velocities and better energy conservation	Calculation more expensive
Predictor-corrector	Predict positions and velocities, calculate force and then correct the prediction.	High accuracy, permit larger time steps	Calculation more expensive and implimetatin is more complex

8.2.4 Choosing an Integration Algorithm for Your Simulation

There are also many ways to step the system through time, and therefore many choices for integrators. Any algorithm you pick should ideally conserve energy (and therefore be time reversible) and also conserve momentum. It should be computationally efficient (minimizing the number of force calculations) and should also permit a long time step for integration. The table below summarizes some commonly used integration algorithms and their benefits as well as drawbacks.

8.2.5 More Complicated Stuff: Modeling Molecules with Molecular Dynamics

Many of the systems we study that are interesting are comprised of molecules and are therefore more complicated than the simple atoms we have discussed so far. In fact, they have internal (intra-) as well as external (inter-) molecular degrees of freedom which must be taken into account. The intramolecular degrees of freedom are illustrated in Figure 8.1a–c.

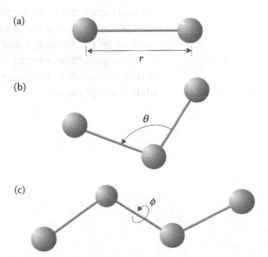

Figure 8.1. Illustration of bond length stretching (a), bond angle bending (b), and dihedral angle torsion (c) degrees of freedom (R. He et al., *Nano Lett.*, 12(5), 2408–2413, 2012, doi: 10.1021/nl300397v.) [25].

Table 8.4. Expressions for the Types of Interaction Potentials Used in Molecular Simulations

Interaction Potential	Type	Formula
Bond stretching	Intra; bonded; 2-body	$u_{stretch} = k(l - l_0)^2$
Bond angle bending	Intra; bonded; 3-body	$u_{bend}(\theta) = k_\theta (\theta - \theta_0)^2$
Dihedral torsion	Intra; bonded; 4-body	$u_{dihed} = k_d \{ 1 + \cos(n\phi_d - \delta) \}$
Van der Waals (modified Lennard-Jones)	Inter; nonbond; 2-body	$u_{LJ}(r_{ij}) = \varepsilon_{ij} \left[\left(\dfrac{r_0}{r_{ij}} \right)^{12} - 2 \left(\dfrac{r_0}{r_{ij}} \right)^6 \right]$
Coulomb (electroscatic)	Inter; nonbond; 2-body	$u_C = \dfrac{k q_i q_j}{r_{ij}}$
Scaled nonbond (Coulomb and van der Waals)	Intra; nonbond	Excluded for 1–3 pairs; scaled for 1–4 pairs, and fully included for 1–5 and more distant pairs.

Source: M.J. Connolly, M.W. Roth, Carlos Wexler, Paul A. Gray, Molecular Dynamics Computer Simulations of Hexane Adlayers on Graphite: Comparing All – Atom and United Atom Models, *53rd Midwest Solid State Physics Conference*, Kansas City, MO, October 7, 2006 [41].
Note: Interactions between atoms on the same molecule are classified as "intra" and those between atoms on different molecules or adatom – substrate interactions are tagged "inter."

Typical interactions and expressions important in molecular simulation are shown in Table 8.4 and we are not showing the *improper* angular degree of freedom because it doesn't apply to molecules having the geometry that we are working with here. I recommend outside reading on the subject; it just wouldn't be *ahem* proper here.

Temperature constraint in molecular simulationis achieved by using velocity rescaling to maintain the center-of-mass, rotational and internal temperatures at the simulated temperature:

$$T_{CM} = \frac{1}{3 N_m k_B} \sum_{i=1}^{N_m} M_i v_{i,CM}^2, \tag{8.16}$$

$$T_{ROT} = \frac{1}{3 N_m k_B} \sum_{i=1}^{N_m} \vec{\omega}_i^{\ T} \vec{I}_i \vec{\omega}_i, \tag{8.17}$$

$$T_{INT} = \frac{1}{2 n_C - 5} \sum_{i=1}^{N_m} \sum_{j=1}^{n_C} m_{ij} (\vec{v}_{ij} - \vec{v}_{i,CM} - \vec{\omega}_{i,CM} \times \vec{r}_{ij,CM})^2. \tag{8.18}$$

So, if you are reading this and are surprised that there are more than one temperature that characterizes the system you are absolutely correct!! Here T_{CM}, T_{ROT}, and T_{INT}

are the respective temperatures of the system and the summation index (i) runs over molecules and the index (j) runs over atoms that comprise the molecule. All variables indexed with (i) are standard and apply to the ith molecule and those indexed with (ij) apply to the jth atom within molecule (i). In many simulations, the internal and external temperatures are kept constant separately and the simulation is allowed to naturally partition energy within them.

8.2.6 There's More to Life than Solid, Liquid, and Gas

Crude oil and its derivatives are compounds which have deep practical and industrial significance. Although of varying composition, even from the same source, crude oils have distribution spectra of predominantly of n – alkanes (abbreviated C_n), which are straight-chained hydrocarbons having the general formula C_nH_{2n+2}. Natural gas is comprised of C_1–C_5 molecules in varying ratios and crude oil generally from C_6 to C_{30} [16–22]. Some basic practical information about alkanes in crude oil can be found in Table 8.5.

Since alkanes comprise a family of compounds they lend themselves quite nicely to comparative studies in theoretical, computational, and experimental arenas. Because of the wide distribution of alkane chain lengths in crude oils, they make ideal systems for studying comparisons, contrasts, and selective behavior within the alkane family. It's kind of like living in a yellow submarine, only it's usually deep in the earth and very dark. Computer simulations can give much insight into the behavior of physical systems from both supplementary as well as predictive standpoints, and can shed light on system dynamics not readily accessible in experiments.

With the hexane/graphite system [23] shown in Figure 8.2, there is clearly a distinct *nematic* phase between the solid and liquid, which is ordered but behaves like a liquid. The behavior of this nematic mesophase seen between the *herringbone* solid and *isotropic* liquid agrees much more nicely with experimental data than does that of the

Table 8.5. Basic Information about Various Alkanes in Crude Oil

Compound Range	Mixture(s) Contained in	Use
C1–C4	Natural gas	Cooking; heating
C5–C10	Naptha and gasoline	Solvent chemicals; car fuel
C10–C16	Kerosene	Heating; jet fuel; lighting
C14–C20	Diesel fuels	Fuel for trucks/trains
C20–C70	Lubricating and fuel oils	Machinery, ships, oils, waxes, polishes
>C70	Residue/asphaltines	Pavement; roofing

Source: Z. Wang, M. Fingas, K. Li, *Journal of Chromatographic Science*, 32(9), 361–366, 1994 [16]; C.A. Hughey, R.P. Rodgers, A.G. Marshall, *Anal. Chem*, 74(16), 4145–41499, 2002 [17]; K. Sugiura et al. *Environ. Sci. Technol.*, 31, 45–51, 1997 [18]; F. Mutelet, G. Ekulu, M. Rogalski *Journal of Chromatography A*, 969, 207–213, September 2002 [19]; Z. Ha, Z. Ring, S. Liu, *Energy & Fuels*, 19(4), 1660–1672, 2005 [20]; N. Thanh, M. Hsieh, R.P. Philp, *Organic Geochemistry*, 30, 119–132, 1999 [21]; D.C. Villalanti, J.C. Raia, J.B. Maynard, *Encyclopedia of Analytical Chemistry*, R.A. Meyers (Ed.) John Wiley & Sons Ltd: Chichester, 2000, 6726–6741 [22].

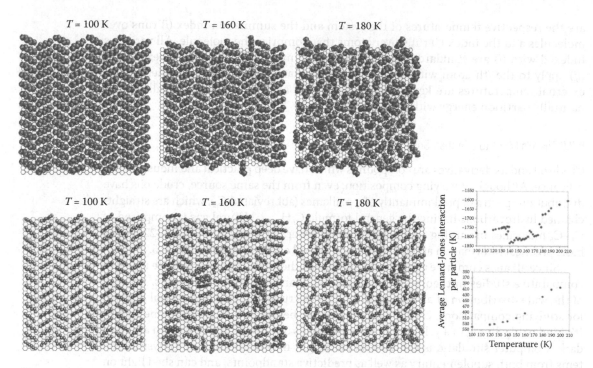

Figure 8.2. Final snapshots of hexane (C6) on graphite at various temperatures in the united atom model (bottom) and with explicit hydrogens (top). With hydrogens in place, molecular rolling agrees well with experiment, and the nematic mesophase (not seen in experiment) is suppressed. Moreover, the artificial drop in the molecular interaction energy (top panel of lower right hand plot) is not seen with explicit hydrogens (bottom panel of the same graph). Atoms and pseudoatoms are colored in the electronic version of the text (M.J. Connolly, M.W. Roth, Carlos Wexler, Paul A. Gray, Molecular Dynamics Computer Simulations of Hexane Adlayers on Graphite: Comparing All – Atom and United Atom Models, *53rd Midwest Solid State Physics Conference*, Kansas City, MO, October 7, 2006.) [41].

United atom model, also shown in Figure 8.2, with hydrogens suppressed to save computational time but unrealistic changes of the energy upon phase transitions.

In the case of tetracosane on graphite [24], Figure 8.3 shows that there is a distinct phase in between the solid and liquid, the *smectic* mesophase, which is seen in experiment. In the smectic, the molecules remain basically straight but the rows, or lamellae begin to shift very slowly. The nematic and smectic phases appear because molecules interact more strongly when they are on a surface, which gives rise to their rich and interesting behavior there.

8.2.7 When Molecular Dynamics Works Out Better Than You had Hoped

In the scientific world, it is great to have friends and collaborators. I completed a Molecular Dynamics project with another research group that was so thrilling to me that I can't help but share it here. Graphite is made up of many sheets of *graphene*, which is just a flat sheet of hexagonal carbon. When you place graphene on another one that has a different periodicity or pattern, you create stresses and strains. You can get the same effect by placing two window screens about a foot apart and looking

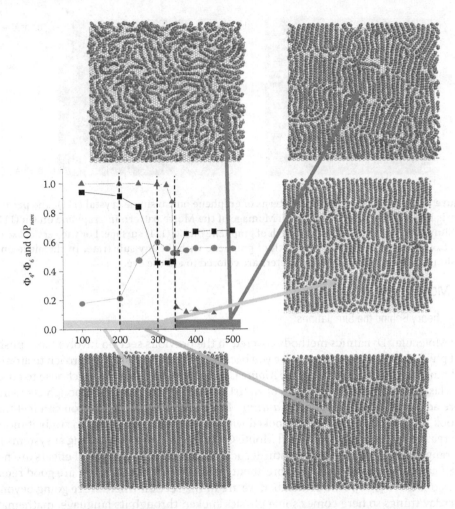

Figure 8.3. Final snapshots of all hydrogen tetracosane (C24) on graphite. Two order parameters change slightly above $T = 200$ K when the system enters the nematic phase, and another order parameter drops at around $T = 350$ K, indicating melting. In the electronic version, snapshots and order parameters are colored to aid the eye (L. Firlej, B. Kuchta, M.W. Roth, Paul A. Gray, Carlos Wexler, Molecular Dynamics study of tetracosane monolayers adsorbed on graphite, *2008 APS March Meeting*, New Orleans, LA, March 10–14, 2008.) [42].

through them at different angles. The experimentalists in our group were placing a graphene sheets on copper and noticing evidence of compressive strain patterns. I repeated the same conditions in our computer simulations and the result is shown in Figure 8.4. The simulations were able to explain not only the effect and presence of compressive strain on the graphene [25], but the actual magnitude of it as well. I will remember that day as being a very good day at work. Another good day at work for me had to do with Monte Carlo simulations in the next section, including an e-mail whose title line was "YAY!"

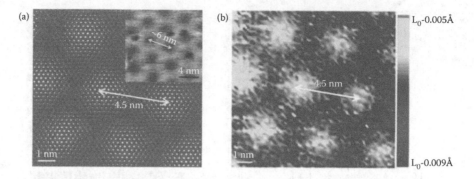

Figure 8.4. (a) Simulated Moire patterns of graphene on Cu single crystal (111). The periodicity is labeled. The inset shows an STM image of the Moire pattern in graphene on Cu (111). (b) Simulated nearest C–C bond length of graphene on Cu (111) surface. For the vertical scale bar, L_0 is the equilibrium bond length of graphene without any substrate. In the electronic version, snapshots and order parameters are colored to aid the eye.

8.3 Monte Carlo Simulations

8.3.1 Theory Behind the Simulations

The Molecular Dynamics methods covered in the previous section involve basic push-pull physics. I would now like to give you background on a statistical approach to atomic and molecular modeling called the *Monte Carlo* method. I am going to choose to cover the classical Monte Carlo method of statistical mechanics [26,27], although there are more advanced methods called *quantum Monte Carlo* methods [28]; you can feel free to look into that if you really get hooked with this stuff and you want to study it more. The reason we're choosing classical Monte Carlo is because we're looking at systems in the range of densities and temperatures such that quantum mechanical effects are not significant. If you want to learn more about quantum mechanics, there are good references out there [4–6]. And remember, we are in the section where we're going beyond everyday things so here comes some physics spoken through its language, mathematics. If you don't have much of a physics background that's OK, This book is still written for you, but do put your seat belt on.

Consider an isolated system. It is a system that does not interact with the outside world; it is isolated and said to be conservative. I know, I know, I shouldn't bring politics into all this but I would also mention a liberal system... if there were any. Anyway, let's say this system has a constant number of particles N, which for example later will be the number of atoms. It also has a constant area or volume V depending upon whether you're working in two or three dimensions, and it also has a constant energy E. We typically call such a collection of particles an ensemble, and this particular ensemble is called the *Gibbs Microcanonical*, or (N,V,E) ensemble.

There is a mathematical construct in statistical mechanics called the *partition function*, defined using calculus as

$$Z = \int_\Gamma W(\vec{p},\vec{r})d\vec{p}\,d\vec{r}$$

(8.19)

Here, we are integrating over the phase space of the system—the range of positions and momenta that are accessible to it [29]. For any ensemble, the partition function can be used to calculate quantities that can be measured in a laboratory. Such quantities are called *ensemble averages,* and the whole idea behind Monte Carlo calculations are to calculate ensemble averages so we can compare those with what is measured in the lab or, even more significantly, to predict laboratory behavior. Now, in Equation 8.19, W is the energy probability distribution as a function of the position and momentum of every particle in the system. For the Gibbs Microcanonical Ensemble, since there is only one accessible energy, the probability of the distribution is a sharp spike. That is to say that there's only one energy accessible mathematically. There is a special function that we use for such spikes and physics, which are called Dirac delta functions, and I recommend reading up on those [30].

$$P(\vec{p},\vec{r}) = \frac{W(\vec{p},\vec{r})}{Z} = \frac{\delta(H(\vec{p},\vec{r})-E)}{Z}. \tag{8.20}$$

Dirac delta functions are essentially infinite at one spot and zero everywhere else. It is very important to think about what your model is doing, however, and it is not realistic for an atomic and molecular system to be isolated and have a constant energy. That means we have to look at a different ensemble.

So consider a system now with a constant number of particles N, constant volume V, and constant temperature T. Sometimes physics can be counterintuitive: just because it has a constant temperature does not mean it has constant energy. One way to understand such a thing is to imagine the system to be connected to an infinitely large heat bath at temperature T, which can input or output energy so that the temperature remains constant. That is to say that if the system experiences any change in its *potential* energy, the heat bath has to adjust the system's *total* energy so its average *kinetic* energy is constant. The system we're talking about here is called the *Gibbs Canonical* ensemble, or (N,V,T) ensemble, and it has an exponential probability distribution that looks like

$$P(\vec{p},\vec{r}) = \frac{e^{-\beta U(\vec{p},\vec{r})}}{Z}. \tag{8.21}$$

Now, in principle the probability distribution contains all the information about the system in question, and knowledge of that distribution allows us to calculate observable quantities. So suppose we want to calculate the ensemble average $<A>$ for some parameter of the system knowing how it is expressed in terms of momentum and position $A = A(\vec{p},\vec{r})$. Then, to calculate the average we have to do the following integral:

$$<A> = \int_{\Gamma} A(\vec{p},\vec{r})P(\vec{p},\vec{r})d\vec{p}\,d\vec{r}. \tag{8.22}$$

One may visualize Equation 8.22 as being similar to calculating the centroid of an object. The expression for centroid actually involves an integral over the system's volume: $\vec{R} = (1/M)\int_V \vec{r}\rho(\vec{r})d\vec{r}$. Taking one more step we can write $\vec{R} = \int_V \vec{r}\rho(\vec{r})d\vec{r} / \int_V \rho(\vec{r})d\vec{r}$ and the expression becomes immediately recognizeable fom the mean value theorem encountered in first-year calculus! The above is really all we're doing in Monte Carlo—only we are calculating the weighted average of A (not $\vec{r}$) over phase space (not volume V). And, to tie the two ideas together the density ρ may be thought of as the mass probability distribution.

In Monte Carlo methods, the integral in Equatin 8.22 involves position and momentum, but the momentum cancels out and doesn't matter and so the actual integral we have to do looks like

$$\langle A \rangle = \frac{\int_\Gamma A(\vec{r})e^{-\beta U(\vec{r})}\, d\vec{r}}{\int_\Gamma e^{-\beta U(\vec{r})}\, d\vec{r}}. \tag{8.23}$$

Now, from past work we know that a computer cannot do integrals directly, so we have to express the integrals as sums over the configurations $\{\vec{r}_i\}$ accessible to the system:

$$\langle A \rangle = \frac{\sum_{i=1}^{M} A\{\vec{r}_i\}e^{-\beta U\{\vec{r}_i\}}}{\sum_{i=1}^{M} e^{-\beta U\{\vec{r}_i\}}}. \tag{8.24}$$

Now, everything we have talked about so far involves fundamental statistical mechanics, but the Monte Carlo method comes into play for calculation of the averages. In principle, we can average over all possible configurations for the system, but actually the system will tend to spend most of its time preferentially in certain configurations. Therefore, instead of blindly averaging over the systems, we use what is called smart sampling. To not use smart sampling woud be like to search for your car in a parking lot blindfolded. We advance the system by a move, thereby generating a trial configuration. Then we look at the difference in potential energies between the initial configuration and the trial configuration. We then accept the trial configuration as a new one if its energy is lower, and we accept it with the probability $e^{-\beta\Delta U}$ if the trial configuration is of higher energy. What the latter part amounts to is comparing $e^{-\beta\Delta U}$ to a random number between 0 and 1 inclusive, and then accepting the trial configuration if the random number is smaller, rejecting if it is larger and conditionally accepting if they are equal. The simulation is advanced for a desired number of steps. Just as with Molecular Dynamics, Monte Carlo simulations are usually broken into two phases. There is an *equilibration* part where the system is transitioning from its initial configuration to reaching its most likely configuration phase space (sometimes called *thermalization*), and a *production* part over which meaningful averages are taken.

Then, the average for the parameter A simply looks like

$$\langle A \rangle = \frac{1}{M}\sum_{h=1}^{M} A(\{\vec{r}_h\}). \tag{8.25}$$

And we just average over the configurations generated in the simulation.

There are also advanced Monte Carlo methods such as *Histogram Monte Carlo* [31]. The idea behind Histogram Monte Carlo is that, as we run a Monte Carlo simulation we know the system's configuration probability distribution as a function of temperature. So, using the idea of smart sampling we can, with some limited success, calculate averages for another temperature T' from a run at a single temperature T, by reweighting the configurations generated by smart sampling:

$$\langle A_{T'} \rangle = \frac{\sum_{i=1}^{M} A\{\vec{r}_i\} e^{-(\beta' - \beta)U\{\vec{r}_i\}}}{\sum_{i=1}^{M} e^{-(\beta' - \beta)U\{\vec{r}_i\}}}. \tag{8.26}$$

The Histogram Monte Carlo method can dramatically reduce calculation time, as you might imagine, but it also has limitations that must be noted. Calculating the sums in Equation 8.26 over regions of phase space, which have not been thoroughly sampled, will result in poor statistics for the probability distribution. One way to mitigate this is to calculate contributions to the thermal average using runs at several strategically-placed different initial temperatures, which significantly reduces the statistical error in Histogram Monte Carlo calculations. Another Histogram Monte Carlo method at multiple temperatures is to calculate curves at new temperatures generated by simulations at various single temperatures, and then overlapping them. There is literature I'll refer you to regarding other ways to improve and optimize Histogram Monte Carlo [32], and I recommend a follow-up if you're interested. Histogram Monte Carlo is best suited for systems near phase transitions where different states of the system are often sampled, or systems at high enough temperatures where energy probability distributions are sufficiently broad. How broad is sufficiently broad? Use your judgement in a homework problem!

8.3.2 Applying Monte Carlo Theory: A Program for Well-Known Magnetic Systems

Just like we constructed the MD program in the previous section around a system for which results are already known so the program can be validated, we can do the same for a Monte Carlo program. Consider a system of magnets placed on a lattice, with an additional external magnetic field present. The magnets can interact with each other, as they can also with the external field. If we look at each interaction separately, there are systems for which we can validate our program. In the case where there is no external field, the magnets interact with each other and we have an Ising Magnet, and their behavior with temperature and magnetic field are very well-known. In the other case, when we have a magnet interacting with an external field only (we use many magnets and shut off their interactions with each other [what's the advantage of this??]) then we have a Lengevin Magnet, whose behavior is also well-characterized.

The provided program MagneticMC.f is related to the following discussion; in FORTRAN we name the program first:

```
ProgramSurRMag
```

Now for the parameters, which are global constants in the program:

```
parameter(B_max=5.0,nB=1)!Magnetic field maximum and incriment
parameter (iLengevin=1,iIsing=0)!Flags for choosing the system
parameter(T_max=20.0,nT=10)!Temperature maximum and incriment
parameter (bohrmu=9.274e-24,boltz=1.3806503e-23)!Bohr magneton and
                                               !Boltzmann's constant
parameter(spin=2.5)!Particle spins
parameter(nmax=10000,nmag=5000,nevery=5000)!Number of steps and pov
                                               !frequency
parameter(tri_spacing=3.0)!Triangular lattice spacing
parameter(nx=10,ny=10)!Number of particles in the x and y dimensions
parameter (vdw=1.0)
```

Next, declare program variables.

```
character*18 fileanim !Character string that holds pov file names
integer i,j,k,ii,isite,iatom,igoto,icount !Loop indices
integer itold,itnew,im,ib !More loop indices
integer ifr,istep,imstep !Step loop indices
integer nBsteps,nTsteps !Number of temperature and magnetic steps
real pi,rnum !Pi and a real nmber placeholder variable
real Sxold,Syold,Szold,Sxnew,Synew,Sznew !Pre, post spin components
real red,green,blue !Colors for pov files
real Upre,Upost ! Pre, post energies
real Ubext ! External magnetic field interaction energy
real sxdum,sydum,szdum !Spin component placeholders
real sdrawx,sdrawy,sdrawz !Placeholders for visualization variables
real theta,dtheta,phi,dphi!Colatitudinal,azimuthal angles,incriments
real Ubextbar,Uexchbar !Avg. external interaction,exchange energies
real RMagxbar,RMagybar,RMagzbar !Average magnetization components
real Sratio,dB !Ratio to find spin components, mag. Field incriment
real Tmag,dT !Temperature for magnetic simulations and incriment
real Splanedum!Dummy plane projectin of spin,
real ranguess ! Random Number for Monte Carlo algorithm
real RMagl! Analytical formula for Lengevin magnetization
```

These common blocks share data across the program and subroutines.

```
COMMON/VECTORS/x(nmax),y(nmax),z(nmax) !Positoin vector
COMMON/VECTORS2/Sx(nmax),Sy(nmax),Sz(nmax),Bext(3) !Spin vector and
                                !z-component of the magnetic field
COMMON/SCALARS/ntot !Total number of particles
COMMON/SCALARS2/xbox,ybox,zbox,Ubext !Dimensions of computational
                     !cell, external magnetic field interaction energy
COMMON/SCALARS3/RMagx,RMagy,RMagz !Magnetization vector
COMMON/SCALARS4/Uexch !Spin-spin interaction energy
```

Open I/O files.

```
open(7,file='magnetic_energies.res')
open(8,file='Magnetization.res')
open(10,file='last_space_spin_config.res')
```

Initialize scalars.

```
fileanim='Interc-0000000.pov'
ifr=6
```

```
      Bext(1)=0.0
      Bext(2)=0.0
      Bext(3)=5.0
      ntot=100
      pi=4.0*atan(1.0)
      xbox=tri_spacing*float(nx)
      ybox=tri_spacing*sin(pi/3.0)*(float(ny))
      nBsteps=nB
      if(nBsteps.eq.1) then
      dB=0.0
      endif
      if(nBsteps.ne.1) then
      dB=2.0*B_max/float(nBsteps-1)
      endif
      nTsteps=nT
      if(nTsteps.eq.1) then
      dT=0.0
      endif
      if(nTsteps.ne.1) then
      dT=T_max/float(nTsteps-1)
      endif
```

Initialize arrays.

```
      do 100 i=1,nmax
      x(i)=0.0
      y(i)=0.0
      z(i)=0.0
      Sx(i)=0.0
      Sy(i)=0.0
      Sz(i)=spin
 100  continue
```

Calculate empty lattice positions.

```
      isite=0
      do 300 j=1,ny
      do 200 i=1,nx
      isite=isite+1
      x(isite)=float(i-1)*tri_spacing+
     1(1.0+(-1.0)**j)*tri_spacing/4.0
      y(isite)=float(j-1)*tri_spacing*sin(pi/3.0)
      z(isite)=0.0
 200  continue
 300  continue
```

Find the upper and lower limits for magnetic field and temperature sequences.

```
      if(nBsteps.eq.1) then
      ibupper=1
      else
      ibupper=2*nBsteps-1
      endif
      if(nTsteps.eq.1) then
      itupper=1
      else
      itupper=2*nTsteps-1
      endif
      do 7776 ib=1,ibupper
```

```
Bext(1)=0.0
Bext(2)=0.0
if(ibupper.ne.1) then
if(ib.le.nBsteps) then
Bext(3)=B_max-float(ib-1)*dB
endif
if(ib.gt.nBsteps) then
Bext(3)=-B_max+float(ib-nBsteps)*dB
endif
else
Bext(3)=B_max
endif
```

Begin the temperature sequence.

```
do 7777 it=1,itupper
if(itupper.ne.1) then
if(it.le.nTsteps) then
Tmag=float(it-1)*dT
endif
if(it.gt.nTsteps) then
Tmag=T_max-float(it-nTsteps)*dT
endif
else
Tmag=T_max
endif
do 7788 im=1,1
```

Calculate analytical Lengevin magnetization.

```
if(Bext(3).ne.0) then
RMagl=1.0/tanh(spin*0.67*Bext(3)/(Tmag))-
1(Tmag)/(Bext(3)*0.67*spin)
else
RMagl=0.0
endif
RMagl=RMagl*float(ntot)*spin

Ubextbar=0.0
Uexchbar=0.0
RMagxbar=0.0
RMagybar=0.0
RMagzbar=0.0
```

Begin the magnetic field sequence.

```
do 7000 imstep=1,nmag
Ubext=0.0
do 3390 i=1,ntot
iatom=i
```

Get the pre-move energies.

```
if(iLengevin.eq.1) then
CALL EMAGEXT
endif
if(iIsing.eq.1) then
CALL LATTICE
endif
CALL MAGNET
```

```
      Upre=Ubext+Uexch
      Sxold=Sx(iatom)
      Syold=Sy(iatom)
      Szold=Sz(iatom)
3333  continue
```

Now a spin move for the trial configuration.

```
      Sratio=Sz(iatom)/spin
      if(Sratio.gt.1.0) then
      Sratio=1.0
      endif
      if(Sratio.lt.-1.0) then
      Sratio=-1.0
      endif
      theta=acos(Sratio)
      phi=atan2(Sy(iatom),Sx(iatom))
      szdum=spin*2.0*(rand()-0.5)
      Splanedum=sqrt(spin*spin-szdum*szdum)
      sxdum=Splanedum*2.0*(rand()-0.5)
      sydum=sqrt(Splanedum**2-sxdum**2)
      ranguess=rand()
      if(ranguess.le.0.5) then
      sydum=-sydum
      endif
      Sx(iatom)=sxdum
      Sy(iatom)=sydum
      Sz(iatom)=szdum
```

Get the post-move energies.

```
      if(iLengevin.eq.1) then
      CALL EMAGEXT
      endif
      if(iIsing.eq.1) then
      CALL LATTICE
      endif
      CALL MAGNET
      Upost=Ubext+Uexch
```

Monte Carlo acceptance/rejection.

```
      rnum=rand()
      if(Upost.gt.Upre) then
      if(exp(-(Upost-Upre)/Tmag).le.rnum) then
      Sx(iatom)=Sxold
      Sy(iatom)=Syold
      Sz(iatom)=Szold
      endif
      endif
3390  continue
      if(iLengevin.eq.1) then
      CALL EMAGEXT
      endif
      if(iIsing.eq.1) then
      CALL LATTICE
      endif
      CALL MAGNET
```

Calculate average magnetization components

```
        Ubextbar=Ubextbar+Ubext
        Uexchbar=Uexchbar+Uexch
        RMagxbar=RMagxbar+RMagx
        RMagybar=RMagybar+RMagy
        RMagzbar=RMagzbar+RMagz
        if (mod(imstep,nevery).eq.0) then
        write(7,8777) imstep,Ubext,Ubextbar/float(imstep)
       1,Uexch,Uexchbar/float(imstep)
 8777 format(I7,4(1X,F8.4))
        endif
```

Now write data to results files.

```
        if (imstep.eq.nmag.and.iIsing.eq.0) then
        write(8,8778) Tmag,Bext(3),RMagzbar/float(imstep),RMagl
 8778 format (4(1X,F8.4))
        endif
        if (imstep.eq.nmag.and.iIsing.eq.1) then
        write(8,7778) Tmag,Bext(3),
       1RMagzbar/float(imstep)
 7778 format (3(1X,F8.4))
        endif
```

Write out visualization files.

```
        if (mod(imstep,nevery).eq.0) then
        ifr=ifr+1
        write(fileanim(8:14),1701) ifr
 1701 format(i7.7)
        open(unit=ifr+6,file=fileanim)
        write(ifr+6,*) '#version 3.6;'
        write(ifr+6,*) '#include "colors.inc"'
        write(ifr+6,*) 'global_settings'
        write(ifr+6,*) '{'
        write(ifr+6,*) 'assumed_gamma 1.0'
        write(ifr+6,*) '}'
        write(ifr+6,*) 'camera'
        write(ifr+6,*) '{'
        write(ifr+6,*) 'location   <',0.05*xbox,',',0.05*ybox,',',
       15.0,'>'
        write(ifr+6,*) 'direction 1.5*z'
        write(ifr+6,*) 'right       -4/3*x'
        write(ifr+6,*) 'sky z'
        write(ifr+6,*) 'look_at    <',0.05*xbox,',',0.05*ybox,',',
       10.00*zbox,'>'
        write(ifr+6,*) '}'
        write(ifr+6,*) 'background { color red 0.5 green 0.5 blue 0.5 }'
        write(ifr+6,*) 'light_source'
        write(ifr+6,*) '{'
        write(ifr+6,*) '0*x'
        write(ifr+6,*) 'color red'
        write(ifr+6,*) '1.0  green 1.0  blue 1.0'
        write(ifr+6,*) 'translate <-30, 30, 30>'
        write(ifr+6,*) '}'
        write(ifr+6,*) 'light_source'
        write(ifr+6,*) '{'
        write(ifr+6,*) '0*x'
        write(ifr+6,*) 'color red'
        write(ifr+6,*) '0.5  green 0.5  blue 0.5'
        write(ifr+6,*) 'translate <30, 30, 30>'
```

```
      write(ifr+6,*) '}'
      write(ifr+6,*) 'light_source'
      write(ifr+6,*) '{'
      write(ifr+6,*) '0*x'
      write(ifr+6,*) 'color red'
      write(ifr+6,*) '0.1  green 0.1  blue 0.1'
      write(ifr+6,*) 'translate <30, -30, 30>'
      write(ifr+6,*) '}'
      DO 7673 i=1,ntot
      red=0.0
      green=0.0
      blue=0.0
      WRITE(ifr+6,*) 'sphere { <',
     1x(i)/10.,',',y(i)/10.,',',z(i)/10.,'>,'
      WRITE(ifr+6,*) vdw/10.
      WRITE(ifr+6,*) 'texture {pigment
     1{color rgb<',red,',',green,','
     2,blue,'>}'
      WRITE(ifr+6,*) 'finish{specular 1}} }'
      sdrawx=Sx(i)*(1.8/2.5)*vdw
      sdrawy=Sy(i)*(1.8/2.5)*vdw
      sdrawz=Sz(i)*(1.8/2.5)*vdw
      if(Sx(i).ne.0.0.or.Sy(i).ne.0.0.or.Sz(i).ne.0.0) then
      WRITE(ifr+6,*) 'cylinder { <',
     1x(i)/10.,',',y(i)/10.,',',z(i)/10.0,'>,',
     2'<',(x(i)+sdrawx)/10.,',',(y(i)+sdrawy)/10.,',',
     3(z(i)+sdrawz)/10.0,'>,'
      WRITE(ifr+6,*) vdw/27.
      WRITE(ifr+6,*) 'texture {pigment {color rgb<1.5,0,0>}'
      WRITE(ifr+6,*) 'finish{specular 1}} }'
      WRITE(ifr+6,*) 'cylinder { <',
     1x(i)/10.,',',y(i)/10.,',',z(i)/10.0,'>,',
     2'<',(x(i)-sdrawx)/10.,',',(y(i)-sdrawy)/10.,',',
     3(z(i)-sdrawz)/10.0,'>,'
      WRITE(ifr+6,*) vdw/27.
      WRITE(ifr+6,*) 'texture {pigment {color rgb<0,0,0>}'
      WRITE(ifr+6,*) 'finish{specular 1}} }'
      endif
      WRITE(ifr+6,*) 'sphere { <',
     1x(i)/10.,',',y(i)/10.,',',z(i)/10.,'>,'
      WRITE(ifr+6,*) vdw/18.
      WRITE(ifr+6,*) 'texture {pigment
     1{color rgb<',1.0,',',1.0,','
     2,1.0,'>}'
      WRITE(ifr+6,*) 'finish{specular 1}} }'
7673 CONTINUE
      close(ifr+6)
      endif
c################################################################
7000 continue
7788 continue
7777 continue
7776 continue

      end
```

This subroutine calculates the magnetization of the system:

$$\vec{M} = \sum_{i-1}^{N} \mu_B g \vec{S} \qquad (8.27)$$

```
SUBROUTINE MAGNET

      parameter (nmax=10000,max_types=2,cutoff=15.00)
      parameter (boltz=1.3806503e-23)
      parameter (CexchJoule=3.0672e-19,Lexch=10.0,rkF=0.667)
      integer ii,jj,i,j,jlist
      real dx,dy,dz,rij,arg,f1,f2,f3,Cexch
      COMMON/VECTORS/x(nmax),y(nmax),z(nmax)
      COMMON/VECTORS2/Sx(nmax),Sy(nmax),Sz(nmax),Bext(3)
      COMMON/SCALARS/ntot
      COMMON/SCALARS2/xbox,ybox,zbox,Ubext
      COMMON/SCALARS3/RMagx,RMagy,RMagz
      COMMON/SCALARS4/Uexch

      RMagx=0.0
      RMagy=0.0
      RMagz=0.0
      do 1000 i=1,ntot
      RMagx=RMagx+Sx(i)
      RMagy=RMagy+Sy(i)
      RMagz=RMagz+Sz(i)
 1500 continue
 1000 continue
      end
```

This subroutine calculates the interaction energy of the system with an external magnetic field:

$$U_{B,\text{ext}} = -\frac{1}{k_B} \sum_{i-1}^{N} \mu_B g \vec{S} \cdot \vec{B} \tag{8.28}$$

```
SUBROUTINE EMAGEXT

      parameter (nmax=10000,max_types=2,cutoff=15.00)
      parameter (bohrmu=9.274e-24,boltz=1.3806503e-23)
      integer ii,jj,i,j,jlist
      real dx,dy,dz,rij
      COMMON/VECTORS/x(nmax),y(nmax),z(nmax)
      COMMON/VECTORS2/Sx(nmax),Sy(nmax),Sz(nmax),Bext(3)
      COMMON/SCALARS/ntot
      COMMON/SCALARS2/xbox,ybox,zbox,Ubext
      COMMON/SCALARS3/RMagx,RMagy,RMagz
      COMMON/SCALARS4/Uexch

      Ubext=0.0
      do 1000 i=1,ntot
      Ubext=Ubext-(Sx(i)*Bext(1)+Sy(i)*Bext(2)+Sz(i)*Bext(3))
     1*bohrmu/boltz
 1000 continue
      end
```

And this subroutine calculates the spin–spin (exchange) interaction energy of the system:

$$U_{B,\text{exch}} = -J \sum_{i-1}^{N} \sum_{j=i+1}^{N} \vec{S}_i \cdot \vec{S}_j \tag{8.29}$$

```
SUBROUTINE LATTICE

parameter (nmax=10000,max_types=2,cutoff=15.00)
parameter (bohrmu=9.274e-24,boltz=1.3806503e-23)
integer ii,jj,i,j,jlist
real dx,dy,dz,rij
real Jconst
COMMON/VECTORS/x(nmax),y(nmax),z(nmax)
COMMON/VECTORS2/Sx(nmax),Sy(nmax),Sz(nmax),Bext(3)
COMMON/SCALARS/ntot
COMMON/SCALARS2/xbox,ybox,zbox,Ubext
COMMON/SCALARS3/RMagx,RMagy,RMagz
COMMON/SCALARS4/Uexch

Jconst=1.0
Uexch=0.0
do 1000 i=1,ntot
do 1500 j=1,ntot
if(j.gt.i) then
Uexch=Uexch-Jconst*(Sx(i)*Sx(j)+Sy(i)*Sy(j)+Sz(i)*Sz(j))
endif
1500 continue
1000 continue
end
```

In Figure 8.5a–d I show a suggested visualization scheme for magnetic spins (as in the code in the above program) illustrating disordering in the Lengevin magnet as temperature increases.

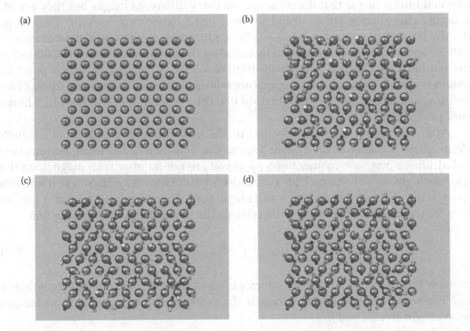

(a)   (b)   (c)   (d)

Figure 8.5. (a–d) Suggested visualization scheme (rendered pov output from the MC program in this chapter) for magnetic spins illustrating disordering in the Lengevin magnet as temperature increases. In the grayscale printing, shading differences help distinguish magnetic N and S poles which are colored red and black, respectively, in the electronic color version.

8.3.3 Learning New Things about How Layered Magnetic Systems Behave

With a validated MC program we can be comfortable taking our calculations into uncharted territory and loking at new research problems. One summer, I worked with an experimental group simulating intercalated layered magnetic systems [33] (say that 10 times fast), where guest atoms (various compositions of Mn and Ti) populate a static lattice with sites that can either be occupied by an atom or can be empty. Intercalation of Mn into the lattice can change its vertical spacing from $c = 5.695$ Å with no Mn to $c = 5.9$ Å for 25% Mn population. You can examine Reference 35 for the computational details.

There is a new twist here though because the atoms interact spatially and also magnetically in a way that depends upon their spatial ordering. So, we first equilibrated the spatial ordering to find the lattice structure, starting out from an initial configuration involving random placement of the correct percentages of Mn and Ti atoms on each layer in the lattice. For the space equilibration, the effective interaction between guest atoms i and j is modeled as being a screened Coulomb interaction [33],

$$u_C = \frac{kq_i q_j}{r_{ij}} e^{-\Gamma r_{ij}}, \tag{8.30}$$

with all parameters for the interaction given in the reference. The reason for the screened interaction is that different species carry different charges but they are in a structure that alters what we would see were just a lone charge present. The charges of (Mn,Ti) are taken to be $(+2e, +4e)$. To equilibrate, we assign a temperature of 1000 K, and then anneal the system by reducing the temperature after 100,000 steps. The temperature reduction continues until $T = 20$ K is reached, at which point the system is said to be in its low-temperature equilibrium state. If you can think of any advantages of attacking the equilibration like this, you can discuss them in a homework problem!

After the spatial Monte Carlo part of the simulation is completed, a *magnetic* Monte Carlo sequence of the system begins. The Ti is modeled as being nonmagnetic and Mn has a magnetic moment of 5 μ_B. If you are not familiar with magnetism theory, I strongly suggest you pick up a second-semester University Physics text and read through the chapters on electricity and magnetism. There are two important interactions for the magnetic behavior of the system. The first is a one-body interaction

$$U_{iB} = -\vec{\mu}_i \cdot \vec{B}_{\text{ext}}, \tag{8.31}$$

between Mn atom i with magnetic moment $\vec{\mu}_i$ and the external applied magnetic field $\vec{B}_{\text{ext}}$. The second magnetic interaction is the two-body RKKY spin–spin interaction between Mn atoms i and j:

$$U_{RKKY} = \sum_{i,j}^{N} J_{ij} \vec{S}_i \cdot \vec{S}_j \tag{8.32}$$

$$J_{ij} = 9\pi \frac{J_{sd}^2}{E_f} Z^2 \left[\frac{\sin 2k_F R_{ij} - 2k_F R_{ij} \cos 2k_F R_{ij}}{(2k_F R_{ij})^4} \right] e^{-R_{ij}/\lambda}, \tag{8.33}$$

The interaction shown in Equation 8.33 is named after the individuals who contributed to the idea and formalism of the interaction, and I recommend more exploration if you are interested [33]. Here we treat k_F, J_{sd}, E_f, and Z as adjustable parameters that depend on carrier charge densities and have been fine tuned to match relevant experimental measurements. The magnetic Monte Carlo sequence runs for 50,000 equilibration steps and then 100,000 steps thereafter over which averages are calculated.

It is useful to understand how the results of the simulations depend on various physical as well as computational effects. Some effects of the former type that were varied are the *physical* boundary conditions for a given layer by changing the number of layers present in the system (anywhere from 1 to 5). Another is the effect of the Ti/Mn charge disparity by assigning a value of the nonmagnetic atom's charge to be in the range of $-4e$ to $8e$. The most significant *computational* effect is artificial stabilization and introduction of false correlations through implementation of the periodic boundary conditions. Therefore, we simulated lattice sizes ranging from 12×12 up to 900×900.

8.3.4 When Monte Carlo Works out Better Than You Had Hoped

As in the section on Molecular Dynamics, Monte Carlo simulations can exhibit surprising correlation with—and insight into—physical systems. One important result of the simulations discussed on layered intercalated magnetic systems is that the cluster size distributions are exponential. That is to say, there are a few large clusters and exponentially more smaller ones. Such a phenomenon is also seen in experiments and the presence of the clusters characterizes the system in a quite fundamental sense. Another e-mail that came to me with a smiley face in the subject line involves a related project where we carefully looked at the topology of the real systems and compared those with what we see in corresponding simulations [34]. As it turns out, there are one dimensional structures seen in the experiment that the simulations were able to reproduce, as shown in Figure 8.6. In the case of the exponential cluster size distributions, as well as the one dimensional ordered structures, the fact that the simulations are reproducing such fine points seen in the experiment attests to their robustness and relavance, as well as to their degree of equilibration. Only well-equilibrated MC simulations can give meaningful results.

8.4 How Do We Choose MD or MC?

Because the two methods discussed in this chapter apply to the same types of systems, the question comes as to which method one should choose when simulating a particular system. Even though both simulation techniques can give meaningful and very helpful information about the systems under study, I will offer some general guidelines that can assist you in making a choice about the simulation technique you utilize.

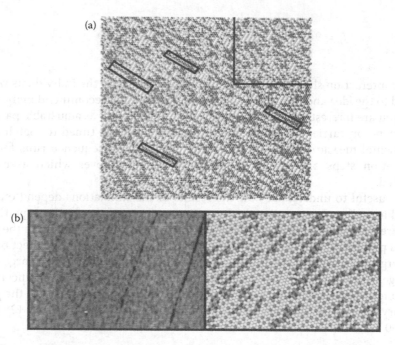

Figure 8.6. Results (a) of equilibrated Monte Carlo simulations of intercalated TiS2 for a
200×200 system showing positions of intercalated Ti ions (lighter shading) and Mn ions
(darker shading) in one layer for an Mn concentration of 9% and a Ti concentration of 10%.
The layer shown is the middle layer in a three-layer system. The inset shows the results for a
72×72 system. In the electronic color version, gold circles represent Ti ions, and purple/blue
circles denote Mn ions. Examples of linear structures formed by the Mn ions are surrounded by
rectangles [33]. (b) Comparison of simulated results to experimental, prompting the smiley face
email (A. Stollenwerk et al., *J. Phys. Chem.* C, 116 (1), 764–769, 2012, doi: 10.1021/jp208422n.) [34].

Molecular Dynamics deals with simulating a dynamical sequence of a given system, so it gives the path through time that the particular system would take. It captures
vibrations, rotations, and more complicated mechanical mechanisms which the system
and its constitutents undergo. MD can also deal with nonequilibrium processes such
as diffusion, desorption, and heating. Monte Carlo, on the other hand is inherently an
equilibrium simulation tool. There is a dynamical interpretation, however, because the
system spends it time jumping between configurations that it is most likely to be in; it's
great at giving averages. There is also a difference in mathematics and coding that is
important: MD deals with forces and MC deals with energies, so to implement MD you
have to take derivatives of the energies with respect to various coordinates, which can
be a highly nontrivial job. It can also make the computing time longer because not only
are the expressions for force longer in general than for energies, but there are also three
force components compared to only one for energy (and that's not really a directional
component, rather just a value). So, at first glance, if you are interested in calculating
averages and doing simulations at various temperatures (or other set physical properties) then MC might be your choice. But if you are looking for detailed dynamics of a

system and how processes within it behave with time then MD will give a much better return on your investment.

8.5 The Dynamics of Planetary and Galactic Systems

Now we zoom out to another lengthscale that we don't deal with (normally) when discussing everyday phenomena... that of the formation of planetary and galactic systems. The topic is fascinating and probes nearly every aspect of our humanity, including faith systems, government support of science, our origins, and destiny.

Nobody fully understands the physical behavior of solar systems and galaxies but there has been much observational, numerical, and analytical research conducted that forwards the body of knowledge. The presence and dynamics of planetary rings are subjects which have attracted a lot of attention in recent decades. With the development of increasingly better computational resources, models have gotten larger and been run longer; they are great tools for students to learn about simulation with (Reference 35, for example) because the physics includes momentum conservation, energy transfer, and electromagnetic forces. Here, we will walk through a FORTRAN program that can simulate, among other things, the coalescing of planetary rings into solar systems. The provided program NewDiscCollide.f is code that is related to the following discussion.

First, name the program and declare global constants: number of steps, system size, and write-out frequencies and clumping tolerance.

```
program rings
parameter (n=2000,nstep=50000)
parameter (neq=0,n6=200,n7=1,n8=100,n9=200,n10=200)
parameter (n11=200,n12=200)
parameter (dclump=1.0)
```

Now declare varaibles including positions, velocities, dummy variables/arrays and distributions.

```
character*19 fileanim
dimension x(n),y(n),z(n),vx(n),vy(n),vz(n)
dimension xhold(n),yhold(n),zhold(n)
dimension r(n),rm(n),q(n)
dimension rhold(n),rmhold(n)
dimension fx(n),fy(n),fz(n),ex(n),ey(n),ez(n)
dimension xvirt(n),yvirt(n),zvirt(n)
dimension vxold(n),vyold(n),vzold(n)
dimension vxhold(n),vyhold(n),vzhold(n)
dimension vxa(n),vya(n),vza(n),vxb(n),vyb(n),vzb(n)
dimension peg(n),peq(n),amx(n),amy(n),amz(n)
dimension bintx(n),binty(n),bintz(n),bextx(n),bexty(n),bextz(n)
dimension rcount(1000),rinst(1000),zcount(1000),zinst(1000)
real ke(n)
integer itime,i,iter,icount
integer j,k
integer ifile,ishorten,ninst
real t,G,PI,epnot,munot
real ei,epi,eki,ep,ek,etot,eknew
real dx,dy,dz,rij,dvx,dvy,dvz,vdotr,rati,ratj
real xcm,ycm,zcm,rmtot,xcmvir,ycmvir,zcmvir
real xlt,ylt,zlt,vxlt,vylt,vzlt
real xrt,yrt,zrt,vxrt,vyrt,vzrt
```

```
      real theta,phi,rpos
      real pet,ket,et,amxt,amyt,amzt,pegt,peqt
      real rin,rout,rmsmax,radmax,vel,tvel,vrad
      real top,bottom,zlow,zhigh,rplane
      real red,green,blue
      real drand
      real xcent,ycent,zcent
      real vxcent,vycent,vzcent
```

Set up I/O.

```
      fileanim='TopView-0000000.pov'
      open(6,file='rf1.res')
      open(7,file='rftotals.res')
      open(9,file='rfdist.res')
      open(11,file='rfrad.res')
      open(12,file='rfz.res')
      ifile=12
      write (7,*) "time amx amy amz ke pe energy"
```

Initialize variables.

```
C*****Define constants for the simulation*********************
      PI=4.*atan(1.)
      t=0.0001
      G=4.*PI**2
      epnot=1.
      munot=1.
C*****Initialze variables/arrays*******************************
      do 4 i=1,1000
      rcount(i)=0.
      zcount(i)=0.
    4 continue
      do 10 i=1,n
      x(i)=0.
      y(i)=0.
      z(i)=0.
      xhold(i)=0.
      yhold(i)=0.
      zhold(i)=0.
      vxhold(i)=0.
      vyhold(i)=0.
      vzhold(i)=0.
      vx(i)=0.
      vy(i)=0.
      vz(i)=0.
      r(i)=0.
      rm(i)=0.
      rhold(i)=0.0
      rmhold(i)=0.0
      q(i)=0.
      xvirt(i)=0.
      yvirt(i)=0.
      zvirt(i)=0.
      bintx(i)=0.
      binty(i)=0.
      bintz(i)=0.
      bextx(i)=0.
      bexty(i)=0.
```

```
          bextz(i)=0.
   10 continue
C*************************************************************
```

Limiting dimensions for initial conditions.

```
          rmsmax=1.0/float(n)
          radmax=0.005/(float(n)**0.333)
          rin=0.5
          rout=2.5
          top=0.0
          bottom=0.0
```

Now begin to define initial conditions: masses, radii, positions, and velocities.

```
c*****Start with central mass*********************************

          rm(1)=1.
          r(1)=0.001
          q(1)=0.
          x(1)=0.
          y(1)=0.
          z(1)=0.
          vx(1)=0.
          vy(1)=0.
          vz(1)=0.
C*************************************************************
c*****assign masses & radii***********************************
          do 4495 i=2,n
          q(i)=0.
          rm(i)=rmsmax
          r(i)=radmax
          phi=rand()
c         rm(i)=0.1*rmsmax+0.9*rmsmax*phi
c         write(6,*) i,r(i),rm(i)
C         phi=rand()
C         r(i)=radmax*phi
 4495 continue
C*************************************************************
          do 789 i=2,n
 2367 continue
          theta=2.*PI*rand()
          drand=rand()
          x(i)=(rin+(rout-rin)*drand)*cos(theta)
          y(i)=(rin+(rout-rin)*drand)*sin(theta)
          z(i)=(bottom+(top-bottom)*rand())
          vel=sqrt(G*rm(1)/sqrt(x(i)**2+y(i)**2+z(i)**2))
          tvel=2.*PI*rand()
          rpos=sqrt(x(i)**2+y(i)**2)
          vx(i)=vel*(-y(i)/rpos)
          vy(i)=vel*(x(i)/rpos)
          vz(i)=0.
          if(i.eq.n) then
c*****include any colliding/brushing masses*******************
C         rm(n)=0.25
C         r(n)=0.001
C         q(n)=0.
C         x(n)=rin+(rout-rin)/2.0
C         x(n)=rin-(rout-rin)/2.0
C         x(n)=(rin+rout)/2.0
C         y(n)=0.0
```

```
C       z(n)=-0.01
C       z(n)=0.00
C       vx(n)=50.0
C       vx(n)=0.00
C       vy(n)=0.
C       vy(n)=sqrt(G*rm(1)/sqrt(x(n)**2+y(n)**2+z(n)**2))
C       vz(n)=0.0
C       vz(n)=0.0
        endif
```

Although nature desn't have to, we need to chack to make sure that any two masses aren't too close; otherwise the simulation will behave unreasonably.

```
C*****Make sure masses aren't too close****************
      do 914 j=2,i
      if(j.ne.i) then
      dx=x(j)-x(i)
      dy=y(j)-y(i)
      dz=z(j)-z(i)
      rij=sqrt(dx**2+dy**2+dz**2)
      if(rij.lt.2.0*(r(i)+r(j))) then
      iter=iter+1
      goto 2367
      endif
      endif
  914 continue
      write(*,*) i,' masses placed in disc'
  789 continue
```

Now start the time loop

```
C!!!!!!Begin Time Loop!!!!!!!!!!!!!!!!!!!!!!!!!!!!!!!!!!!!!!!!!!!!!!!!!!!
      do 556 i=1,n
      write(6,*) i,r(i),rm(i)
  556 continue
C*********************************************************************
      ninst=n
      do 100 itime=1,nstep
```

Here is an inelastic collision

```
C*******Combine Particles for Collisions*********************
      if(itime.gt.neq) then
      do 1717 i=1,ninst
      if(r(i).ne.0.0) then
      do 1718 j=i+1,ninst
      if(r(j).ne.0.0) then
      if(j.gt.i) then
      dx=x(j)-x(i)
      dy=y(j)-y(i)
      dz=z(j)-z(i)
      rij=sqrt(dx**2+dy**2+dz**2)
      if(rij.le.dclump*(r(i)+r(j))
     1.and.rm(j).ne.0.0.and.rm(i).ne.0.0) then
      xcent=(x(i)*rm(i)+x(j)*rm(j))/(rm(i)+rm(j))
      ycent=(y(i)*rm(i)+y(j)*rm(j))/(rm(i)+rm(j))
      zcent=(z(i)*rm(i)+z(j)*rm(j))/(rm(i)+rm(j))
      vxcent=(vx(i)*rm(i)+vx(j)*rm(j))/(rm(i)+rm(j))
      vycent=(vy(i)*rm(i)+vy(j)*rm(j))/(rm(i)+rm(j))
```

```
        vzcent=(vz(i)*rm(i)+vz(j)*rm(j))/(rm(i)+rm(j))
        x(i)=xcent
        y(i)=ycent
        z(i)=zcent
        vx(i)=vxcent
        vy(i)=vycent
        vz(i)=vzcent
        rm(i)=rm(i)+rm(j)
        r(i)=(r(i)**3+r(j)**3)**0.33
        r(j)=0.0
        rm(j)=0.0
        endif
        endif
        endif
1718 continue
        endif
1717 continue
```

Now shorten the particle arrays by one to accound for the absorbed mass.

```
        ishorten=0
        do 2117 i=1,ninst
        if(rm(i).ne.0.0) then
        ishorten=ishorten+1
        xhold(ishorten)=x(i)
        yhold(ishorten)=y(i)
        zhold(ishorten)=z(i)
        rmhold(ishorten)=rm(i)
        rhold(ishorten)=r(i)
        vxhold(ishorten)=vx(i)
        vyhold(ishorten)=vy(i)
        vzhold(ishorten)=vz(i)
        endif
2117 continue
        ninst=ishorten
        do 3317 i=1,ninst
        x(i)=xhold(i)
        y(i)=yhold(i)
        z(i)=zhold(i)
        rm(i)=rmhold(i)
        r(i)=rhold(i)
        vx(i)=vxhold(i)
        vy(i)=vyhold(i)
        vz(i)=vzhold(i)
3317 continue
        endif
cccccccccccccccccccccccccccccccccccccccccccccccccccccccccccccccccccc
        do 34 i=1,1000
        rinst(i)=0.0
        zinst(i)=0.0
  34 continue
```

Now, we use the following formulae to calculate potential and kinetic energies

$$V = -\sum_{i=1}^{N}\sum_{j=i+1}^{N}\frac{Gm_im_j}{r_{ij}},$$ (8.34)

$$T = \frac{1}{2}\sum_{i=1}^{N} m_i v_i^2,$$ (8.35)

And the total mechanical energy is just their sum.

```
C*****Calcualte initial potential, total and mechanical energies*****
      if(itime.eq.1) then
      epi=0.
      eki=0.
      do 5688 i=1,n
      do 5689 j=i+1,n
      if(i.ne.j) then
      dx=x(j)-x(i)
      dy=y(j)-y(i)
      dz=z(j)-z(i)
      rij=sqrt(dx**2+dy**2+dz**2)
      epi=epi-G*rm(i)*rm(j)/rij+q(i)*q(j)/(4.*PI*epnot*rij)
      endif
 5689 continue
      eki=eki+0.5*rm(i)*(vx(i)**2+vy(i)**2+vz(i)**2)
 5688 continue
      ei=eki+epi
      endif
C********************************************************************
      do 1117 i=1,n
      fx(i)=0.0
      fy(i)=0.0
      fz(i)=0.0
 1117 continue
```

Particle (i) experiences a gravitational force

$$\vec{F}_i = \sum_{j=1}^{N} \frac{Gm_i m_j}{r_{ij}^2} \hat{r}_{ij}(1-\delta_{i,j}).$$ (8.36)

Here, G is Newton's universal gravitational constant and the m_i are the masses of any objects in the simulation. The particles composing the planets may also experience forces from external and internal electric fields

$$\vec{F}_e^i = q_i \vec{E}_{\text{ext}} - \sum_{j=1}^{N} \frac{kq_i q_j}{r_{ij}^2} \hat{r}_{ij}(1-\delta_{i,j})$$ (8.37)

as well as from external and internal magnetic fields

$$\vec{F}_m^i = q_i \vec{v}_i \times \left(\vec{B}_{\text{ext}} + \sum_{j=1}^{N} \frac{\mu_0}{4\pi} \frac{q_j \vec{v}_j \times \vec{r}_{ij}}{r_{ij}^3} (1-\delta_{i,j}) \right).$$ (8.38)

Here, F_e and F_m are the electric and magnetic forces, respectively; the q_i are the particles' charges, k is Coulomb's constant and μ_0 is the permeability of free space.

The delta functions are present to ensure we are not counting any particle's interaction with itself.

```
C*****Advance System according to Newton's 2nd Law*******************
      if(itime.gt.neq) then
      do 759 i=1,ninst
      ex(i)=0.
      ey(i)=0.
      ez(i)=0.
      bextx(i)=0.
      bexty(i)=0.
      bextz(i)=0.
  759 continue
      DO 90 i=1,ninst
      fx(i)=0.
      fy(i)=0.
      fz(i)=0.
      bintx(i)=0.
      binty(i)=0.
      bintz(i)=0.
      DO 91 j=1,ninst
      if(j.ne.i) then
      dx=x(j)-x(i)
      dy=y(j)-y(i)
      dz=z(j)-z(i)
      dvx=vx(j)-vx(i)
      dvy=vy(j)-vy(i)
      dvz=vz(j)-vz(i)
      rij=sqrt(dx**2+dy**2+dz**2)
      xdumg=fx(i)+(G*rm(i)*rm(j)/rij**2)*dx/rij
      xdumq=(q(i)*q(j)/(4.*PI*epnot*rij**2))*dx/rij
      ydumg=fy(i)+(G*rm(i)*rm(j)/rij**2)*dy/rij
      ydumq=(q(i)*q(j)/(4.*PI*epnot*rij**2))*dy/rij
      zdumg=fz(i)+(G*rm(i)*rm(j)/rij**2)*dz/rij
      zdumq=(q(i)*q(j)/(4.*PI*epnot*rij**2))*dz/rij
      fx(i)=xdumg+xdumq
      fy(i)=ydumg+ydumq
      fz(i)=zdumg+zdumq
      bintx(i)=(munot*q(j)/(4.*PI*rij**3))*(dvy*dz-dvz*dy)
      binty(i)=(munot*q(j)/(4.*PI*rij**3))*(dvz*dx-dvx*dz)
      bintz(i)=(munot*q(j)/(4.*PI*rij**3))*(dvx*dy-dvy*dx)
      if(rij.lt.0.5*rmsmax) then
      write(*,*) itime,i,j,rij
      endif
      endif
   91 continue
C     fx(i)=fx(i)+q(i)*
C    1(ex(i)+vy(i)*(bintz(i)*bextz(i))-vz(i)*(binty(i)*bexty(i)))
C     fy(i)=fy(i)+q(i)*
C    1(ey(i)+vz(i)*(bintx(i)*bextx(i))-vx(i)*(bintz(i)*bextz(i)))
C     fz(i)=fz(i)+q(i)*
C    1(ez(i)+vx(i)*(binty(i)*bexty(i))-vy(i)*(bintx(i)*bextx(i)))
   90 continue
      endif
```

After having calculated the forces and hence accelerations, the system is advanced in time:

$$\vec{r}_i(t+\Delta t) = \vec{r}_i(t) + \vec{v}_i(t)\Delta t + \frac{1}{2}\frac{\vec{F}_i}{m_i}(\Delta t)^2 \qquad (8.39)$$

$$\vec{v}_i(t+\Delta t) = \vec{v}_i(t)\Delta t + \frac{\vec{F}_i}{m_i}\Delta t. \qquad (8.40)$$

```
      do 309 i=1,ninst
      x(i)=x(i)+vx(i)*t+0.5*(fx(i)/rm(i))*t*t
      y(i)=y(i)+vy(i)*t+0.5*(fy(i)/rm(i))*t*t
      z(i)=z(i)+vz(i)*t+0.5*(fz(i)/rm(i))*t*t
  309 continue
      do 310 i=1,ninst
      vx(i)=vx(i)+(fx(i)/rm(i))*t
      vy(i)=vy(i)+(fy(i)/rm(i))*t
      vz(i)=vz(i)+(fz(i)/rm(i))*t
  310 continue
C*************************************************************
```

Now we calculate angular momenta as well as potential and kinetic energies
and write them out with desired frequencies

```
C*********** Calculate various quantities ****************

      amxt=0.0
      amyt=0.0
      amzt=0.0
      ket=0.0
      pegt=0.0
      peqt=0.0
      pet=0.0

      do 6000 i=1,ninst
       amx(i)=(y(i)*vz(i)-z(i)*vy(i))*rm(i)
       amy(i)=(z(i)*vx(i)-x(i)*vz(i))*rm(i)
       amz(i)=(x(i)*vy(i)-y(i)*vx(i))*rm(i)
      amxt=amxt+amx(i)
      amyt=amyt+amy(i)
      amzt=amzt+amz(i)

      ke(i)=0.5*rm(i)*(vx(i)**2+vy(i)**2+vz(i)**2)
      ket = ket+ke(i)
      peg(i)=0.0
      peq(i)=0.0
      do 6001 j=1,ninst
      if (i.ne.j) then
      dx=x(j)-x(i)
      dy=y(j)-y(i)
      dz=z(j)-z(i)
      rij=sqrt(dx**2+dy**2+dz**2)
      peg(i)=peg(i)-G*rm(i)*rm(j)/rij
      peq(i)=peq(i)+q(i)*q(j)/(4.*PI*epnot*rij)
      end if
 6001 continue
      pegt=pegt+0.5*peg(i)
      peqt=peqt+0.5*peq(i)
      pet=pet+0.5*(peg(i)+peq(i))
```

```
6000   continue
       et=pet+ket
       if (float(itime/n7).eq.float(itime)/float(n7)) then
         write (7,*) float(itime)*t,amxt,amyt,amzt,ket,pegt,peqt,pet,et,
      1numtg
       endif
C*************************************************************
C*****

C*************************************************************
```

Keep track of the simulation progress.

```
       write(*,*) itime-neq,'out of ',nstep-neq,' run steps'
```

Write out the classical state (positions and velocities of all particles) with desired frequency.

```
       if (float(itime/n6).eq.float(itime)/float(n6)) then
       write(6,*) itime
       do 7889 i=1,ninst
       write(6,*)  float(itime)*t,x(i),y(i),z(i),vx(i),vy(i),vz(i)
7889   continue
       endif
```

Calculate and write out various particle distributions that may be helpful to have.

```
       do 4485 i=1,ninst
       rdist=sqrt((x(i)-x(1))**2+(y(i)-y(1))**2)
       do 4486 j=1,1000
       rlow=10.*float(j-1)/999.
       rhigh=10.*float(j)/999.
       zlow=-0.01+0.02*float(j-1)/999.
       zhigh=-0.01+0.02*float(j)/999.
       if(rdist.gt.rlow.and.rdist.le.rhigh) then
       rcount(j)=rcount(j)+1.
       rinst(j)=rinst(j)+1.
       endif
       if(z(i).gt.zlow.and.z(i).le.zhigh) then
       zcount(j)=zcount(j)+1.
       zinst(j)=zinst(j)+1.
       endif
4486   continue
4485   continue
       if (float(itime/n11).eq.float(itime)/float(n11)) then
       write(11,*) itime
       do 7295 i=1,1000
       rlow=10.*float(i-1)/999.
       rhigh=10.*float(i)/999.
       write(11,*) 0.5*(rlow+rhigh),rcount(i)/float(itime),rinst(i)
7295   continue
        endif
       if (float(itime/n12).eq.float(itime)/float(n12)) then
       write(12,*) itime
       do 7296 i=1,1000
       zlow=-0.01+0.02*float(i-1)/999.
       zhigh=-0.01+0.02*float(i)/999.
       write(12,*) 0.5*(zlow+zhigh),zcount(i)/float(itime),zinst(i)
```

```
7296   continue
       endif
```

Write out a POV raytracer file with desired frequency.

```
       if (float(itime/n8).eq.float(itime)/float(n8)) then
       ifile=ifile+1
       write(fileanim(9:15),1011) ifile
 1011  format(i7.7)
       open(unit=ifile,file=fileanim)
       write(ifile,*) '#version 3'
       write(ifile,*) '#include "colors.inc"'
       write(ifile,*) 'global_settings'
       write(ifile,*) '{'
       write(ifile,*) 'assumed_gamma 1.0'
       write(ifile,*) '}'
       write(ifile,*) 'background { color red 0.5 green 0.5 blue 0.5 }'
       write(ifile,*) 'camera'
       write(ifile,*) '{'
       write(ifile,*) 'location   <',0.0,',',0.0,',',' 8.0>'
       write(ifile,*) 'direction 1.5*z'
       write(ifile,*) 'right      4/3*x'
       write(ifile,*) 'look_at    <0,0,0>'
       write(ifile,*) '}'
       write(ifile,*) 'light_source'
       write(ifile,*) '{'
       write(ifile,*) '0*x'
       write(ifile,*) '1.0  green 1.0  blue 1.0'
       write(ifile,*) 'translate <-30,30,-30>'
       write(ifile,*) '}'
       write(ifile,*) 'light_source'
       write(ifile,*) '{'
       write(ifile,*) '0*x'
       write(ifile,*) '1.0  green 1.0  blue 1.0'
       write(ifile,*) 'translate <30,-30,30>'
       write (ifile,*) '}'

       do 300 i=1,ninst
       if(i.eq.1) then
       red=1.0
       green=0.0
       blue=0.0
       write(ifile,*) 'sphere { <',
      1x(i),',',y(i),',',z(i),'>,'
       write(ifile,*) 50.0*r(i)
       write(ifile,*) 'texture {pigment
      1{color rgb<',red,',',green,',',blue,'>}'
       write(ifile,*) 'finish{specular 1}} }'
       endif
       if(i.eq.ninst) then
       red=0.0
       green=0.0
       blue=1.0
       write(ifile,*) 'sphere { <',
      1x(i),',',y(i),',',z(i),'>,'
       write(ifile,*) 50.0*r(i)
       write(ifile,*) 'texture{pigment
      1{color rgb<',red,',',green,',',blue,'>}'
       write(ifile,*) 'finish{specular 1}} }'
       endif
```

```
       if(i.gt.1.and.i.lt.ninst) then
       red=0.0
       green=0.0
       blue=1.0
       write(ifile,*) 'sphere { <',
      1x(i),',',y(i),',',z(i),'>,'
       write(ifile,*) 50.0*r(i)
       write(ifile,*) 'texture {pigment
      1{color rgb<',red,',',green,',',blue,'>}'
       write(ifile,*) 'finish{specular 1}} }'
       endif
  300 continue
       close(ifile)
       endif

  100 continue
       write(6,*) ' '
       icount=0
       do 5579 i=1,ninst
       rdist=sqrt((x(i)-x(1))**2+(y(i)-y(1))**2+(z(i)-z(1))**2)
       if(rdist.le.100.) then
       icount-icount+1
       write(6,*) x(i),y(i),z(i),r(i),rm(i)
       endif
 5579 continue
       write(6,*) icount
C!!!!!!End Time Loop!!!!!!!!!!!!!!!!!!!!!!!!!!!!!!!!!!!!!!!!!!!!!!!!!!!!!
```

Wrap it up.

```
       end
```

8.6 Advanced Planetary Dynamics Methods Designed to Save Time: Go Climb a Tree

The methods covered in the previous section work well and are great teaching tools, but when it comes time to do very long runs for larger systems we would have to wait way too long for results. This is in part because the simulation scales like N^2–N, where N is the number of particles. There has been much effort placed in writing more efficient algorithms [36–40]. Although they always sacrifice accuracy to some degree, such methods can provide dramatic speedup and still allow reasonable physical interpretation of the results. One such method is the Barnes–Hut tree algorithm, which is described in detail elsewhere complete with sample code [40]. The algorithm basically uses a leaf/tree setup to approximate the mass distribution that a particle "sees" in terms of *moments* of the mass distribution that it interacts with. A student I worked with, Christopher Stark, implemented a Barnes–Hut algorithm and studied it. Comparisons of the tree code method to the "brute force" method discussed in the last section are shown in Table 8.6.

The Barnes–Hut tree code scales like $N \log(N)$—a remarkable improvement over the brute force method! Simulations of 20,000 initial planetesimals bodies over 98,289 time steps (approximately 1000 Earth years), resulted in significant accretion, with only 159 particles remaining.

Table 8.6. Comparisons of the "Brute Force" and Barnes–Hut Algorithms Using a Pentium III 1.0 GHz with 512 MB RAM under a Linux OS

Number of Particles	Tree Code Method (s)	Brute Force Method (s)
250	<1	<1
500	<1	1
1000	<1	3
2000	1	11
4000	2	44
8000	3	177 (2 min 57 sec)
16000	7	724 (12 min 24 sec)
32000	16	2904 (48 min 22 sec)

Source: Courtesy of Christopher Stark.

So, were I transitioning from learning about simulating solar system formation or planetary collision dynamics to doing some long or large-scale runs, I suppose I would go climb a tree [40].

PROBLEMS

8.1. Justify that the reduced units shown in Table 8.1 make sense and have reasonable numerical values in the Molecular Dynamics world.

8.2. Consult Table 8.2 and calculate values of reduced units for all the noble gases. Comment on trends you observe in each parameter and tie your observation to physical properties of the systems.

8.3. Using some arbitrary numerical values for N_{max} and ρ, calculate the box size as specified in Equation 8.3 as well as using the pow function in C++. Discuss any differences in the results and which method you would prefer for applications that come to mind.

8.4. Rewrite Equation 8.6 using an identical double sum, and discuss why this doesn't apply to the force sum.

8.5. Show that Equations 8.8 (a–c) come from Equation 8.7.

8.6. Run the MD program provided in this chapter. Look up MD results in Lennard-Jones systems and make sure that your program validates agains accepted results. You may need to change some of the units depending on the examples you find.

8.7. Run the MD program provided in this chapter. Now, rescale the temperature at every 1, 10, and 100 steps and track the results. Discuss your findings.

8.8. Do some research into Binder's fourth cumulant and discuss an application that is of interest to you vis-à-vis some system you are interested in studying.

8.9. Add code to MD program provided in this chapter to allow writing to pov files every so many steps as opposed to every step.

8.10. Run the MD program provided in this chapter. Impliment three different thermostats and discuss the differentces in results that you observe. Discuss which thermostat is the best one for your application.

8.11. Run the MD program provided in this chapter using three different integration algorithms. Discuss the differences in results that you observe, including which integrator you feel is the best one for your application.

8.12. We all feel good when we are validated. So, run the MC program provided in this chapter and make sure it validates against the Lengivin magnet for various temperature and magnetic field runs. Be sure to do at least one hysteresis run in T and B, and discuss your results. You will likely have to research Lengevin magnets a bit for this.

8.13. In the MC program provided in this chapter, we use many magnets and shut off their interactions with each other in order to model the Lengevin magnet. Discuss what the advantage of doing such a thing is.

8.14. Now run the MC program provided in this chapter and make sure it is consistent with an Ising magnet for various temperature and magnetic field runs. Be sure to do at least one hysteresis run in T, and discuss your results including what aspect of the system gives hysteresis. It is a very good idea to research Ising magnets for this.

8.15. Histogram Monte Carlo is a reasonably powerful tool which has its limits. Discuss specifics of what you would expect to get when you have poor statistics and are on the edges of a Gaussian-shaped energy distribution. This problem will require some research on your part.

8.16. How broad is "sufficiently broad" for multiple temperature Histogram MC? Discuss and justify.

8.17. In the work we did on the intercalated layered magnetic systems, discuss why cooling the sample from 1000K to 20K useful.

8.18. In the FORTRAN planetary ring program, discuss under what condition you would use a different time integrator (Equations 8.39 and 8.40). Select one, (I suggest Euler-Cromer), code up the integrator, run your code, and compare the results with those of the program provided here.

8.19. In any of the programs we don't want particles to start out too close. Remove that protection in a program of your choice and discuss what happens. Impliment a mitigation technique other than the protection statements in the initial conditions. Hint: Think about particle–particle interactions or the time step.

8.20. Reference 35 has a different collision algorithm implemented from the one in the code provided here. Review this reference and implement the collision algorithm or something similar to it. The coding for tracking clumping is fairly advanced, and this problem is easily on the order of a semester project.

8.21. For any of the programs provided in this chapter, construct and execute validation runs—special cases for which you know what the results should be. How much do you have to see before you are convinced of validation?

References

1. R. Ferraro, *Einstein's Space-Time*, 2000, Springer, New York, NY, Hardback, 310p, ISBN 03876994732.

2. R. Rucker, *Geometry, Relativity and the Fourth Dimension*, 1977, Dover Publications, Mineola, NY, Ebook, 133p, ISBN 0486140334.

3. R. Skinner, *Relativity for Scientists and Engineers*, 1982, Dover Publications, Mineola, NY, Ebook, 372p, ISBN 0486793672.

4. R.J. Scherrer, *Quantum Mechanics: An Accessible Introduction*, 2006, Benjamin Cummings, San Francisco, CA, Paperback, 334p, ISBN 0805387161.

5. D.J. Griffiths, *Introduction to Quantum Mechanics*, Cambridge University Press, Cambridge, United Kingdom, 2017.

6. R.L. Liboff, *Introductory Quantum Mechanics*, Addison-Wesley, Boston, MA, 2002.

7. M. Finnis, *Interatomic Forces in Condensed Matter*, 2004, Oxford University Press, Oxford, Ebook, 304p, ISBN 0191545295.

8. M.K. Balasubramanya, M.W. Roth, P. Tilton, B. Suchy, Molecular dynamics simulations of noble gas release from endohedral fullerene aggregates due to cage disintegration, *J. Comput. Theor. Nanosci.* 5, 627–634, 2008.

9. P. Attard, *Thermodynamics and Statistical Mechanics: Equilibrium by Entropy Maximisation*, 2002, Elsevier Science, London; San Diego, Ebook, 450p, ISBN 0080519180.

10. S.T. Thornton, J.B. Marion, *Classical Dynamics of Particles and Systems*, Cengage Learning, Belmont, CA, ISBN 13: 9780534408961, 2003.

11. A.L. Fetter, J.D. Walecka, *Theoretical Mechanics of Particles and Continua*, Dover Books on Physics, Mineola, NY, ISBN-13: 978-0486432618, 2003.

12. H. Goldstein, C.P. Poole Jr. J.L. Safko, *Classical Mechanics*, Addison-Wesley, Essex, England, Hardcover, 3rd Edition, 2001.

13. M.P. Allen, D.J. Tildesley, *Computer Simulation of Liquids*, Clarendon Press, Oxford, England, ISBN 0-19-855375-7, 1991.

14. P.H. Hünenberger, Thermostat algorithms for molecular dynamics simulations, *Adv. Polym. Sci.* 173, 105–149, 2005, doi:10.1007/b99427.

15. N. Bou-Rabee, Time integrators for molecular dynamics, *Entropy* 16, 138–162, 2014, doi:10.3390/e16010138.

16. Z. Wang, M. Fingas, K. Li, Fractionation of a light crude oil and identification and quantitation of aliphatic, aromatic, and biomarker coumpounds by GC-FID and GC-MS.I, *Journal of Chromatographic Science*, 32(9), 361–366, 1994.

17. C.A. Hughey, R.P. Rodgers, A.G. Marshall, Resolution of 11,000 compositionally distinct components in a single electrospray ionization Fourier transform ion cyclotron resonance mass spectrum of crude oil. *Anal. Chem.*, 74(16), 4145–41499, 2002.

18. K. Sugiura, M. Ishihara, T. Shimauchi, S. Harayama, Physicochemical properties and biodegradability of crude oil, *Environ. Sci. Technol.*, 31, 45–51, 1997.

19. F. Mutelet, G. Ekulu, M. Rogalski, Characterization of crude oils by inverse gas chromatography, *Journal of Chromatography A*, 969, 207–213, September 2002.

20. Z. Ha, Z. Ring, S. Liu, Estimation of isomeric distributions in petroleum fractions, *Energy & Fuels*, 19(4), 1660–1672, 2005.

21. N. Thanh, M. Hsieh, R.P. Philp, Waxes and asphaltenes in crude oils. *Organic Geochemistry*, 30, 119–132, 1999.

22. D.C. Villalanti, J.C. Raia, J.B. Maynard, High-temperature simulated distillation applications in petroleum characterization, *Encyclopedia of Analytical Chemistry*, R.A. Meyers (Ed.) John Wiley & Sons Ltd, Chichester, 2000, 6726–6741.

23. M.J. Connolly, M.W. Roth, C. Wexler, P.A. Gray, Explicit hydrogen molecular dynamics simulations of hexane deposited onto graphite at various coverages, *Langmuir*, 24, 3228–3234, 2008.

24. L. Firlej, B. Kuchta, M.W. Roth, M.J. Connolly, C. Wexler, Structural and phase properties of tetracosane (C24H50) monolayers adsorbed on graphite: An explicit hydrogen molecular dynamics study, *Langmuir*, 24(21), 12392–12397, 2008.

25. R. He, L. Zhao, N. Petrone, K.S. Kim, M. Roth, J. Hone, P. Kim, A. Pasupathy, A. Pinczuk, Large physisorption strain in CVD graphene on copper substrates, *Nano Lett.*, 12(5), 2408–2413, 2012, doi: 10.1021/nl300397v

26. K. Binder, D. Heermann, Lyle Roelofs, A. John Mallinckrodt, Susan McKay, *Monte Carlo Simulation in Statistical Physics: An Introduction [Book]*, Springer, Berlin Heidelberg, Paperback, 202p, ISBN 3642264468.

27. N. Metropolis, A.W. Rosenbluth, M.N. Rosenbluth, A.H. Teller, E. Teller, Equations of state calculations by fast computing machines, *Journal of Chemical Physics*, 21(6), 1087–1092, 1953.

28. M.P. Nightingale, C.J. Umrigar, *Quantum Monte Carlo Methods in Physics and Chemistry (Nato Science Series C:)*, Springer, Netherlands, Paperback, December 31, 1998.

29. L.D. Landau, E.M. Lifshitz, *Statistical Physics, Band 5 (Google eBook)*, Elsevier, Oxford, England, October 22, 2013, 544 pages.

30. G.B. Arfken, H.J. Weber, *Mathematical Methods for Physicists*, A Comprehensive Guide Hardcover, 7th Edition, January 31, 2012.

31. A.M. Ferrenberg, R.H. Swendsen, New Monte Carlo technique for studying phase transitions, *Physical Review Letters*, 63, 1658–1661, Published October 9, 1989.

32. A.M. Ferrenberg, R.H. Swendsen, Optimized Monte Carlo data analysis, *Physical Review Letters*, 61(23), 2635–2638, December 5, 1988. doi:10.1103/physrevlett.61.2635

33. P. Shand, A.L. Meyer, M. Streicher, A. Wilson, T. Rash, M.W. Roth, T.E. Kidd, L.H. Strauss, Coulomb-driven cluster-glass behavior in Mn-intercalated $Ti_{1+y}S_2$, *Phys. Rev. B*, 85, 144432, 2012, doi: 10.1103/PhysRevB.85.144432

34. A. Stollenwerk, A. O'Shea, E. Wolter, M.W. Roth, T. Kidd, L. Strauss, Emergence of long range one: Dimensional nanostructures in a disordered two: Dimensional system: Mn doped $Ti_{1+\delta}S_2$, *The Journal of Physical Chemistry*, 116 (1), 764–769, 2012, doi: 10.1021/jp208422n

35. W. Even, M.W. Roth, A new method for simulating the effects of collisions on planets, *American Journal of Undergraduate Research* 1(3), 1, 2002.

36. J. Wisdom, S. Tremaine, Local simulations of planetary rings, *Astrophysical Journal*, 95, 925, 1988.

37. J. Barnes, P. Hut, A hierarchical O(N log N) force-calculation algorithm, *Nature* 324, 446, 1986.

38. D.C. Richardson, T. Quinn, J. Stadel, G. Lake, Direct Large-scale N-body simulations of planetesimal dynamics, *Icarus*, 143(1), 45–49, January 2000.

39. D.C. Richardson, A new tree code method for simulation of planetesimal dynamics, *Monthly Notices of the Royal Astronomical Society*, 261, 396, 1993.

40. Barnes, *Treecode Guide* 2001. http://www.ifa.hawaii.edu/faculty/barnes/treecode/treeguide.html

41. M.J. Connolly, M.W. Roth, Carlos Wexler, Paul A. Gray, Molecular Dynamics Computer Simulations of Hexane Adlayers on Graphite: Comparing All – Atom and United Atom Models, *53rd Midwest Solid State Physics Conference*, Kansas City, MO, October 7, 2006.

42. L. Firlej, B. Kuchta, M.W. Roth, Paul A. Gray, Carlos Wexler, Molecular Dynamics study of tetracosane monolayers adsorbed on graphite, *2008 APS March Meeting*, New Orleans, LA, March 10–14, 2008.

SECTION IV

A Glimpse into More Advanced Computing

CHAPTER 9

//

Parallel Computing, Scripting and GPU's

9.1 Introductory Thoughts

The way I see it, computer modeling and simulation entails a beautiful confluence of applied physics insight and computer skills. We've talked a lot about everyday physics and computing, and we've gone beyond everyday phenomena, and so it is only fair that we talk about going beyond everyday computing. So, this chapter will deal with elements of computing that allow us to obtain significant gains on the work we get done.

One such emerging area is *parallel computing*, where more than one processor handle the (parts of) the computational problem at the same time. There is normally a *head* or *master node* that reads input, farms calculations out to a number of *compute nodes* that do calculations simultaneously and then pass the result back to the head node for organization and further I/O. Congruent with my philosophy throughout this book, I will refer you to the larger body of references out there on how to build computer clusters, how to keep them cool, what hardware to select, what programs to select and so forth. Since the purpose of this book is to be able to help you learn coding and algorithms, I want to talk about construction of parallel computing simulations, and of course provide examples.

A fundamental consideration, when one is constructing and running a parallel computer simulation, is how the simulation is going to divide the calculations among the processors. I will walk through some thoughts that will help you decide such things.

You might have a series of commands or a great number of runs to set up and submit. In those cases, UNIX *scripting* can be of huge benefit. A script is essentially a series of commands automatically issued. The commands can live inside loops and so if you need to do a comparative study and look at 100 different values of a parameter in your simulations a script can be of amazing help. Scripts can compile, run and edit files, as well as create and delete files and directories.

The processing unit is central in determining the size, quality, and number of simulations you can do. As gaming has become a central focus of computers and computing, industry has had to meet demands with fast and efficient processing units. Some of the nicest ones around are Graphics Processing Units (GPU's), which I will briefly discuss later.

9.2 Decompositions: Breaking Up is Easy to Do

There are many ways to break a system up—*decompositions*—when doing parallel processing but there are three that stand in the forefront and have been thoroughly studied [1], and those are the ones I want to discuss with you. One method is called *spatial decomposition*. In spatial decomposition, the system being simulated is divided up into different parts and each of the processors is responsible for its own part of the system. My analogy to this is that if you live in a house and you're doing some computing with the house, you divide space up into blocks (possibly equally sized ones) and each processor gets a box. *Domain decomposition* is similar, only now each processor gets some part of a system that it has its own unique type of region. For example, using the house analogy, one type of domain decomposition would be to give each processor a certain room or type of room regardless of size. So, one processor gets a bedroom another processor the guest room, another one the living room and so on and so forth. The last type of decomposition is called *force decomposition*, where each processor gets part of the forces that need to be calculated. In this case, imagine people in the house and they are holding strings that connect them to each other, which represent person–person interactions. Each processor would take a certain number of strings that people are holding with each other. As you might imagine there are different scenarios where each of the unique types of decomposition work better than others. If you consider N to be the number of particles and P to be the number of processors, Plimpton [1] figured out the benefits and drawbacks to each type of decomposition. I thoroughly recommend that you check the reference out.

Careful inspection of Plimpton's results shows that there are some situations where more processors actually give less efficiency, and this is because *communication* between the processors costs time, and construction of the hardware is such that you get sweet spots for certain numbers of processors (many times, in multiples of 4). In the end, when constructing a parallel simulation you would carefully look at the situation in front of you, move forward with a good first guess of your project, and refine based on the results you obtain.

9.3 Example Parallel Programs

Now let's start out by looking at some example programs and how parallel processing happens in a practical sense. The hardware that you choose and setup will not work unless there is software that supports parallel processing, the communication between nodes and so forth. Such a protocol is *Message Passing Interface* (MPI) and there are copious references for you to learn more about it [2,3]. There is also a very useful repository [4] of MPI examples in C++ and FORTRAN, which I strongly recommend you make use of. As I mention in the Foreword to this text, the point *is* to get you to learn to construct, run, and analyze computer simulations and it *is not* about the choice of programming languages. With that said, I think you will notice that, as you go more "off road" into parallel processing you will notice a lot more programming in C++ and FORTRAN, the staple languages of the examples in this book.

9.3.1 An Integral in C++ with MPI

Consider an integral from calculus as in Chapter 4. The value of the integral is simply the area between a curve $y = f(x)$ and the x axis, and we broke that area up into trapezoids or rectangles to actually compute the integral. Now let's look at an example of spatial decomposition, where we assign each processor a certain region to integrate over. Although my students and I wrote nearly all the programs in this text, I am taking this one from John Burkardt's vast repository [4] and re-narrating it. He has included some very nice "bells and whistles." I want to emphasize that, when you use it you will likely have to make minor compile and run adjustments so the code works on your particular system. I am also preserving the existing indention in the program—sometimes called dovetail programming—which you may adopt in your own practice as a matter of code organization if you like.

Looking at Intervals_mpi.cpp will be most relevant here, which can be run with the Intervals.sh script provided, if you so choose. Here are the include statements—note the presence of `mpi.h`.

```
# include<cmath>
# include<cstdlib>
# include<ctime>
# include<iomanip>
# include<iostream>
# include<mpi.h>

using namespace std;

int main ( int argc, char *argv[] );
double f ( double x );
void timestamp ( );
```

Program Header.

```
int main ( int argc, char *argv[] )

{
```

Declare Variables.

```
  double end_time;
  double h;
  int i;
  int id;
  int ierr;
  int m;
  int n;
  int p;
  double pi = 3.141592653589793238462643;
  int process;
  double q_global;
  double q_local;
  int received;
  int source;
  double start_time;
  MPI_Status status;
```

```
int tag;
int target;
double x;
double xb[2];
double x_max = 1.0;
double x_min = 0.0;
```

Here we set up the MPI environment and protocol.

```
ierr = MPI_Init ( &argc, &argv );
```

Now, we determine the rank of the processor.

```
ierr = MPI_Comm_rank ( MPI_COMM_WORLD, &id );
```

How many processors are there?

```
ierr = MPI_Comm_size ( MPI_COMM_WORLD, &p );
```

Now, give some basic information to the user.

```
if ( id == 0 )
{
  timestamp ( );
  cout << "\n";
  cout << "INTERVALS - Master process:\n";
  cout << "  C++ version\n";
  cout << "\n";
  cout << "  An MPI example program,\n";
  cout << "  A quadrature over an interval is done by\n";
  cout << "  assigning subintervals to processes.\n";
  cout << "\n";
  cout << "  The number of processes is " << p << "\n";
```

Get the start time for efficiency comparisons.

```
  start_time = MPI_Wtime ( );
```

If we don't have enough processors for parallel computing, let the user know and shut the program down.

```
  if ( p <= 1 )
  {
    cout << "\n";
    cout << "INTERVALS - Master process:\n";
    cout << "  Need at least 2 processes!\n";
    MPI_Finalize ( );
    cout << "\n";
    cout << "INTERVALS - Master process:\n";
    cout << "  Abnormal end of execution.\n";
    exit ( 1 );
  }
}
cout << "\n";
cout << "Process " << id << ": Active!\n";
```

Now the head (master) node does some things:

```
if ( id == 0 )
{
```

It calculates the intervals over which each compute note is going to do a partial integral over

```
for ( process = 1; process <= p-1; process++ )
{
  xb[0] = ( ( double ) ( p - process       ) * x_min
        + ( double ) (       process - 1 ) * x_max )
        / ( double ) ( p             - 1 );

  xb[1] = ( ( double ) ( p - process - 1 ) * x_min
        + ( double ) (       process     ) * x_max )
        / ( double ) ( p             - 1 );

  target = process;
  tag = 1;
```

and it send them the information. Burkhardt points out that the compute nodes could figure this out themselves, but we are illustrating communication here.

```
    ierr = MPI_Send ( xb, 2, MPI_DOUBLE, target, tag, MPI_COMM_WORLD );
  }
}
else
{
  source = 0;
  tag = 1;
```

If you are a compute node... you then receive the interval information:

```
  ierr = MPI_Recv ( xb, 2, MPI_DOUBLE, source, tag, MPI_COMM_WORLD, &status );
}
```

Now, we exercise manners and wait until everyone has gotten the interval information.

```
  ierr = MPI_Barrier ( MPI_COMM_WORLD );

if ( id == 0 )
{
  cout << "\n";
  cout << "INTERVALS - Master process:\n";
  cout << "  Subintervals have been assigned.\n";
}
```

Now, we broadcast the number of points to be used in each interval:

```
m = 100;
source = 0;

ierr = MPI_Bcast ( &m, 1, MPI_INT, source, MPI_COMM_WORLD );
```

If you are not the head (master) node, do the calculation on your interval:

```
if ( id != 0 )
{
  q_local = 0.0;

  for ( i = 1; i <= m; i++ )
  {
    x = ( ( double ) ( 2 * m - 2 * i + 1 ) * xb[0]
        + ( double ) (         2 * i - 1 ) * xb[1] )
        / ( double ) ( 2 * m               );

    q_local = q_local + f ( x );
  }

  q_local = q_local * ( xb[1] - xb[0] ) / ( double ) ( m );

  target = 0;
  tag = 2;
```

Send the partial integral results to the head node:

```
  ierr = MPI_Send ( &q_local, 1, MPI_DOUBLE, target, tag, MPI_COMM_WORLD );
}

else
{
  received = 0;
  q_global = 0.0;
```

The master node receives the partial integral results and adds them up to the full integral.

```
  while ( received < p - 1 )
  {
    source = MPI_ANY_SOURCE;
    tag = 2;

    ierr = MPI_Recv ( &q_local, 1, MPI_DOUBLE, source, tag, MPI_COMM_WORLD,
      &status );

    q_global = q_global + q_local;
    received = received + 1;
  }
}
```

The head (master) node communicates the result with some I/O.

```
if ( id == 0 )
{
  cout << "\n";
  cout << "INTERVALS - Master process:\n";
  cout << "  Estimate for PI is " << q_global      << "\n";
  cout << "  Error is           " << q_global - pi << "\n";

  end_time = MPI_Wtime ( );

  cout << "\n";
```

```
    cout << "  Elapsed wall clock seconds = "
         << end_time - start_time << "\n";
  }
```

Wrap MPI up.

```
  MPI_Finalize ( );
```

And shut down the program.

```
  if ( id == 0 )
  {
    cout << "\n";
    cout << "INTERVALS - Master process:\n";
    cout << "  Normal end of execution.\n";
    cout << "\n";
    timestamp ( );
  }
  return 0;
}
```

9.3.2 Now the Integral in FORTRAN with MPI

Consider the same integral from Section 9.3.1 but in FORTRAN with MPI. I chose to re-narrate the program as I did for the C++ version because the calls to MPI look different enough and the program is the only full MPI FORTRAN code I'm covering. The same caveat applies, namely that you might have to tweak the code as well as compile/run statements to get it to run on your particular platform/machine. The most relevant code you can find here is intervals_mpi.f.

Program Name.

```
      program main
```

It's MPI, after all.

```
      include 'mpif.h'
```

Declare Variables.

```
      double precision end_time
      double precision f
      double precision h
      integer i
      integer ierr
      integer m
      integer master
      parameter ( master = 0 )
      integer n
      double precision pi
      parameter ( pi = 3.141592653589793238462643D+00 )
      integer process
      integer process_id
      integer process_num
      double precision q_global
      double precision q_local
      integer received
```

```
      integer source
      double precision start_time
      integer status(MPI_Status_size)
      integer tag
      integer target
      double precision x
      double precision xb(2)
      double precision x_max
      parameter ( x_max = 1.0D+00 )
      double precision x_min
      parameter ( x_min = 0.0D+00 )
```

Set up the MPI environment and protocol.

```
      call MPI_Init ( ierr )
```

Now we determine the rank of the ("this") processor.

```
      call MPI_Comm_rank ( MPI_COMM_WORLD, process_id, ierr )
```

How many processors are there?

```
      call MPI_Comm_size ( MPI_COMM_WORLD, process_num, ierr )
```

Now, the head (master) node gives some basic information I/O to the user.

```
      if ( process_id .eq. master ) then
        call timestamp ( )
        write ( *, '(a)' ) ' '
        write ( *, '(a)' ) 'INTERVALS - Master process:'
        write ( *, '(a)' ) '  FORTRAN77 version'
        write ( *, '(a)' ) ' '
        write ( *, '(a)' ) '  An MPI example program.'
        write ( *, '(a)' ) '  A quadrature over an interval is done by'
        write ( *, '(a)' ) '  assigning subintervals to processes.'
        write ( *, '(a,i8)' )
     &    '  The number of processes is ', process_num
```

Get the start time for efficiency comparisons.

```
      start_time = MPI_Wtime ( )
```

If we don't have enough processors for parallel computing, let the user know and shut the program down.

```
      if ( process_num .le. 1 ) then
        write ( *, '(a)' ) ' '
        write ( *, '(a)' ) 'INTERVALS - Master process:'
        write ( *, '(a)' ) '  Need at least 2 processes.'
        call MPI_Finalize ( ierr )
        write ( *, '(a)' ) ' '
        write ( *, '(a)' ) 'INTERVALS - Master process:'
        write ( *, '(a)' ) '  Abnormal end of execution.'
        stop
      end if
```

```
      end if

      write ( *, '(a)' ) ' '
      write ( *, '(a,i8,a)' ) 'Process ', process_id, ': Active.'
```

Now, the head (master) node does some things:

```
      if ( process_id .eq. master ) then
```

It calculates the intervals over which each compute note is going to do a partial integral over.

```
        do process = 1, process_num-1

          xb(1) = ( dble ( process_num - process        ) * x_min
     &            + dble (                     process - 1 ) * x_max )
     &            / dble ( process_num             - 1 )

          xb(2) = ( dble ( process_num - process - 1 ) * x_min
     &            + dble (                     process     ) * x_max )
     &            / dble ( process_num             - 1 )

          target = process
          tag = 1
```

Then it sends them the information. See Burkhardt's note at this point in the C++ program above.

```
          call MPI_Send ( xb, 2, MPI_DOUBLE_PRECISION, target, tag,
     &        MPI_COMM_WORLD, ierr )

        end do

      else

        source = master
        tag = 1
```

If you are a compute node... you then receive the interval information:

```
        call MPI_Recv ( xb, 2, MPI_DOUBLE_PRECISION, source, tag,
     &      MPI_COMM_WORLD, status, ierr )
      end if
```

Now, wait until every node has gotten the interval information.

```
      call MPI_Barrier ( MPI_COMM_WORLD, ierr )

      if ( process_id .eq. master ) then
        write ( *, '(a)' ) ' '
        write ( *, '(a)' ) 'INTERVALS - Master process:'
        write ( *, '(a)' ) '  Subintervals have been assigned.'
      end if
```

Now we broadcast the number of points to be used in each interval.

```
m = 100
source = master

call MPI_Bcast ( m, 1, MPI_INTEGER, source, MPI_COMM_WORLD,
& ierr )
```

If you are not the head (master) node, do the calculation on your interval.

```
if ( process_id .ne. master ) then

  q_local = 0.0D+00

  do i = 1, m
    x = ( dble ( 2 * m - 2 * i + 1 ) * xb(1)
&        + dble (         2 * i - 1 ) * xb(2) )
&        / dble ( 2 * m               )

    q_local = q_local + f ( x )

  end do

  q_local = q_local * ( xb(2) - xb(1) ) / dble ( m )

  target = master
  tag = 2
```

Send the partial integral results to the head node:

```
      call MPI_Send ( q_local, 1, MPI_DOUBLE_PRECISION, target, tag,
&        MPI_COMM_WORLD, ierr )
c
c   Process 0 expects to receive N-1 partial results.
c
    else

      received = 0
      q_global = 0.0D+00

10      continue
```

The master node receives the partial integral results and adds them up to the full integral.

```
      if ( received < process_num - 1 ) then

        source = MPI_ANY_SOURCE
        tag = 2

        call MPI_Recv ( q_local, 1, MPI_DOUBLE_PRECISION, source, tag,
&          MPI_COMM_WORLD, status, ierr )

        q_global = q_global + q_local
        received = received + 1

        go to 10

      end if

    end if
```

The head (master) node communicates the result with some I/O.

```
if ( process_id .eq. master ) then

   write ( *, '(a)' ) ' '
   write ( *, '(a)' ) 'INTERVALS - Master process:'
   write ( *, '(a,g14.6)' )
&     '  Estimate for PI is ', q_global
   write ( *, '(a,g14.6)' )
&     '  Error is            ', q_global - pi

   end_time = MPI_Wtime ( )

   write ( *, '(a)' ) ' '
   write ( *, '(a,f14.6)' ) '  Elapsed wall clock seconds = ',
&       end_time - start_time

end if
```

Wrap MPI up.

```
call MPI_Finalize ( ierr )
```

And shut down the program.

```
if ( process_id == master ) then
   write ( *, '(a)' ) ' '
   write ( *, '(a)' ) 'INTERVALS - Master process:'
   write ( *, '(a)' ) '  Normal end of execution.'
   write ( *, '(a)' ) ' '
   call timestamp ( )
end if

stop
end
```

Here is the function routine that calculates the value of the integrand (isn't it grand??):

```
double f ( double x )

{
  double value;

  value = 4.0 / ( 1.0 + x * x );

  return value;
}
```

There are other C++ and FORTRAN examples provided of heat transfer and integration by quadrature with MPI and I strongly encourage you to check them out.

9.3.3 Planetary Dynamics in Parallel

Another example of parallel processing with MPI is our planetary simulation code, courtesy of Wes Even. Because the code is at nearly the same as in the nonparallel case, I present an abbreviated version below with omitted (implicit) sections from the nonparallel code delimited by dashed lines "====" and explicit excerpts that illustrate the use of MPI, and hence the difference between it and its non-MPI counterpart.

Program name

```
program coll3
```

As before, we are doing MPI so include it

```
include 'mpif.h'
```

```
===================================
```
Here we declared parameters, variables and arrays.
```
===================================
```

Now start MPI as we did in the examples above

```
call MPI_INIT(ierr)
call MPI_COMM_RANK(MPI_COMM_WORLD, myid, ierr)
call MPI_COMM_SIZE(MPI_COMM_WORLD, numprocs, ierr)

strtime = MPI_WTIME()
```

Have the head node open files and set the stage for I/O

```
if (myid.eq.0) then
open(unit=50,file='incon1996.res',status='old')
open(unit=60,file='coll3d.res',status='new')
endif
```

```
=========================================
```
Here we initialized variables and set the initial condition.
```
=========================================
```

Now we broadcast the initial condition to the compute nodes:

```
call MPI_BCAST(x,n,MPI_REAL,0,MPI_COMM_WORLD,ierr)
call MPI_BCAST(y,n,MPI_REAL,0,MPI_COMM_WORLD,ierr)
call MPI_BCAST(z,n,MPI_REAL,0,MPI_COMM_WORLD,ierr)
call MPI_BCAST(vx,n,MPI_REAL,0,MPI_COMM_WORLD,ierr)
call MPI_BCAST(vy,n,MPI_REAL,0,MPI_COMM_WORLD,ierr)
call MPI_BCAST(vz,n,MPI_REAL,0,MPI_COMM_WORLD,ierr)

btime = MPI_WTIME()
```

Now start the time loop.

```
do 100 itime=beginstep,nstep
dO 90 i=1,n
 if (i.le.(n/(numprocs)*(myid+1))) then
 if (i.gt.(n/(numprocs)*(myid))) then
```

```
==================
```
Here we calculated forces.
```
==================
```

```
endif
endif
 90  continue
```

```
      do 310 i=1,n
      if (i.le.(n/(numprocs)*(myid+1))) then
      if (i.gt.(n/numprocs*myid)) then
```

====================================
Here each processor updated positions:
====================================

```
      endif
      endif
310   continue
```

Broadcast the updated results back to the head node.

```
      do 888 i=1,numprocs
         bottom = INT((i-1)*n/numprocs)+1
         top = INT(i*n/numprocs)
         tem = top-bottom+1
         call MPI_BCAST(x(bottom),tem,MPI_REAL,i-1,MPI_COMM_WORLD,ierr)
         call MPI_BCAST(y(bottom),tem,MPI_REAL,i-1,MPI_COMM_WORLD,ierr)
         call MPI_BCAST(z(bottom),tem,MPI_REAL,i-1,MPI_COMM_WORLD,ierr)
         call MPI_BCAST(vx(bottom),tem,MPI_REAL,i-1,MPI_COMM_WORLD,ierr)
         call MPI_BCAST(vy(bottom),tem,MPI_REAL,i-1,MPI_COMM_WORLD,ierr)
         call MPI_BCAST(vz(bottom),tem,MPI_REAL,i-1,MPI_COMM_WORLD,ierr)
889      continue
888   continue
C****************************************************************
```

====================================
Here the compute nodes dealt with collisions.
====================================

Broadcast the updated results back to the head node.

```
      do 2999 i=0,(numprocs-1)
         ricollp = REAL(icolllp)
         rjcmarkp = REAL(jcmarkp)
         ricmarkp = REAL(icmarkp)
         call MPI_BCAST(ricollp,1,MPI_REAL,i,MPI_COMM_WORLD,ierr)
         call MPI_BCAST(rjcmarkp,1,MPI_REAL,i,MPI_COMM_WORLD,ierr)
         call MPI_BCAST(ricmarkp,1,MPI_REAL,i,MPI_COMM_WORLD,ierr)
         if (icoll.eq.0) then
         icoll = INT(ricollp)
         jcmark = INT(rjcmarkp)
         icmark = INT(ricmarkp)
         endif
2999  continue
```

Calculate ending times for efficiency studies.

```
      if (myid.eq.0) then
      forcetime = forcetime + etime-btime
      endif

      btime = MPI_WTIME()
```

```
=====================
```
One more collision algorithm.
```
=====================
```

Calculate more ending times for efficiency studies.

```
etime = MPI_WTIME()
if (myid.eq.0) then
touchtime = touchtime + etime - btime
endif
```

Final I/O from the head node.

```
if (myid.eq.0) then
calculate and write out
stptime = MPI_WTIME()
if (myid.eq.0) then
write(6,*) 'Time:',(stptime-strtime)/nstep
write(6,*) 'Force Time:', forcetime/nstep
write(6,*) 'Touching Group Time:', touchtime/nstep
endif
```

Wrap up MPI.

```
call MPI_FINALIZE(ierr)
```

End the program.

```
end
```

9.4 Compiling and Executing MPI Codes

There can be multiple statements and flags when compiling and running MPI code, just as is true with non-MPI applications and I encourage you to consult Reference 5, for example. Typical compile statements look like

```
mpif77 filename.f
```

for FORTRAN and

```
mpiCC filename.f, mpicxx filename.f, mpic++ filename.f
```

for various cases of C++.

A standard execution statement looks like

```
mpirun -np 4 ./filename > output.txt ,
```

where in this case we are using four processors to do the calculations. At the end of the day, it is important for you to know the details of your hardware, software, and operating system and be able to tweak your code and commands so as to deliver what you need, preferably optimally.

9.5 UNIX Scripting

As mentioned above, UNIX scripting is a wonderful tool for automating commands. The method is often called *shell scripting* because the commands are run by the UNIX shell. There are a great many references and tutorials out there on scripting [6] and I encourage you to search through them. For this book, I am going to narrate typical scripts that I utilize and have provided as sample code, each of which saves me a considerable amount of work.

Here we begin discussion relevant to the Prepareitall script provided and see the beginning of loops for indices i, j, k, and m. In the version of the script provided there are many commands commented out because the required files are lacking but you can get an idea of how powerful it is.

```
for i in 1  2  3
do
for j in 4  5  6
do
for k in 7  8
do
for m in 10 100 1000
do
```

Now, we create a directory whose name reflect the value of the loop indices.

```
mkdir Bc$i"Su$j""Ac$k""Size$m"
```

Next, we copy needed files (FORTRAN code as well as a qsub script) into the new directory.

```
cp  MPM_3D_08_01_17.f Bc$i"Su$j""Ac$k""Size$m"
cp qsub.antipodal Bc$i"Su$j""Ac$k""Size$m"
```

We now go into the new directory

```
cd Bc$i"Su$j""Ac$k""Size$m"
```

and we edit the FORTRAN code so that the text strings in CAPS are replaced by numbers having the values of the loop indices (the code will not compile until the text is replaced with numbers):

```
vi -c :%s/\COLOR\/$i/ -c :x  MPM_3D_08_01_17.f
vi -c :%s/\SURFACE\/$j/ -c :x  MPM_3D_08_01_17.f
vi -c :%s/\ANTI\/$k/ -c :x  MPM_3D_08_01_17.f
vi -c :%s/\SIZE\/$m/ -c :x  MPM_3D_08_01_17.f
```

Compile with an optimizer

```
gfortran -O3   MPM_3D_08_01_17.f
```

and execute (commented out for now with the # sign)

```
#./a.out
```

or instead of executing, submit the job to the machine's queue (this command will vary from machine to machine).

```
qsub -q eternity qsub.antipodal
```

Back up a directory and close the loops.

```
cd ..
done
done
done
done
```

Here is a simple script that finds files whose names start with *Antipod* and renders them in POV (see Chapter 4). In this simulation, the only file names starting with *Antipod* also have a pov extension and so this script won't run with any files that aren't POV files. The Povscript file is called and run within the Render script.

```
for k in ./Antipod*
do
              povray -D -I $k -o- >$k.jpg -W4000 -H3000
done
```

There are also shell scripts I have provided that apply to the parallel programing applications discussed earlier.

9.6 Graphics Processing Units (GPU's)

There are a wealth of references dealing with GPU's [7]. With the warning that a Wikipedia page can be edited without review or fact check, Reference 7 conveys the rich and interesting background behind GPU's as well as the context of their use. Essentially, GPU's are made up of a large number of parallel processors while a CPU (Central Processing Unit) is made up of only a few. Such construction allows a GPU to be very efficient and powerful, often giving processing gains of several hundred times over a CPU. If you are constructing a machine that will be used for weather animation or heavy duty computing in either the memory or time direction, I suggest you have a serious look into GPU's.

PROBLEMS

9.1. The examples of integrals previously covered involve giving $m = 100$ intervals for integrals done on *each processor*. Another way to decompose space is to keep the total number of intervals for the integral constant and divide those out among the processors. Rewrite one of the integral examples to keep the total number of intervals constant. Run your code for a few values of the number of processors and compare results to the example given. Discuss advantages, disadvantages, and situations where each method would be a preferred choice. Hint: Watch out for remainders when dividing the total number of intervals up!

9.2. Consider one of Burkhardt's examples of integrals that I discussed earlier in this chapter. Run the code for various numbers of processors and numbers of intervals for the partial integrals; discuss your results. Can you identify scenarios where each type of decompositions would be best?

9.3.   Consider one of Burkhardt's examples of integrals that I discussed earlier in this chapter. Given that the examples here utilize spatial decomposition, rewrite the codes to utilize domain decomposition. Run your code for various numbers of processors and numbers of intervals for the partial integrals; discuss your results. Can you identify scenarios where each type of decompositions would be best?

9.4.   Modify one of Burkhardt's examples of integrals covered earlier in this chapter so that the compute nodes themselves determine their integration intervals. Run for a few values of processor number and integration intervals and compare results.

9.5.   Run both of Burkhardt's examples discussed here for a few values of processor number and number of integration intervals and compare the C++ to FORTRAN results.

9.6.   Consider the MPI version of the solar system dynamics program I outlined in this chapter. Ignoring collisions, flesh out the program and run for a few different values of particle and processor number.

9.7.   Consider the MPI version of the solar system dynamics program I outlined in this chapter. Including collisions not allowing absorption and with varying coefficients of restitution e, flesh out the program and run for a few different values of particle and processor number. I suggest starting with some limiting case of e. Please feel free to consult the section of this book dealing with the pool game simulation or contact me if you want to discuss the general collision algorithm.

9.8.   Consider the MPI version of the solar system dynamics program I outlined earlier in this chapter. Including perfectly inelastic collisions allowing absorption, flesh out the program and run for a few different values of particle and processor number.

9.9.   The shell scripts provided work with the FORTRAN version of the intervals code. (a) Modify them so that they could work with the C++ versions. (b) After you get the script working, test it out on one of the example programs provided but not covered in the text.

9.10.   Play with some UNIX scripting but play it safe, so that you aren't removing any files or directories. Write scripts that create directories, print working directories, edit files, and such. I suggest you start with the example scripts I provided, and if you compile/move/copy files you'll have to make sure they exist in the proper directory.

9.11.   Some areas of this book covers develop so quickly that updates soon after publication are almost certainly necessary. Look up the current state of GPU's. Invite a friend to coffee, hot chocolate, or any other soft drink and discuss. Report out to your class or colleagues as required.

References

1. S. Plimpton, Fast parallel algorithms for short–range molecular dynamics. *Journal of Computational Physics*, 117, pp. 1–19, 1995—originally Sandia Technical Report SAND91–1144 (May 1993, June 1994).

2. http://mpi-forum.org/

3. W. Gropp, E. Lusk, A. Skjellum, *Using MPI: Portable Parallel Programming with the Message-Passing Interface*, MIT Press, Cambridge, MA, ISBN: 0262571323, Second edition, 1999.

4. http://people.sc.fsu.edu/~jburkardt/index.html

5. https://docs.oracle.com/cd/E19356-01/820-3176-10/compiling.html

6. https://www.shellscript.sh/

7. https://en.wikipedia.org/wiki/Graphics_processing_unit

//

Integrated C++ / Python Simulation of Guitar Sounds*

Here is the Python (main) part of the simulation.
Generously provided by Matt Karl

```
# Matt Karl
# Modeling and Simulation of Physical Systems

# guitar_string_sim.py
# Strings: D'Addario EXL110 strings
# Guitar: 25.5" scale, 24 frets

import audiere, numpy, scipy, time, os, array
from math import cos,sin,pi,sqrt,exp
import matplotlib.pyplot as plt
from matplotlib.pyplot import plot,show,draw,ylabel,xlabel,title,axis,subpl
ot,figure,savefig,semilogy
from scipy import fft
from threading import Thread

recordString=0
```

_____

* Used with gracious permission of Matt Karl.

Parameter and Variable Definitions:

```
# strings in order from low to high [E,A,D,G,B,e]
# sting info

X=None # muted string
chords={'A':[X,0,2,2,2,0],'B':[X,2,4,4,4,2],'C':[X,3,2,0,1,0],'D':[
X,X,0,2,3,2],

'E':[0,2,2,1,0,0],'F':[X,X,3,2,1,1],'G':[3,2,0,0,0,3],'Am':[X,0,2,2,1,0],

'Bm':[X,2,4,4,3,2],'Dm':[X,X,0,2,3,1],'Em':[0,2,2,0,0,0],'O':[0,0,0,0,0,0],

'oE':[0,X,X,X,X,X],'oA':[X,0,X,X,X,X],'oD':[X,X,0,X,X,X],'oG':[X,X,X,0,
X,X],
'oB':[X,X,X,X,0,X],'oe':[X,X,X,X,X,0]}

strum=chords['oA']
duration=2.0
img_interval=8000
```

Strum custom chord, guitar tab style, list needs reversed.

```
##strum = [0,
##          1,
##          0,
##          2,
##          3,
##          X]
##strum.reverse() # reverse list order for correct indexing

pickupNum=2
pickL=0.105
pickD=0.002
```

Now the physical properties of the system.

```
L = 0.6477 # (meters), = 25.5"
tension = [77.875,86.775,81.880,73.870,68.530,72.090] # (Newtons)
mu = [0.006824600783,0.004279483284,0.002262783037,
      0.001143267067,0.000668602296,0.000395553976] # (kg/m)
vel=[sqrt(elms[0]/elms[1]) for elms in zip(tension,mu)] # wave velocities
fund_freqs = [82.4,110.0,146.8,196.0,246.9,329.6] # (Hz)
fret=[L,0.611378,0.577088,0.544576,0.514096,0.48514,0.457962,
      0.432308,0.407924,0.385064,0.363474,0.343154,0.32385,
      0.305562,0.288544,0.272288,0.257048,0.24257,0.229108,
      0.216154,0.203962,0.192532,0.181864,0.17145,0.162052] # string length
at fret, (meters)
fName=["Estring","Astring","Dstring","Gstring","Bstring","e_string"]

pickup=[0.035,0.050,0.090,0.125,0.140]
sf = 44100 # sampling frequency
```

Open sound output device.

```
snd_dev=audiere.open_device()
```

Now here are some function definitions.

```
def soundsc(data,sf):
m=max([abs(i) for i in data])
scaled_data=[]
for i in data:
scaled_data.append(i/m)
snd=snd_dev.open_array(scaled_data,sf)
snd.play()
time.sleep(duration)

def
sim(L,fretNum,vel,sf,duration,pickL,pickupL,pickD,ofilename,recordString,
img_interval):
opts=' '.join([str(i) for i in [L,fretNum,vel,sf,duration,pickL,pickupL,pi
ckD,
                                  ofilename,recordString,img_interval]])
  os.system("guitar_sim "+opts)

def plotfft(data,freq_range=1000,semilog=False):
r=int(duration*freq_range)
Y=fft(data)
Y=abs(Y*Y.conj())
plt.clf()
if semilog:
semilogy([i/duration for i in xrange(r)],Y[:r])
else:
plot([i/duration for i in xrange(r)],Y[:r])
show()
```

Main part of the program:

```
if __name__=='__main__':
time.clock()
print "Building",strum
threads=[]
for i,fretNum in enumerate(strum):
if fretNum != None:
threads.append(Thread(target=sim, args=(L,fretNum,vel[i],sf,duration,pickL,
pickup[pickupNum],

pickD,fName[i],recordString,img_interval)))
for thread in threads:
thread.start()
for thread in threads:
thread.join()
```

Now some I/O:

```
print "Done simulating in", time.clock()

print "Reading in waveforms"
strings=[]
for i,fretNum in enumerate(strum):
if fretNum != None:
with open(fName[i]) as ifile:
strings.append([float(line.strip()) for line in ifile])
# Add individual waveforms to generate full chord waveform
chord_wave=[sum(elms) for elms in [[strings[j][i]
```

```
for j in range(len(strings))]
for i in range(len(strings[0])))]]
print "Playing single notes and full chord"
for note in strings:
soundsc(note,sf)
if len(strings)>1: soundsc(chord_wave,sf)

if recordString:
stringForms=[]
for i,fretNum in enumerate(strum):
if fretNum != None:
with open(fName[i]+"_time") as ifile:
stringForms.append([[float(i) for i in line.split()] for line in ifile])
mgNum=0
for s in stringForms:
for waveform in s:
#_=figure()
_=plot(waveform)
title("waveform "+str(imgNum))
xlabel("x")
ylabel("y")
axis([0,len(waveform),-pickD,pickD])
savefig("vid/"+str(imgNum)+".png",format='png')
plt.clf()
#plt.close()
imgNum+=1
```

Wrap it up.

```
plt.close()
```

Here is the C++ part of the simulation that the Python part calls.
First the headers:

```
#include<cstdlib>
#include<iostream>
#include<math.h>
#include<fstream>
#include<string>

using namespace std;
```

Declare and initialize variables.

```
int main(int argc, char* argv[])
{
//args:
progName,L,fretNum,vel,sf,duration,pickL,pickupL,pickD,ofilename,recordStri
ng,img_interval
float L = atof(argv[1]); //string length
// fret[x] is the string length at fret number x
float fret[]={L,0.611378,0.577088,0.544576,0.514096,0.48514,0.457962,
              0.432308,0.407924,0.385064,0.363474,0.343154,0.32385,
              0.305562,0.288544,0.272288,0.257048,0.24257,0.229108,
              0.216154,0.203962,0.192532,0.181864,0.17145,0.162052};
int fretNum = atoi(argv[2]);
float frettedL = fret[fretNum]; //pick out string length according to
//given fret number
```

```
float vel = atof(argv[3]); //wave velocity
float f0 = vel/(2*frettedL); // fundamental frequency f=v/lambda
int sf = atoi(argv[4]); //sampling frequency
float duration = atof(argv[5]); //duration of simulation
int Nt = int(duration*sf); //Total number of time steps
double dt = duration/float(Nt);
//int skip = int(Nt/sf); // variable to pick out sf samples/sec from final
//array
int Nx = int(sf/2.0/f0);//atoi(argv[6]); //total number of string pieces
float dx = frettedL/float(Nx);
float r = (vel*vel*dt*dt/(dx*dx)); //simplified variable for main loop
float pickL = atof(argv[6]); // picking/plucking position on string
int pickNode = int(pickL/dx); // string piece that is plucked
float pickupL = atof(argv[7]); // pickup position on string to record from
int pickupNode = int(pickupL/dx); // string piece that is recorded
float pickD = atof(argv[8]); //distance string is plucked
float x[Nx+1]; //array of string piece positions
float ** y; // variable for wave array, position and time dimensions
// Create and initialize 2D array for y
y = new float *[Nx+1];
for (int j=0; j<Nx+1; j++)
y[j] = new float[Nt];
}
for (int j=0; j<Nx+1; j++) {
for (int i=0; i<Nt; i++) {
y[j][i] = 0.0;}
}

string ofilename = argv[9]; //output data filename
string ofilename_time = ofilename+"_time"; //output data filename for
string shape through time
bool recordString = (bool)atoi(argv[10]);
int img_interval = atoi(argv[11]);
```

Now some I/O:

```
ofstream ofile,ofile_t;

cout<<"Simulating "<<ofilename<<" held down on fret "<<fretNum<<endl;
```

Now initialize like crazy.

```
for (int j=0;j<Nx+1;j++) {
x[j] = j*dx;
}

float upslope = pickD/pickL;
float downslope = pickD/(frettedL-pickL);
for(int j=1;j<Nx;j++) {
if (j<=pickNode) {
y[j][0] = upslope*x[j];
y[j][1] = y[j][0];
}
else {
y[j][0] = downslope*(frettedL-x[j]);
y[j][1] = y[j][0];
}
}
```

Boundary conditions:

```
y[0][0] = 0.0;
y[0][1] = 0.0;
y[Nx][0] = 0.0;
y[Nx][1] = 0.0;
```

More I/O: string shape recording and initial profile.

```
if (recordString) {
ofile_t.open(ofilename_time.c_str());
for (int j=0; j<Nx+1; j++) {
ofile_t << y[j][0] << " ";
}
ofile_t << endl;
}
```

Now start the main simulation loop.

```
for (int i=1; i<Nt-1; i++) {
y[0][i+1] = 0.0; // boundary conditions
y[Nx][i+1] = 0.0;
for (int j=1; j<Nx; j++) {
y[j][i+1] = 2.0*y[j][i]-y[j][i-1] + r*(y[j+1][i]-2.0*y[j][i]+y[j-1][i]);
}
if(i%img_interval==0) {
if (recordString) {
for (int j=0; j<Nx+1; j++) {
ofile_t << y[j][i] << " ";
}
ofile_t << endl;
}
}
}
```

Write it out and wrap it up.

```
ofile.open(ofilename.c_str());

for (int i=0; i<Nt; i++) {
ofile << y[pickupNode][i] << endl;
}

ofile.close();
if (recordString) {ofile_t.close();}
return(0);
}
```

Index